Lecture Notes in Mathematics

Edited by A. Dold and B. Eckmann



761

Klaus Johannson

# Homotopy Equivalences of 3-Manifolds with Boundaries

Springer-Verlag
Berlin Heidelberg New York 1979

**Author**

Klaus Johannson
Fakultät für Mathematik
der Universität
Universitätsstr. 1
D-4800 Bielefeld 1

AMS Subject Classifications (1970): 55 A 99

ISBN 3-540-09714-7 Springer-Verlag Berlin Heidelberg New York
ISBN 0-387-09714-7 Springer-Verlag New York Heidelberg Berlin

Library of Congress Cataloging in Publication Data
Johannson, Klaus, 1948-
Homotopy equivalence of 3-manifolds with boundaries.
(Lecture notes in mathematics; 761)
Bibliography: p.
Includes index.
1. Manifolds (Mathematics) 2. Homotopy equivalences. I. Title. II. Series: Lecture notes in mathematics (Berlin); 761.
QA3.L28 no. 761 [QA613] [514'.2] 510'.8s 79-23603
ISBN 0-387-09714-7



Printing and binding: Beltz Offsetdruck, Hemsbach/Bergstr.
2141/3140-543210

Contents.

## Introduction

The main object of this book is the study of homotopy equivalences between 3-manifolds. Here a 3-manifold $M$ is always compact, orientable and irreducible. Moreover, we suppose that $M$ is boundary-irreducible, that is, for any component $G$ of $\partial M$, $\pi_1 G \to \pi_1 M$ is injective. However, we do not always insist that the boundary is non-empty. Examples of such 3-manifolds are the knot spaces (of non-trivial knots).

If $M_1$ and $M_2$ are such 3-manifolds with non-empty boundaries, and $f: M_1 \to M_2$ a homotopy equivalence, one knows by [Wa 4] that $f$ can be deformed into a homeomorphism, provided it preserves the peripheral structure, that is, $f|\partial M_1$ can be deformed into $\partial M_2$.

We are interested here in homotopy equivalences whose restrictions to the boundary cannot be deformed into the boundary. Such homotopy equivalences will be called exotic. Our main result is a classification theorem for exotic homotopy equivalences.

Before describing this theorem let us briefly recall the situation in the 2-dimensional case. Many exotic homotopy equivalences can be found between surfaces $F_1$, $F_2$ with boundaries. For example, there is of course at least one such homotopy equivalence between the torus with one hole and the 2-sphere with three holes. Since a surface with boundary is a $K(\pi,1)$-space whose fundamental group is free, the exotic homotopy equivalences of surfaces can be analyzed by using the presentations of the outer automorphism groups of free groups [Ni 1]. In particular, it follows that they are (finitely) generated by Dehn flips (along arcs). Here a Dehn flip means an exotic homotopy equivalence $f: F_1 \to F_2$ for which there is an arc $k$ in $F_2$, $k \cap \partial F_2 = \partial k$, such that $f^{-1}k$ is again an arc and that $f$ is the identity outside of regular neighborhoods of these arcs. If, on the other hand, a homotopy equivalence $f: F_1 \to F_2$ is boundary preserving, i.e. not exotic, it can be deformed into a homeomorphism (Nielsen's theorem). Furthermore, one knows that there is a finite set of Dehn twists which generate a normal subgroup of index two in the whole mapping class group of a surface (orientable or not) (see [De 1] [Li 1, 2, 4].) Here a Dehn twist is a homeomorphism which is the identity outside of a regular neighborhood of a closed curve. Hence, altogether, this shows that the homotopy equivalences of surfaces are built up of locally defined maps (neglecting orientation-phenomena). However, one does not know in general the relations between homotopy equivalences (but see the recent work of Hatcher and Thurston on the mapping class group).

In order to switch to dimension three, take the product of a surface with the 1-sphere. In this way we get our first examples of exotic homotopy equivalences of 3-manifolds. Indeed these homotopy equivalences are generated by Dehn flips along annuli. Similarly, $S^1$-bundles over a surface with boundary contain a lot of exotic homotopy equivalences. Still more can be found in Seifert fibre spaces with boundaries. Here the exceptional fibres give rise to additional phenomena. Although every homotopy equivalence of a sufficiently large Seifert fibre space can be deformed into a fibre preserving map which maps exceptional fibres to exceptional fibres (see 28.4), the restriction to the complement of the exceptional fibres is in general <u>not</u> a homotopy equivalence. Hence exotic homotopy equivalences of Seifert fibre spaces might be rather complicated, and they are, in fact, not yet completely understood.

The examples considered so far are found in very special 3-manifolds and one might expect in other 3-manifolds still more freedom for constructing exotic homotopy equivalences. In contrast to that, it is very difficult to find them. Hence one turns hindsight into foresight and conjectures that the above examples are the only ones (up to modifications). This idea turns out to be correct and therefrom it emerges what we call the <u>characteristic submanifold</u>. Since the concept of the characteristic submanifold plays a crucial role throughout the whole book we present here an explicit definition of it--at least in the absolute case:

<u>Definition</u>. Let $M$ be a 3-manifold (with or without boundary). A codim zero submanifold $V$ of $M$ is called a <u>characteristic submanifold</u> if the following holds:

1. Each component $X$ of $V$ admits a structure as Seifert fibre space, with fibre projection, $p: X \to B$, such that

$$X \cap \partial M = p^{-1}p(X \cap \partial M),$$

or as I-bundle, with fibre projection, $p\colon X \to B$, such that

$$X \cap \partial M = (\partial X - p^{-1}\partial B)^{-}.$$

2. If $W$ is a non-empty codim zero submanifold of $M$ which consists of components of $(M - V)^{-}$, then $V \cup W$ is not a submanifold satisfying 1.
3. If $W'$ is a submanifold of $M$ satisfying 1 and 2, then $W'$ can be deformed into $V$, by using a proper isotopy.

A codim zero submanifold $W$ of $M$ is called an <u>essential</u> <u>F-manifold</u> ($F$ = fibered) if 1 holds and if every component of $(\partial W - \partial M)^{-}$ is incompressible (a surface $G$ in $M$, $G \cap \partial M = \partial G$, is called incompressible if it is not a 2-sphere and $\pi_1 G \to \pi_1 M$ is injective).

Some work is required to show that the characteristic submanifold for sufficiently large 3-manifolds (in the sense of [Wa 4], e.g. if $\partial M \neq \emptyset$) indeed exists (i.e. that it is well-defined--of course we do not assert that it is always non-empty) and that it is unique, up to ambient isotopy.

<u>Example</u>. The characteristic submanifold of a Seifert fibre space $M$ is equal to $M$.

It turns out that the characteristic submanifold is a very useful geometric structure in $M$. In particular, our classification of exotic homotopy equivalences can be given within this concept. Indeed, we shall prove the following (see 24.2).

<u>Classification theorem</u>. Let $M_1$, $M_2$ be 3-manifold (irreducible etc.) with non-empty boundaries, and $V_1$, $V_2$ resp. their characteristic submanifolds. Let $f\colon M_1 \to M_2$ be any homotopy equivalence. Then $f$ can be deformed so that afterwards

1. $f(V_1) \subset V_2$ and $f(\overline{M_1 - V_1}) \subset \overline{M_2 - V_2}$,

2. $f|V_1 \colon V_1 \to V_2$ is a homotopy equivalence,

3. $f|\overline{M_1 - V_1} \colon \overline{M_1 - V_1} \to \overline{M_2 - V_2}$ is a homeomorphism.

The proof of this theorem takes up a large part of these notes. An outline of the proof will be given below. But first we would like to mention some of its consequences, and also we will describe some results, obtained in the course of its proof, which are of interest in their own right.

To begin, recall from [Wa 4] that a homotopy equivalence between 3-manifolds which is a homeomorphism on the boundaries can be deformed into a homeomorphism by using a homotopy which is constant on the boundary. That means--in our context--that in the classification theorem above, $V_i$, $i = 1,2$, can be replaced by the submanifold $W_i$ which consists of all those components of $V_i$ which do meet $\partial M_i$. This in turn implies

<u>Corollary</u>. If there are no essential annuli, there are no exotic homotopy equivalences.

Here an essential annulus $A$ in $M$, $A \cap \partial M = \partial A$, means an annulus which is incompressible and not boundary-parallel.

To consider a concrete example, let $M$ be the <u>knot space</u> of a non-trivial knot. If the submanifold $W$ is not trivial (i.e. not a regular neighborhood of $\partial M$) one knows a priori that the knot is either non-prime, or a cable knot, or a torus knot (see 14.8). In any other case, the classification theorem says that the homotopy type of $M$ contains just one 3-manifold, up to homeomorphism.

Recall that, in general, 3-manifolds can be homotopic without being homeomorphic. However, with the help of the classification theorem, we shall see that the homotopy type of a 3-manifold (irreducible, sufficiently large, etc.) contains only finitely many 3-manifolds (§29).

The last remark leads us to the <u>isomorphism problem</u> for the fundamental groups of sufficiently large 3-manifolds. This problem asks for an algorithm for deciding whether or not two such fundamental groups are isomorphic. Since a sufficiently large 3-manifold is a $K(\pi,1)$-space (recall the restrictions in the beginning) one knows that every isomorphism between their fundamental groups is induced by a homotopy equivalence. Using this fact, together with the classification theorem, the above isomorphism problem can be reduced to the homeomorphism problem for sufficiently large 3-manifolds (§29). But the latter problem was completely solved recently [Ha 2] [He 1] [Th 1]. Hence, in particular, the isomorphism problem for knot groups is solved.

Having established the classification theorem we can push the study of homotopy equivalences a bit further still. Two directions of this study are conceivable. One is to describe one given exotic homotopy equivalence more fully, the other is to study the (exotic) homotopy equivalences all at once.

To describe a result towards the first direction, let a homotopy equivalence $f: M_1 \to M_2$ be given. Furthermore let us assume that the 3-manifolds $M_1$, $M_2$ contain no Klein bottles and no essential annuli which separate a solid torus (e.g. no exceptional fibre). Then we find an essential F-manifold $O_f$ of $M_2$ which is unique, up to ambient isotopy, and which has the following properties:

1. $f$ can be deformed such that afterwards

   $f|f^{-1}O_f: f^{-1}O_f \to O_f$ is a homotopy equivalence, and

   $f|(M_1 - f^{-1}O_f)^-: (M_1 - f^{-1}O_f)^- \to (M_2 - O_f)^-$ is a homeomorphism.

2. $O_f$ can be properly isotoped into all essential F-manifolds which satisfy 1.

$O_f$ will be called an <u>obstruction submanifold for</u> $f$, because $f$ is exotic if and only if $O_f \neq \emptyset$. Some work is required to establish the obstruction submanifold for homotopy equivalences of surfaces (see 30.15). After this the forementioned result can be deduced with the help of the classification theorem (see §28).

As a first attack in the second direction we investigate the <u>mapping class group</u> $H(M)$ of sufficiently large 3-manifolds. Our approach to this is the following. First observe that, by the uniqueness of the characteristic submanifold, V, the computation of H(M) can be split into that of $H_{\overline{M-V}}(V)$ (= isotopy classes of homeomorphisms h: V → V which extend to M) and that of $H_V(\overline{M-V})$ (= isotopy classes of homeomorphisms h: $\overline{M-V} \to \overline{M-V}$ which extend to M). For $H_{\overline{M-V}}(V)$ one can give a fairly explicit computation (see §25), using the recent presentation of the mapping class group of surfaces [HT 1]. Furthermore, we shall prove (§27) that the mapping class group of a simple 3-manifold is finite (and so $H_V(\overline{M-V})$). Here a simple 3-manifold means a sufficiently large 3-manifold whose characteristic submanifold is trivial. To prove this, we use the theory of characteristic submanifolds and Haken's finiteness theorem for surfaces [Ha 1] in order to reduce the problem to the conjugacy problem for the mapping class group of surfaces. The latter was recently solved [He 1][Th 1]. Altogether, we obtain in any sufficiently large 3-manifold a finite set of Dehn twists (along annuli and tori) which generate a normal subgroup in the mapping class group of finite index (see §27, and cf. the 2-dimensional case mentioned in the beginning). As a consequence of this property of the mapping class group we obtain infinitely many examples of surface-homeomorphisms which cannot be extended to any 3-manifold (see 27.10).

Observe that the definition of the characteristic submanifold does not depend on the presence of any homotopy equivalence. In fact, the characteristic submanifold has still other very nice properties besides the above relationship with exotic homotopy equivalences. The most important one is that one can prove a certain enclosing theorem. To describe this we first have to give a definition of an essential singular annulus and torus.

Definition. Let T be an annulus or torus. Then a map f: (T,∂T) → (M,∂M) will be called an essential singular annulus or torus if f induces an injection of the fundamental groups and if it cannot be deformed into ∂M.

Note that by the above definition of the characteristic submanifold, the characteristic submanifold contains all essential (non-singular) annuli or tori of a sufficiently large 3-manifold, up to proper isotopy. In addition to this we shall prove (see §12):

Enclosing theorem. If M is a sufficiently large 3-manifold (with or without boundary), then every essential singular annulus or torus in M can be deformed into the characteristic submanifold of M.

Now recall that the characteristic submanifold consists of I-bundles and Seifert fibre spaces only. Of course, one finds many essential non-singular annuli and tori in such 3-manifolds. Hence, as an immediate consequence of the enclosing theorem, we obtain the "annulus" and "torus theorems" [Wa 6]: the existence of an essential singular annulus implies the existence of an essential non-singular one. The same is true for the torus, except in the very special case of a Seifert fibre space over the 2-sphere with holes, where the sum of boundary components and exceptional fibres is at most three. Furthermore, essential singular annuli and tori can be fairly explicitely classified in I-bundles and Seifert fibre spaces, and so it follows from the enclosing theorem that any such map can be deformed into the composition of a covering map and an immersion without triple points.

Working in a suitable relative framework and using the notion of an essential map, we also obtain a version of the enclosing theorem for essential maps of I-bundles and Seifert fibre spaces (see §13). As an immediate corollary we get: if M has a finite covering which is a Seifert fibre space, then M must be a Seifert-fibre space itself (see 12.11).

In an appendix, we finally apply the enclosing theorem to questions about the fundamental groups of sufficiently large 3-manifolds. There we give a geometric characterization of sufficiently large 3-manifolds whose fundamental group is an R-group (an R-group is a group, where $x^n = y^n$ implies $x = y$), and we apply this to give another proof of Shalen's result [Sh 1] that no element of a 3-manifold group is infinitely divisible. An easy consequence of this and the enclosing theorem is that the centralizer of any element of a sufficiently large 3-manifold group is always carried by an embedded

Seifert fibre space (or a 2-sheeted covering of such a submanifold).

We now give a more detailed description of parts I - IV of this paper in the form of a Leitfaden.

## Part I: The concepts of characteristic submanifolds and manifolds with boundary-patterns.

Part 1 consists of three chapters.

Chapter I. As a first motivation (others will become obvious later) for this chapter the reader should keep in mind the following observation. On the one hand, the "ball" and the "cube" are topologically the same, but on the other hand, the notion of a "cube" does involve much more information. For example, we may distinguish corners, edges, and faces of the cube. That is, we have more information about the boundary. Since we would like to obtain results about manifolds whose boundaries are non-empty, it would be wise to preserve as much information about the boundary as possible. From the effort to present this information in a more organized way the concept of "manifolds with boundary-patterns" emerges.

A boundary-pattern of an n-manifold $M$, $n \geq 1$, is a collection of (n-1)-manifolds in the boundary of $M$ which meet nicely (see Def. 1.1). An admissible map is a map which preserves this structure (see Def. 1.2)--the same with admissible deformations and admissible homotopy equivalences.

It appears that the ensueing formalism is of some interest in its own right, e.g. there is a relative version of the loop-theorem for manifolds with boundary-patterns (see 2.1), and this is equivalent (or at least implies) the main technical result of[Wa 5]. Furthermore, we introduce the notions of "useful boundary-patterns" and "essential maps" (see Def. 2.2 and Def. 3.1). These notions are relativized versions of "boundary irreducible" and "maps which induce an injection of the fundamental groups". A first advantage of these new notions is immediate from their definitions: While, in general, 3-manifolds do not stay boundary-irreducible after splitting at incompressible surfaces, this is true for 3-manifolds with useful boundary-patterns after splitting at essential surfaces.

After having reproduced the proof of Waldhausen's theorem [Wa 4] for 3-manifolds with non-empty boundaries (respectively with boundary-patterns) (3.4) we finally conclude chapter I by establishing some general position theorems (see e.g. 4.4 and 4.5).

Note. Throughout the whole book we have to work entirely within the framework of manifolds with boundary-patterns. However, in this introduction we mostly ignore the boundary-patterns, for convenience. Here the reader should keep in mind that "deformation", "homotopy equivalence" etc. always mean "admissible deformation", "admissible homotopy equivalence" etc.

Chapter II. In this chapter we study singular essential annuli and tori in I-bundles, Seifert fibre spaces (§5), generalized Stallings fibrations (§6) and generalized Seifert fibre spaces (§7). In particular, the proof of the annulus- and torus-theorem for Stallings fibrations is contained here (§6). The 3-manifolds considered in this chapter are fairly special. But on the other hand, information about them is very important for us, because many questions on 3-manifolds can be reduced, via characteristic submanifolds, to questions on these special 3-manifolds.

Chapter III. Here the concept of characteristic submanifolds will be developed. Several definitions of characteristic submanifolds will be given--the most convenient one was already mentioned in the beginning (at least in the absolute case). Using the finiteness theorem for surfaces [Kn 1], [Ha 3], we prove the existence of a characteristic submanifold for sufficiently large 3-manifolds with useful boundary-patterns (§9). Later on (§10) we prove some useful facts about these submanifolds (including the equivalence of all their definitions (10.1)), and we end up with the proof of the uniqueness of characteristic submanifolds, up to ambient isotopy.

Part II: The enclosing theorems.

Part II consists of two chapters.

Chapter IV. In this chapter the first enclosing theorem will be proved, asserting that every essential singular torus, annulus, or square in a sufficiently large 3-manifold $M$ with useful boundary-pattern can be admissibly deformed into the characteristic submanifold of $M$.

Here we give a short indication of the proof. It uses induction on a hierarchy. Recall from [Wa 4] that a hierarchy is a sequence $M = M_0, M_1, \ldots, M_n$ of 3-manifolds, where $M_i$ is given by splitting $M_{i-1}$ along an incompressible surface. The boundary-pattern of each $M_i$ is the "trace" of the previous splittings.

Since the first step of this induction is trivial, we turn at once to the inductive step from $M_{i+1}$ to $M_i$. For this denote by $S$ the surface which splits $M_i$ to $M_{i+1}$, and identify a regular neighborhood of $S$ with $S \times I$.

Let $f_i$ be any essential singular torus, annulus, or square in $M_i$. Then with the help of chapter I, we may assume that $f_i$ is deformed so that firstly $f_{i+1} = f_i | f_i^{-1} M_{i+1}$ consists also of such singular surfaces in $M_{i+1}$, and so, by our induction assumption, that secondly $f_{i+1}$ is contained in the characteristic submanifold $V_{i+1}$ of $M_{i+1}$. Observe that, by the very definition, every essential F-manifold in $M_i$ can be isotoped into the characteristic submanifold $V_i$ of $M_i$. Hence it suffices to prove that $f_i$ can be deformed into an essential F-manifold. This is comparatively easy if $S$ is either a torus, annulus, or square. So let us assume that $S$ cannot be chosen to be one of these surfaces. Then, in general, the surfaces $V_{i+1} \cap (S \times 0)$ and $V_{i+1} \cap (S \times 1)$ do not correspond via $S \times I$.

In this situation we use a combing process. Similar sorts of this process will be used also in chapter VIII and in the proof of the finiteness of the mapping class group. Generally speaking, a combing process is an organized way (by means of the characteristic submanifold) to extend results which are true for a 2-dimensional submanifold to the whole 3-manifold. In the case at hand, the corresponding 2-manifold result is the following

Lemma (see §11). Let $F$ be a surface which is not a torus, annulus or square. Let $F_0$, $F_1$ be two essential surfaces in $F$ which are in a "very good position" (this position can always be obtained by an isotopy of $F_0$, see §11). Then every essential singular curve (closed or not) which can be deformed into both $F_0$ and $F_1$ can be deformed into $F_0 \cap F_1$.

By inductive application of this lemma, we find in $V_{i+1}$ an essential F-manifold $W_{i+1}$ with the properties that $f_{i+1}$ can be deformed into $W_{i+1}$ and that, moreover, $W_{i+1} \cap (S \times 0)$ and $W_{i+1} \cap (S \times 1)$ correspond via $S \times I$. Thus the components of $W_{i+1}$ can now be fitted together, across $S \times I$, and the outcome is a submanifold in $M_i$ such that $f_i$ can be deformed into one of its components, X.

If X is an essential I-bundle or Seifert fibre space, we are done. If not, it is a generalized Stallings fibration. Then, applying the results of Chapter II, the existence of f implies the existence of an essential non-singular torus, annulus or square in $M_i$. But this contradicts our choice of S.

Chapter V. As a corollary of the above enclosing theorem on essential singular tori, annuli, and squares, we prove in §13 an enclosing theorem for essential maps of I-bundles and Seifert fibre spaces.

With the help of these enclosing theorems we are in the position to present the classification theorem in the special case of homotopy equivalences $f: M_1 \to M_2$, where $M_2$ is a 3-manifold whose boundary consists of tori. This will be proved in §15: first we observe that $\partial M_1$ necessarily consists of tori as well. Hence $f|\partial M_1$ is a system of essential singular tori in $M_2$. Recalling Waldhausen's theorem [Wa 4] it now follows immediately from the first enclosing theorem that f is homotopic to a homeomorphism, provided the characteristic submanifold of $M_2$ is a regular neighborhood of $\partial M_2$. If it is not, a bit more work is required, using also the second enclosing theorem.

However, in the general case, i.e. if the boundaries are arbitrary, the proof of the classification theorem is much more complicated. The idea is not to consider the restriction $f|\partial M_1$ as suggested by the Waldhausen theorem, but instead to use induction on a great hierarchy--a concept which will be described later. To make this idea work we first have to prove certain splitting theorems.

Part III: The splitting theorems.

Part III consists of two chapters, in which two splitting theorems will be proved. The first of these splitting theorems (see 18.3) says that every homotopy equivalence $f: M_1 \to M_2$ can be deformed so that afterwards $f|V_1: V_1 \to V_2$ and $f|\overline{M_1 - V_1}: \overline{M_1 - V_1} \to \overline{M_2 - V_2}$ are homotopy equivalences. Here $V_i$, $i = 1,2$, denotes the characteristic submanifold of $M_i$. This theorem will be proved in chapter VI (the use of boundary-patterns is here crucial, both in the proof and in the correct formulation of the theorem, see 18.3).

Chapter VI. In §15 and §16, we prove the existence of two homotopies, one which deforms $f$ so that afterwards $f^{-1}V_2$ is an essential F-manifold, and one which deforms $f$ so that afterwards $f(V_1) \subset V_2$. In §17 we see that these homotopies can be chosen independently from each other. Indeed, we find a homotopy $f_t$ such that $f_1^{-1}V_2$ is an essential F-manifold and such that every component of $V_1$ is a component of $f_1^{-1}V_2$. The properties of the characteristic submanifolds then ensure that $V_1$ is necessarily equal to $f_1^{-1}V_2$.

In §18 it will be shown that the characteristic submanifold is very rigid with respect to homotopies. More precisely, we prove that any given homotopy $h_t$ of a 3-manifold $M$ with $h_i^{-1}V = V$, $i = 0,1$, can be deformed (relative the ends) into a homotopy $g_t$ with $g_t^{-1}V = V$, for all $t \in I$. This completes the proof of the first splitting theorem.

Chapter VII. Having established the above splitting theorem we are led to consider the behavior of homotopy equivalences in the complement of the characteristic submanifolds. The proper setting for this problem is to study homotopy equivalences between simple 3-manifolds. This will be done in chapter VII:

A simple 3-manifold is a sufficiently large 3-manifold whose characteristic submanifold is trivial, i.e. either empty or a regular neighborhood (of parts) of the boundary. For example, the complement of a characteristic submanifold consists of simple 3-manifolds, together with product I- or $S^1$-bundles over the square or annulus. In §19 we give some technical results concerning surfaces in simple 3-manifolds. This information is needed for the proof of the <u>second splitting theorem</u>. This theorem (see 21.3) asserts that there <u>exists</u> a non-separating, essential, connected surface $F_2$ in $M_2$, $F_2 \cap \partial M_2 = \partial F_2$, such that any homotopy equivalence $f: M_1 \to M_2$ can be deformed so that afterwards

1. $f^{-1}F_2$ is connected,
2. $\underline{f|U(f^{-1}F_2)}$: $\underline{U(f^{-1}F_2) \to U(F_2)}$ $\underline{\text{and}}$
   $f|M_1-U(f^{-1}F_2)$: $M_1-U(f^{-1}F_2) \to M_2-U(F_2)$ are homotopy equivalences, where $U(F_2)$ is a regular neighborhood of $U(F_2)$

(again the use of boundary-patterns is needed for a correct statement of the theorem, see 21.3). The theorem is definitely wrong for homotopy equivalences between 3-manifolds which are not simple. Counterexamples can easily be constructed for $S^1$-bundles over surfaces with boundaries. To find surfaces in simple 3-manifold along which homotopy equivalences are splittable we introduce, for technical reasons, a complexity for surfaces. With the help of the results of §19 on surfaces with minimal complexity we are able to prove in §20, that every connected and non-separating surface $F_2 \subset M_2$, $F_2 \cap \partial M_2 = \partial F_2$, with minimal complexity has the following property: any homotopy equivalence $f: M_1 \to M_2$ and any homotopy inverse g of f can be deformed so that afterwards

$$f^{-1}F_2 = F_1 \text{ is a connected surface and } g^{-1}F_1 = F_2.$$

To complete the proof of the second splitting theorem we show in §21 that homotopies can always be split along $F_2$ and $f^{-1}F_2$.

<u>Part IV:</u> <u>The conclusion of the proof of the classification theorem.</u>

Chapter VIII. The existence of the two splitting theorems suggests the construction of a great hierarchy. A great hierarchy for a 3-manifold $M$ is a sequence $M = M_0, M_1, \ldots, M_n$ of 3-manifolds such that

> $M_{2i-1} = (M_{2i-2} - V_{2i-2})^-$, $1 \leq i \leq \frac{n+1}{2}$, where $V_{2i-2}$ is the characteristic submanifold of $M_{2i-1}$, and that
>
> $M_{2i} = (M_{2i-1} - U(F_{2i-1}))^-$, $1 \leq i \leq \frac{n-1}{2}$, where $F_{2i-1}$ is a surface in $M_{2i-1}$ which satisfies the conclusion of the second splitting theorem and where $U(F_{2i-1})$ denotes a regular neighborhood of $F_{2i-1}$. Finally, the boundary-patterns of the $M_i$'s are given by the "traces" of the splittings, and $M_n$ consists of simple 3-balls.

Now, let us be given a homotopy equivalence $f: N \to M$. Then, with the help of the two splitting theorems, we get a hierarchy $N = N_0, N_1, \ldots, N_n$, and we find that $f$ induces a sequence of homotopy equivalences $f_i: N_i \to M_i$, $0 \leq i \leq n$.

$f_n: N_n \to M_n$ is an admissible homotopy equivalence between simple 3-balls with boundary-patterns! As a consequence of the Jordan curve theorem we obtain in §22, that such a homotopy equivalence always can be admissibly deformed into a homeomorphism. This establishes the first step of an induction.

Finally, we have to work up the hierarchy again. This refers to the inductive application of the following glueing theorem:

Glueing theorem (see 23.1). If $f_{2i+1}$ is homotopic to a homeomorphism, then also $f_{2i-1}$.

To prove this we use a combing process again. To describe the corresponding 2-manifold result, denote by $F$, $F'$ two surfaces. Let $F_i$ and $F_i'$, $i = 0,1$, be surfaces in $F$ and $F'$, respectively, which are in very good position. Moreover, let $g_0, g_1: F \to F'$ be homotopy equivalences with $g_i^{-1}F_i' = F_i$ and such that $g|F_i: F_i \to F_i'$ is a homeomorphism. Then we will prove

<u>Lemma</u> (see 31.1. for a more precise statement). If $g_0$ and $g_1$ are homotopic, then there is a homotopy $h_t$, $t \in I$, of $g_0$ with

1. $h_t^{-1}F_i' = F_i$, and
2. $h_1|F_0 \cup F_1: F_0 \cup F_1 \to F_0' \cup F_1'$ is a homeomorphism.

The proof of this lemma is rather involved. Since it is a 2-dimensional result we defer the proof to chapter XI of the appendix. This lemma also leads to the existence of <u>obstruction surfaces</u> for homotopy equivalences of surfaces (see §30). In §23, however, we apply the lemma inductively to prove the glueing theorem as follows: Starting with the fact that $f_{2i+1}$ can be deformed into a homeomorphism and using a combing process, we obtain submanifolds $W_{2i-1}$ and $W_{2i-1}'$ in $N_{2i-1}$ and $M_{2i-1}$, respectively, with the following properties:

1. $W_{2i-1}'$ is either empty, or it is equal to $M_{2i-1}$ and a generalized Stallings fibration
2. $f_{2i-1}$ can be deformed so that afterwards $f_{2i-1}|\overline{N_{2i-1} - W_{2i-1}}: \overline{N_{2i-1} - W_{2i-1}} \to \overline{M_{2i-1} - W_{2i-1}'}$ is a homeomorphism.

We are done if $W_{2i-1}'$ is empty. If it is not empty, $\partial M_{2i-1}$ consists of tori, and the glueing theorem follows from the results obtained in chapter V. This completes the outline of the proof of the classification theorem.

The author wants to thank F. Waldhausen who has initiated this work. Indeed, this book can be considered as a continuation of his work on homotopy equivalences in [Wa 4] [Wa 6]. The main results have been announced in [Jo 1], [Jo 5], [Wa 7]. The papers [Jo 2] and [Jo 3], which constitute a first presentation of the main results of the present paper, were widely distributed in early 1976 and 1977 respectively. Results concerning the enclosing theorem and its applications have also been proved by Feustel, Jaco, Shalen, Scott and others. Recently Swarup [Sw 1] gave a different approach to the classification theorem.

The final draft of this manuscript was completed while I held a visiting position at Columbia University. I would like to thank the Mathematics Department of Columbia for its hospitality, and Mrs. Kate March for her very good job of typing.

Universität Bielefeld

and

Columbia University, New York

# Part I. THE CONCEPTS OF CHARACTERISTIC SUBMANIFOLDS AND MANIFOLDS WITH BOUNDARY PATTERNS.

## Chapter I: General theory.

Throughout this book we work in the PL-category (unless otherwise stated, e.g. in §6). More precisely, as indicated in the introduction, we work in the category of PL-manifolds with boundary-patterns and admissible (i.e. pattern-preserving) PL-maps. This concept will turn out to be very convenient for our purposes, especially for induction proofs on hierarchies. In this chapter we translate for later use some of the well-known theorems of 3-manifold-theory into this language.

### §1. Definitions and preliminaries.

1.1 Definition. Let $M$ be a compact n-manifold, $n \geq 1$. A boundary-pattern for $M$ consists of a set $\underline{\underline{m}}$ of compact, connected (n-1)-manifolds in $\partial M$, such that the intersection of any $i$, $i = 1,2,\ldots,n+1$, of them consists of (n-i)-manifolds.

The elements of $\underline{\underline{m}}$ are called the bound sides of $(M,\underline{\underline{m}})$, and $\underline{\underline{m}}$ is complete if $\partial M = \cup_{G\in\underline{\underline{m}}} G$. In general, $\underline{\underline{m}}$ is not complete and the components of $(\partial M - \cup_{G\in\underline{\underline{m}}} G)^-$ are called the free sides. The completed boundary-pattern of $(M,\underline{\underline{m}})$ will be denoted by $\bar{\underline{\underline{m}}}$ and is defined to be the union of the set of all bound sides and the set of all free sides. Hence, by the very definition, $(M,\bar{\underline{\underline{m}}})$ is always a manifold with complete boundary-pattern. Observe that the boundary-pattern of $(M,\underline{\underline{m}})$ induces a boundary-pattern of the sides (bound or free) of $(M,\underline{\underline{m}})$ in the obvious way. For free sides this induced boundary-pattern is complete.

1.2 Definition. Let $(M,\underline{\underline{m}})$ and $(N,\underline{\underline{n}})$ be manifolds with boundary-patterns, not necessarily of the same dimension. An admissible map from $(M,\underline{\underline{m}})$ to $(N,\underline{\underline{n}})$ is a map $f: M \to N$ satisfying

$$\underline{\underline{n}} = \dot{\bigcup}_{G \in \underline{\underline{m}}} \{\text{components of } f^{-1}G\} \quad (\dot{\cup} = \text{disjoint union})$$

As a consequence of this definition, every bound side is mapped into precisely one bound side (e.g. every end-point of an admissible arc in a surface with complete boundary-pattern must lie in the interior of some bound side). Every two neighboring bound sides are mapped into neighboring ones. Furthermore the composition of two admissible maps is of course again an admissible map.

An <u>admissible homotopy</u> is a continuous family of admissible maps. Alternatively, an admissible homotopy can be described as an admissible map $f\colon (M \times I, \underline{\underline{m}} \times I) \to (N, \underline{\underline{n}})$, where $I$ denotes the closed interval $[0,1]$, $\underline{\underline{m}}$ is the boundary-pattern of $M$, and $\underline{\underline{m}} \times I = \{G \times I \mid G \in \underline{\underline{m}}\}$. The terms "admissible isotopy" and "admissible ambient isotopy" are defined analogously. We sometimes say that $g\colon (M\ \underline{\underline{m}}) \to (N, \underline{\underline{n}})$ can be admissibly deformed (or pulled, or moved) or isotoped in $(N, \underline{\underline{n}})$ <u>into</u> $Z$, if there is an admissible homotopy or isotopy $f\colon M \times I \to N$ with $f|M \times 0 = g$ and $f(M \times 1) \subset Z$.

An admissible map $f\colon (M, \underline{\underline{m}}) \to (N, \underline{\underline{n}})$ is an <u>admissible homotopy equivalence</u>, if there is an admissible map $g\colon (N, \underline{\underline{n}}) \to (M, \underline{\underline{m}})$ such that $g \circ f$ and $f \circ g$ are admissibly homotopic to the identities ($g$ is sometimes called an admissible homotopy inverse of $f$). Similarly, we define an <u>admissible homeomorphism</u>.

In this book , a <u>manifold</u> will always mean a compact manifold with boundary-pattern (of course, the empty set is admitted as boundary-pattern). On occasion the boundary-pattern will be left out of the notation, especially when it is clear from the context.

A 3-manifold will mean an orientable 3-manifold (not necessarily connected). 2-manifolds (surfaces) are not generally required to be orientable or connected. However, when we are dealing with maps of 2-manifolds into 3-manifolds, the 2-manifolds will be orientable unless otherwise noted. In particular, a 2-manifold properly embedded in a 3-manifold will be 2-sided. Whenever the notion of an <u>annulus</u> or <u>Möbius band</u> appears, it is to be understood that the boundary-pattern is the collection of its boundary curves. A <u>square</u> is a disc with complete boundary-pattern and precisely four sides.

We will not repeat here the definitions of such terms as irreducible, boundary-irreducible, incompressible, parallel and so on. These can be found in [Wa 4]. In particular, the term "sufficiently large" is used in the sense of [Wa 4]. However, an irreducible 3-manifold which is sufficiently large is sometimes called a Haken 3-manifold. We will also use without further comment the material given in the preliminary section of Waldhausen's paper [Wa 4]. In particular, by Baer's and Nielsen's theorem we mean the theorems mentioned there.

Let $(M,\underline{\underline{m}})$ be a 3-manifold with boundary-pattern. The graph $J = \cup_{G\in\underline{\underline{m}}} \partial G$ is called the graph of $(M,\underline{\underline{m}})$. Note that every point of the graph $J$ must have order three.

Now let $(N,\underline{\underline{n}})$ be either a 2- or a 3-manifold with boundary-pattern. By an admissible singular curve in $(N,\underline{\underline{n}})$ we mean an admissible map $f\colon (k,\underline{\underline{k}}) \to (N,\underline{\underline{n}})$, such that $k$ is $I$ or $S^1$ (where $S^n$ denotes the n-sphere) and $\underline{\underline{k}}$ consists of the end-points of $k$. By an admissible singular surface in $(N,\underline{\underline{n}})$ we mean an admissible map $h\colon (F,\underline{\underline{f}}) \to (N,\underline{\underline{n}})$ such that $(F,\underline{\underline{f}})$ is some 2-manifold with boundary-pattern.

Finally, we mention that we shall often have to consider splitting situations. These can be described as follows: let $(N,\underline{\underline{n}})$ be an admissible submanifold in a manifold $(M,\underline{\underline{m}})$ of codimension 0. Then define

$$\tilde{M} = (M - N)^-, \quad \text{and} \quad \tilde{\underline{\underline{m}}} = \underline{\underline{m}} \cup \{\text{components of } (\partial N - \partial M)^-\}.$$

We call $\tilde{\underline{\underline{m}}}$ the proper boundary-pattern of $\tilde{M}$ induced by $N$. An important case of this situation is when $N$ is a regular neighborhood of a properly embedded codim 1 submanifold $F$ in a 2- or 3-manifold $M$. In this case we sometimes have to use $(\tilde{M},\tilde{\underline{\underline{m}}})$ as defined alternatively by the following properties: $\partial\tilde{M}$ contains (connected) submanifolds $F_1$ and $F_2$ which are copies of $F$, and identification of $F_1$ and $F_2$ gives a natural projection $p\colon (\tilde{M}, F_1 \cup F_2) \to (M,F)$. $\tilde{\underline{\underline{m}}} = \{F_1,F_2\} \cup \{\text{components of } p^{-1}G,\ G \in \underline{\underline{m}}\}$.

In any case we say that $(\tilde{M},\tilde{\underline{\underline{m}}})$ is the manifold obtained from $(M,\underline{\underline{m}})$ by splitting at $N$ (resp. $F$).

## §2. Useful boundary-patterns

Let an *i-faced disc*, $i \geq 1$, denote a disc with complete boundary-pattern and precisely $i$ sides. For $1 \leq i \leq 4$, i-faced discs will be of great technical importance throughout this paper. In particular, a 4-faced disc is the same as a square.

*2.1 Proposition. Let* $(M,\underline{\underline{m}})$ *be a 3-manifold, and* $J$ *its graph. Then the following is equivalent:*

1. *The boundary curve of any admissible i-faced disc,* $1 \leq i \leq 3$, *in* $(M,\underline{\underline{m}})$ *bounds a disc,* $D$, *in* $\partial M$ *such that* $J \cap D$ *is the cone on* $J \cap \partial D$.
2. *For any admissible singular i-faced disc* $f\colon (D,\underline{\underline{d}}) \to (M,\underline{\underline{m}})$, $1 \leq i \leq 3$, *there exists a map* $g\colon D \to M$ *such that*
   (a) $g(D) \subset \partial M$ *and* $g|\partial D = f|\partial D$, *and*
   (b) $g^{-1}J$ *is the cone on* $g^{-1}J \cap \partial D$.

*Remark.* This proposition is very closely related to the main technical result of [Wa 5]. Indeed, using the proposition, one can simplify somewhat the argument leading to the algorithm of [Wa 5].

*2.2 Definition. Let* $(M,\underline{\underline{m}})$ *be a 3-manifold. If* $\underline{\underline{m}}$ *satisfies* 1 *or* 2 *of* 2.1, *then* $\underline{\underline{m}}$ *is called a "useful boundary-pattern of* $M$".

*Proof of 2.1.*

1 *implies* 2.

Let $f$ be given as in 2 of 2.1. $f|\partial D$ is in general position with respect to the graph $J$, and, without loss of generality, $f|\partial D$ is in general position itself. In particular, a regular neighborhood $U(f\partial D)$ in $\partial M$ is not a disc. For later use fix also a regular neighborhood, $U(J)$, of $J$ in $\partial M$. It remains to construct a map $g\colon D \to M$ with (a) and (b) of 2.1.2. This is easy if $f(\partial D)$ is entirely contained in a disc which lies in $\partial M$, and so we suppose the converse. Then the following can be deduced which is the key to our construction.

2.3 Assertion. There is a map $\hat{g}: D \to \partial M$ with the following properties:

1. $\hat{g}|\partial D = f|\partial D$ and $\hat{g}|\hat{g}^{-1}U(J)$ is an immersion, i.e. locally a homeomorphism.
2. Every knot point of the graph $\hat{g}^{-1}J$ is mapped under $\hat{g}$ to a component of $(\partial M - U(f\partial D))^{-}$ which itself is a disc.

To prove this result recall from 1 of 2.1 that the boundary of every admissible i-faced disc, $1 \leq i \leq 3$, in $(M,\underline{\underline{m}})$ is contractible in $\partial M$. Hence it follows from the proof of the loop-theorem [St 1] that $f|\partial D$ is contractible in $\partial M$. Let $h: D \to \partial M$ be any contraction of $f|\partial D$. After a small general position deformation of $h$ (rel $\partial D$) we have that $S = h^{-1}\partial U(h\partial D) = h^{-1}\partial U(f\partial D)$ is a system of simple closed curves. This system splits $D$ into connected surfaces. Let $F$ be that one of them which contains the boundary of $D$. Now we suppose that $h$ has been chosen so that the number of components of $(D - F)^{-}$ is as small as possible.

Let $D_1$ be a component of $(D - F)^{-}$, and let $k_1$ be the component of $\partial U(h\partial D)$ which contains $h(\partial D_1)$. Then, by our minimality condition on $(D - F)^{-}$, the restriction $h|\partial D_1$ cannot be contractible in $k_1$. This implies that $k_1$ bounds a disc, $H_1$, in $\partial M$ [Ep 2]. Moreover, it follows from our suppositions on $f|\partial D$ that this disc must be a component of $(\partial M - U(f\partial D))^{-}$. Since, without loss of generality, $h|\partial D_1: \partial D_1 \to \partial H_1$ is a covering map, we easily find (using a cone construction) an extension of $h|\partial D_1$ to a map $h_1: D_1 \to H_1$ such that $h_1|h^{-1}U(J)$ is an immersion.

Replacing all $h|D_1$ by $h_1$, we get a map $\hat{g}: D \to \partial M$ from $h$ so that $\hat{g}|F = h|F$, $\hat{g}(D - F)^{-} \subset (\partial M - U(h\partial D))^{-}$, and $\hat{g}|(D-F)^{-} \cap \hat{g}^{-1}U(J)$ is an immersion. Furthermore, $\hat{g}|(\partial F - \partial D)^{-}: (\partial F - \partial D)^{-} \to \partial U(h\partial D)$ is an immersion.

Now we consider $\hat{g}|F: F \to U(h\partial D)$. $J \cap U(h\partial D)$ consists of arcs and $\hat{g}|\partial F$ is transverse with respect to $J$. Hence, by the transversality lemma [Wa 3, p. 60], $\hat{g}$ can be deformed, by a homotopy which is constant on $D - F^{\circ}$, so that afterwards $F \cap \hat{g}^{-1}J$ is a system of simple arcs and simple closed curves which are non-contractible in F. Then, by our minimality condition on $(D - F)^-$, $F \cap \hat{g}^{-1}J$ has to be a system of simple arcs. Let $b_1$ be one of them. $b_1$ is mapped under $\hat{g}$ into an arc $b_1'$ of $J \cap U(h\partial D)$.

Assume the end-points of $b_1$ are mapped under $\hat{g}$ to one point. Then $\partial b_1$ must lie in $(\partial F - \partial D)^-$. Since $U(h\partial D)$ is not a disc, no boundary curve of $U(h\partial D)$ can be contracted in $U(h\partial D)$, and so it follows that $b_1$ cannot be boundary-parallel in F. Now either $\partial b_1$ lies in one component of $(\partial F - \partial D)^-$, or $b_1$ joins two different components of $(\partial F - \partial D)^-$. In either case, contracting $\hat{g}|b_1: b_1 \to b_1'$ into a point (rel $\partial b_1$) and extending this deformation to a homotopy of $\hat{g}$ (rel $\partial D$), the number of components of $(D - F)^-$ can be diminished (surgery). This is a contradiction.

Thus $b_1$ is an arc whose end-points are mapped under $\hat{g}$ to different points of $b_1'$. This means that $g|b_1: b_1 \to b_1'$ can be deformed (rel $\partial b_1$) into an embedding. Hence we may suppose that $\hat{g}|F: F \to U(h\partial D)$ is deformed (rel $\partial F$) so that $\hat{g}|F \cap \hat{g}^{-1}U(J): F \cap \hat{g}^{-1}U(J) \to U(J)$ is an immersion. Since $\hat{g}(F) \subset U(h\partial D)$ and $\hat{g}(D - F)^- \subset (\partial M - U(h\partial D))^-$, we conclude that $\hat{g}$ is a map which satisfies 1 of 2.3. That it also satisfies 2 of 2.3 follows directly from its construction. This proves the assertion.

To continue the proof, fix a map $\hat{g}: D \to \partial M$ as described in 2.3. Assume that the graph $\hat{g}^{-1}J$ has strictly more than one knot point. Since $\hat{g}|\hat{g}^{-1}U(J)$ is an immersion and since $\partial D \cap \hat{g}^{-1}J$ consists of at most three points, it follows the existence of a component $J_1$ of $\hat{g}^{-1}J$ with at least two knot points mapped under $\hat{g}$ to two different knot points $x_1', x_2'$ of J. By our choice of $\hat{g}$, the points $x_1'$, $x_2'$ lie in components $H_1$, resp. $H_2$, of $(\partial M - U(f\partial D))^-$ which are discs. $J \cap \partial H_1$ consists of at most three points. To see this recall that $J \cap U(f\partial D)$ consists of at most three arcs and that the intersection number of $f|\partial D$ with any simple closed curve in $\partial M$ has to be zero, for $f|\partial D$ is contractible in $\partial M$. By 1 of 2.1, $J \cap H_1$ is the cone on $J \cap \partial H_1$. In particular, $H_1 \neq H_2$. Moreover, it follows that $x_1'$ and $x_2'$ both lie in a component, $J_1'$, of J whose only knot points are $x_1'$, $x_2'$ and which has three edges, all joining $x_1'$ with $x_2'$. Now let

$U_1$ be the component of $\hat{g}^{-1}U(J)$ which contains $J_1$. By our choice of $J_1$, $U_1$ must have a boundary curve which does not meet $\partial D$. Since $\hat{g}|\hat{g}^{-1}U(J)$ is an immersion, this boundary curve is mapped under $\hat{g}$ to a contractible boundary curve k' of $U(J_1')$. This means that k' is the boundary of an admissible 1-faced disc in $(M,\underline{\underline{m}})$ (loop-theorem). Hence $\partial M$ must be a 2-sphere. To see this fix also an appropriate admissible 2-faced disc in $(M,\underline{\underline{m}})$ whose boundary lies near k', and apply 1 of 2.1 twice. So $f|\partial D$ is contained entirely in a disc which lies in $\partial M$, but this was excluded. This shows that $\hat{g}^{-1}J$ has at most one knot point.

By what we have seen so far, $\hat{g}^{-1}J$ must be the disjoint union of a cone on $\hat{g}^{-1}J \cap \partial D$ and components of $\hat{g}^{-1}J$ which do not meet $\partial D$. We still have to excise the latter. For this let $D_1$ be any disc in the interior of D whose boundary lies on $\partial\hat{g}^{-1}U(J)$. Since $\hat{g}|\hat{g}^{-1}U(J)$ is an immersion, $\hat{g}|\partial D_1$ is mapped onto a contractible boundary curve of $\partial U(J)$ which itself bounds a disc $H_1$ in $\partial M$ which does not meet J (see 1 of 2.1). Hence we may replace $\hat{g}|D_1$ by any map of $D_1$ into $H_1$ whose restriction to $\partial D_1$ is equal to $\hat{g}|\partial D_1$. After finitely many such steps we finally obtain from $\hat{g}$ the required map $g: D \to M$ with (a) and (b) of 2.1.2.

<u>2 implies 1</u>.

Let $(D,\underline{\underline{d}})$ be an admissible i-faced disc, $1 \leq i \leq 3$, in $(M,\underline{\underline{m}})$. Then, by 2 of 2.1, there is a map $g: D \to M$ with $g|\partial D = id|\partial D$, $g(D) \subset \partial M$ and $g^{-1}J$ is the cone on $g^{-1}J \cap \partial D$.

Now suppose first that $(D,\underline{\underline{d}})$ is a 3-faced disc. Let $k_1$, $k_2$ be two edges of the graph $g^{-1}J$. Then $k_1 \cup k_2$ splits D into two discs and let $D_1$ be that one of them which does not contain the third edge of $g^{-1}J$. Let $s_1 = D_1 \cap \partial D$ and denote by $G_1$ the side of $(M,\underline{\underline{m}})$ containing $s_1$. Observe that g maps $k_1$ and $k_2$ into edges of J, different from closed curves, and that J splits $\partial D$ into arcs. Hence, applying an excision procedure if necessary (see the first part of this proof), we may suppose that $D_1 \cap g^{-1}\partial D$ is a system of arcs joining $k_1$ with $k_2$. This system splits $D_1$ into discs. Let D* be that one of them whose intersection with $k_1 \cup k_2$ is connected (possibly $D_1 = D^*$). Now recall that the disc is the only surface whose boundary curves are contractible. Hence the existence of the map $g|D^*$ shows that the arc $s_1$ is parallel in $G_1$ to an arc, t, of J which contains precisely one knot point of J. By our suppositions on D, we find an admissible 2-faced disc in $(M,\underline{\underline{m}})$ whose boundary lies near $(\partial D - s_1) \cup t$. Putting the last two facts together, we find that it remains to consider admissible 1- or 2-faced discs. In fact, using the same argument, we see that we may restrict ourselves to the case that $(D,\underline{\underline{d}})$ is a 1-faced disc. Then the existence of a disc, D', in $\partial M$ with $\partial D' = \partial D$ and $D' \cap J = \emptyset$ follows from the existence of g and the above characterization of discs. q.e.d.

## §3. Essential maps

Let $(X,\underline{\underline{x}})$ be a 2- or 3-manifold. An admissible singular curve $h: (k,\underline{\underline{k}}) \to (X,\underline{\underline{x}})$ is called inessential, if it can be admissibly deformed near a point.

To be more precise, $h$ is *inessential* if there is a disc $D$, with boundary-pattern $\underline{\underline{d}}$, and an admissible map $g: (D,\underline{\underline{d}}) \to (X,\underline{\underline{x}})$ such that

1. $k = (\partial D - \bigcup_{t \in \underline{\underline{d}}} t)^-$,
2. $(D,\bar{\underline{\underline{d}}})$ is an i-faced disc, $1 \leq i \leq 3$ (recall, $\bar{\underline{\underline{d}}}$ denotes the complete boundary-pattern associated to $(D,\underline{\underline{d}})$),
3. $g|k = h$.

*3.1 Definition. Let $(X,\underline{\underline{x}})$, $(Y,\underline{\underline{y}})$ be 2- or 3-manifolds (not necessarily of the same dimension). An admissible map $f: (X,\underline{\underline{x}}) \to (Y,\underline{\underline{y}})$ is called "essential" if for any essential curve $h: (k,\underline{\underline{k}}) \to (X,\underline{\underline{x}})$ the composed map $f \cdot h: (k,\underline{\underline{k}}) \to (Y,\underline{\underline{y}})$ is also an essential singular curve.*

*Remark.* The composition of essential maps is certainly again an essential map, also, if $X$, $Y$ are connected, an essential map $f: (X,\underline{\underline{x}}) \to (Y,\underline{\underline{y}})$ induces a monomorphism of the fundamental groups.

*3.2 Examples.* Let $(X,\underline{\underline{x}})$, $(Y,\underline{\underline{y}})$ be 2- or 3-manifolds and $f: (X,\underline{\underline{x}}) \to (Y,\underline{\underline{y}})$ be either an admissible covering map or an admissible homotopy equivalence. Then $f$ is an essential map.

The following two propositions are translations of well-known theorems of Nielsen and Waldhausen (see [Wa 4]) into our language.

*3.3 Proposition. Let $(F,\underline{\underline{f}})$, $(G,\underline{\underline{g}})$ be connected surfaces with complete boundary-patterns. Suppose that $(G,\underline{\underline{g}})$ is neither the disc with exactly one side nor the 2-sphere, and that $F$ is not the projective plane.*

*Then any essential map $f: (G,\underline{\underline{g}}) \to (F,\underline{\underline{f}})$ can be admissibly deformed* *into a covering map.*

*If* $f|\partial G$ *is locally homeomorphic, the homotopy may be chosen constant on* $\partial G$.

*3.4 Proposition.* *Let* $(M,\underline{\underline{m}})$, $(N,\underline{\underline{n}})$ *be connected, irreducible* 3-*manifolds with useful and complete boundary-patterns. Suppose that* $N$ *has non-empty boundary and* $(N,\underline{\underline{n}})$ *is not a ball with one or two sides.*

*Then any essential map* $f\colon (N,\underline{\underline{n}}) \to (M,\underline{\underline{m}})$ *can be admissibly deformed into a covering map*

*If* $f|\partial N$ *is locally homeomorphic, the homotopy may be chosen constant on* $\partial N$.

*Remark.* For closed and sufficiently large 3-manifolds the reader is referred to [Wa 4].

We only give the proof of 3.4 (following [Wa 4]), for the proof of 3.3 is similar.

*Proof of 3.4.* Let $(R,\underline{\underline{r}})$ be any surface of $\underline{\underline{n}}$; it is a surface with *complete* boundary-pattern. Since $f$ is an admissible map, there is a surface, $(S,\underline{\underline{s}})$ of $\underline{\underline{m}}$ such that $f|R\colon (R,\underline{\underline{r}}) \to (S,\underline{\underline{s}})$ is admissible. By our supposition on $(N,\underline{\underline{n}})$, $R$ cannot be the 2-sphere. Moreover, $(R,\underline{\underline{r}})$ cannot be the disc with exactly one side. Otherwise we choose an admissible 1-faced disc in $(N,\underline{\underline{n}})$ near $R$, and, since $\underline{\underline{n}}$ is a useful boundary-pattern of $N$, we find that $(N,\underline{\underline{n}})$ must be a ball with exactly two sides. But this case is excluded. Now let $k$ be any admissible singular curve in $(R,\underline{\underline{r}})$ such that $f \circ k$ is inessential in $(S,\underline{\underline{s}})$. $k$ is not admissible in $(N,\underline{\underline{n}})$, and therefore, by a small deformation, we push $k$ out of $R$ and to an admissible singular curve, $k'$, in $(N,\underline{\underline{n}})$. $f \circ k'$ is inessential in $(M,\underline{\underline{m}})$, and, since $f$ is an essential map, this implies that $k'$ is inessential in $(N,\underline{\underline{n}})$. Therefore $k$ is a side of an admissible singular $i$-faced disc, $1 \leq i \leq 3$, in $(N,\underline{\underline{n}})$. Since $\underline{\underline{n}}$ is a useful boundary-pattern of $N$, we conclude that $k$ is inessential in $(R,\underline{\underline{r}})$. Thus $f|R\colon (R,\underline{\underline{r}}) \to (S,\underline{\underline{s}})$

is essential, and so, by 3.3, $f|R$ can be admissibly deformed into a covering map. Therefore $f$ can be admissibly deformed so that $f|\partial N$ is locally homeomorphic (note that $f$ maps adjacent sides of $(N,\underline{\underline{n}})$ into adjacent sides of $(M,\underline{\underline{m}})$).

The remainder of the proof follows rather closely the proof of 6.1 of [Wa 4]; indeed, that argument seems to be tailor made to apply in the present context.

Let $\{F_i\}$, $1 \leq i \leq n$ be a hierarchy for $M$ in the sense of [Wa 4]. This exists since the boundary of $M$ is non-empty cf. [Wa 3]. Let $\{(M_i,\underline{\underline{m}}_i)\}$, $1 \leq i \leq n$, be the sequence of 3-manifolds associated with the hierarchy, with $(M_1,\underline{\underline{m}}_1) = (M,\underline{\underline{m}})$ and $F_i \subset M_i$. Here $(M_{i+1},\underline{\underline{m}}_{i+1})$ denotes that manifold which we obtain from $(M_i,\underline{\underline{m}}_i)$ by splitting at $F_i$ ($F_i$ can clearly be chosen as an admissible surface). Let $f_1 = f$ and suppose $f_i\colon (N_i,\underline{\underline{n}}_i) \to (M_i,\underline{\underline{m}}_i)$ is already defined. Then applying the transversality lemmas [Wa 3, p. 60], $f_i$ can be admissibly deformed so that $f_i$ is transversal and $f_i^{-1}F_i$ is an admissible incompressible surface in $(N_i,\underline{\underline{n}}_i)$. Moreover, if $f_i|\partial N_i$ is locally homeomorphic, this deformation may be chosen constant on $\partial N_i$. Then define $(N_{i+1},\underline{\underline{n}}_{i+1})$ to be that manifold which we obtain from $(N_i,\underline{\underline{n}}_i)$ by splitting at $f_i^{-1}F_i$, and also define $f_{i+1}$ as the restriction $f_i|N_{i+1}$. We prove easily, $\ker(f_{i+1})_* = 0$ if $\ker(f_i)_* = 0$.

We suppose $f_r$ is admissibly deformed so that $f_r|\partial N_r\colon \partial N_r \to \partial M_r$ is locally homeomorphic. As proved above, this holds for $r = 1$. We assert that $f_{r+1}\colon (N_{r+1},\underline{\underline{n}}_{r+1}) \to (M_{r+1},\underline{\underline{m}}_{r+1})$ can be admissibly deformed so that $f_{r+1}|\partial N_{r+1}$ is locally homeomorphic. Let $G$ be a component of $f_r^{-1}F_r$. Then we have to show that $f|G\colon G \to F_r$ can be deformed (rel $\partial G$) into a covering map. For this we want to apply 3.3.

First $G$ is not a 2-sphere since $N$ is irreducible and $G$ is incompressible. Assume $G$ is a 1-faced disc in $(N_r,\underline{\underline{n}}_r)$. Then it follows that also $F_r$ is a 1-faced disc in $(M_r,\underline{\underline{m}}_r)$ since $f|\partial N_r$ is locally homeomorphic. $\partial F_r$ either lies in a surface of $\underline{\underline{m}}$ or in a copy of a component of $F_p$, for some $1 \leq p \leq r-1$. Hence, in either case, there is an integer $q$, $1 \leq q \leq r-1$, such that $\partial F_r$ lies in a surface, $B$, in $\underline{\underline{m}}_q$ and bounds a disc in $B$. Hence $F_r$ is boundary-parallel in $M_q$ since $M_q$ is irreducible, and so boundary-parallel in $M_r$, which contradicts the fact that $F_r$ belongs to the hierarchy. Thus $G$ is neither a 2-sphere nor a 1-faced disc in $(N_r, \underline{\underline{n}}_r)$.

In order to apply 3.3 we still have to show that $f_r|G: G \to F$ is essential. For this let $t$ be any essential singular curve in $G$ and assume $f \circ t$ is inessential in $F_r$. Since $G$ is incompressible and since $\ker(f_r)_* = 0$, $t$ must be a singular arc. We may suppose that $t$ is chosen so that $f(\partial t)$ is one point. Combining $t$, if necessary, with two suitable arcs (e.g. obtained by lifting an arc which joins $f(\partial t)$ inside $(M - M_r)^-$ to $\partial M$) we find an admissible singular arc, $k$, in $(N\ \underline{\underline{n}})$ such that $k$ joins two different points $p_1, p_2 \in \partial N$ and such that $f \circ k$ is contractible in $M$ (in particular $f(p_1) = f(p_2)$). That means that $f \circ k$ is inessential in $(M, \underline{\underline{m}})$. Therefore $k$ is inessential in $(N, \underline{\underline{n}})$ since $f$ is essential. Let $R_i$ be that surface of $\underline{\underline{n}}$ which contains $p_i$, $i = 1,2$, and $S$ that surface of $\underline{\underline{m}}$ which contains $f(p_i)$. Then, in particular, $R_1 = R_2$ and $k$ is homotopic (rel $\partial k$) to a singular arc, $\tilde{k}$, in $R_1$. $f \circ \tilde{k}$ defines a loop in $S$, which is not contained in the subgroup $(f|R_1)_*(\pi_1 R_1)$ since $\tilde{k}$ is a singular arc. On the other hand, $f \circ \tilde{k}$ is homotopic (rel $\partial\tilde{k}$) in $M$ to the loop $f \circ k$ which is contractible. Since $\underline{\underline{m}}$ is a useful boundary-pattern of $M$, it follows that $f \circ \tilde{k}$ is contained in every subgroup of $\pi_1 S$. Hence we have the contradiction, and so $f|G: G \to F_r$ must be essential.

Thus, by 3.3, $f|G$ can be deformed (rel $\partial G$) into a covering map. Therefore $f_{r+1}$ can be admissibly deformed so that $f_{r+1}|\partial N_{r+1}$ is locally homeomorphic.

Inductively, $f_{n+1}|\partial N_{n+1}$ is locally homeomorphic. $M_{n+1}$ is a system of balls. Since every covering map onto the 2-sphere is a homeomorphism, the restriction of $f_{n+1}$ to any component of $N_{n+1}$ can be deformed (rel $\partial N_{n+1}$) into a homeomorphism. q.e.d.

Applying 3.2 and 3.4 we obtain

3.5 Corollary. Let $(N, \underline{\underline{n}})$, $(M, \underline{\underline{m}})$ be as in 3.4. Then any admissible homotopy equivalence $f: (N, \underline{\underline{n}}) \to (M, \underline{\underline{m}})$ can be admissibly deformed into a homeomorphism.

Recall from the introduction that 3.5 is no longer true if the manifolds are allowed to have free sides. For this case the reader is referred to 24.2.

We end this paragraph with the observation that essential maps can be "split" along surfaces. Later on we shall consider still other splitting-situations (see Part III).

3.6 Corollary. *Let* $(N,\underline{\underline{n}})$, $(M,\underline{\underline{m}})$ *be as in* 3.4. *Let* $F$ *be an admissible surface in* $(M,\underline{\underline{m}})$, *and* $f\colon (N,\underline{\underline{n}}) \to (M,\underline{\underline{m}})$ *be an essential map.*

*Then* $f$ *can be admissibly deformed so that* $f^{-1}F$ *is an admissible surface in* $(N,\underline{\underline{n}})$ *and that* $f|\tilde{N}\colon (\tilde{N},\tilde{\underline{\underline{n}}}) \to (\tilde{M},\tilde{\underline{\underline{m}}})$ *is an essential map, where* $(\tilde{N},\tilde{\underline{\underline{n}}})$, $(\tilde{M},\tilde{\underline{\underline{m}}})$ *is the manifold obtained from* $(N,\underline{\underline{n}})$, $(M,\underline{\underline{m}})$ *by splitting at* $f^{-1}F$, $F$, *respectively.*

Proof. By 3.4, $f$ can be admissibly deformed into a covering map. In particular, $f|\tilde{N}\colon \tilde{N} \to \tilde{M}$ is locally homeomorphic. Then the corollary follows from 3.2. q.e.d.

## §4. Essential surfaces and useful boundary-patterns

By an *essential singular surface* we will mean an essential map $f\colon (F,\underline{\underline{f}}) \to (X,\underline{\underline{x}})$, where $F$ is a 2-manifold and $X$ may be either a 2- or 3-manifold. This term will be applied only if $(F,\underline{\underline{f}})$ is *not* a 2-sphere or an i-faced disc, $1 \leq i \leq 3$. But in order to avoid a conflict of notation we will not make this part of the definition.

The reader should keep in mind that *essential* and *useful boundary-pattern* are the appropriate general concepts that in special cases reduce to such notions as "incompressible", "boundary-incompressible", and "boundary-irreducible".

*4.1 Lemma. Let* $(M,\underline{\underline{m}})$ *be a* 3-*manifold, and* $F$ *an admissible surface in* $(M,\underline{\underline{m}})$ *with* $F \cap \partial M = \partial F$. *Then* $F$ *is inessential in* $(M,\underline{\underline{m}})$, *if and only if there is an admissible singular disc* $g\colon (D,\underline{\underline{d}}) \to (M,\underline{\underline{m}})$ *such that* $(D,\bar{\underline{\underline{d}}})$ *is an i-faced disc,* $1 \leq i \leq 3$, $g^{-1}F$ *is equal to one side* $k$, *of* $(D,\bar{\underline{\underline{d}}})$, *and* $g|k$ *is essential in* $F$.

*Proof.* One direction is obvious. Therefore let $F$ be inessential in $(M,\underline{\underline{m}})$. Then, by definition, there is an admissible singular disc $g\colon (D,\underline{\underline{d}}) \to (M,\underline{\underline{m}})$ such that $(D,\bar{\underline{\underline{d}}})$ is an i-faced disc, $1 \leq i \leq 3$, and that the restriction of $g$ to one side, $k$, of $(D,\bar{\underline{\underline{d}}})$ is an essential singular curve in $F$. After a small general position deformation of $g$ (rel $k$), $g^{-1}F - k$ is a system of admissible curves in $(D,\underline{\underline{d}})$. We suppose $g$ is chosen so that it has the above properties and that, in addition, the number of components of $g^{-1}F$ is as small as possible.

Assume $g^{-1}F \neq k$. Then there exists at least one curve, $k_1$, of $g^{-1}F$ which is disjoint to $k$. $k_1$ separates an admissible disc, $(D_1,\underline{\underline{d}}_1)$, from $(D,\underline{\underline{d}})$ such that $(D_1,\bar{\underline{\underline{d}}}_1)$ is an i-faced disc, $1 \leq i \leq 3$. $g|k_1$ cannot be essential in $F$, for otherwise we could choose $g|D_1$ instead of $g$, and we have a contradiction to our minimality condition on $g^{-1}F$. Thus there is also an admissible map $\bar{g}\colon (D_1,\underline{\underline{d}}_1) \to (F,\underline{\underline{f}})$ with $\bar{g}|k_1 = g|k_1$. Then, replacing $g|D_1$ by $\bar{g}$ and applying a small admissible general position deformation (rel $k$), we get an admissible map $\hat{g}$: $(D,\underline{\underline{d}}) \to (M,\underline{\underline{m}})$ from $g$ such that the number of curves of $\hat{g}^{-1}F$ is smaller than that of $g^{-1}F$. This again contradicts our minimality condition on $g^{-1}F$. q.e.d.

4.2 Lemma. *Let $F$ and $(M,\underline{\underline{m}})$ be given as in 4.1. Suppose that $\underline{\underline{m}}$ is a useful boundary-pattern of $M$, and that no component of $F$ is an admissible i-faced disc, $1 \leq i \leq 3$, in $(M,\underline{\underline{m}})$. Then $F$ is inessential in $(M,\underline{\underline{m}})$ if and only if there is an admissible (non-singular) disc $(D,\underline{\underline{d}})$ in $(M,\underline{\underline{m}})$ such that $(D,\bar{\underline{\underline{d}}})$ is an i-faced disc, $1 \leq i \leq 3$, and that $D \cap F$ is a side of $(D,\bar{\underline{\underline{d}}})$ which is an essential curve in $F$.*

*Proof.* One direction is obvious. Therefore let $F$ be inessential in $(M,\underline{\underline{m}})$. Let $g$: $(D,\underline{\underline{d}}) \to (M,\underline{\underline{m}})$ be an admissible singular disc as given in 4.1. Define $(\tilde{M},\tilde{\underline{\underline{m}}})$ to be that manifold obtained from $(M,\underline{\underline{m}})$ by splitting at $F$. Then $g$ can be considered to be an admissible singular i-faced disc, $1 \leq i \leq 3$, in $(\tilde{M},\tilde{\underline{\underline{m}}})$. The existence of $g$ shows that $\tilde{\underline{\underline{m}}}$ is not a useful boundary-pattern of $\tilde{M}$. $(M,\underline{\underline{m}})$ is a manifold with useful boundary-pattern and $F$ not an i-faced disc, $1 \leq i \leq 3$. Thus the existence of the required disc $(D,\underline{\underline{d}})$ follows easily from 2.1. q.e.d.

4.3 Lemma. *Let $(M,\underline{\underline{m}})$ be a 3-manifold with useful boundary-pattern. Suppose the boundary is not empty; and $M$ is not a ball.*

*Then there exists a non-separating essential surface, $F$, in $(M,\underline{\underline{m}})$, with $F \cap \partial M = \partial F$.*

*Proof*: By [Wa 3], there exists at least one admissible, connected surface, $F$, in $(M,\underline{\underline{m}})$, $F \cap \partial M = \partial F$, which is non-separating. Suppose $F$ is chosen so that, in addition, $(\beta_1(F), \eta(F))$ is minimal with respect to the lexicographical order, where $\beta_1(F)$ denotes the first Betti number of $F$ and $\eta(F)$ the number of intersection points of $\partial F$ with the graph, $J$, of $(M,\underline{\underline{m}})$.

Assume $F$ is inessential in $(M,\underline{\underline{m}})$. We may apply 4.2. Let $(D,\underline{\underline{d}})$ be the admissible disc in $(M,\underline{\underline{m}})$ as given in 4.2. Then $D \cap F$ is an essential curve in $F$ which is a side of $(D,\underline{\underline{d}})$. Let $U(D)$ be a regular neighborhood in $M$, and $D_1$, $D_2$ the two copies of $D$ in $\partial U(D)$. Define $\tilde{F} = (F - U(D))^{-} \cup D_1 \cup D_2$ and let $\tilde{F}_1$, $\tilde{F}_2$ (which are possibly equal) be the components of $\tilde{F}$. Since $D \cap F$ is an essential curve in $F$, it is easily checked that $(\beta_1(\tilde{F}_i), \eta(\tilde{F}_i)) < (\beta_1(F), \eta(F))$, $i = 1,2$. Hence, by our choice of $F$, both $\tilde{F}_1$ and $\tilde{F}_2$ must be separating. But this contradicts the fact that $F$ is non-separating. q.e.d.

4.4 Proposition. Let $(M_i,\underline{\underline{m}}_i)$, $i = 1,2$, be irreducible and aspherical 3-manifolds with useful boundary-patterns. Let $F$ be an essential surface in $(M_2,\underline{\underline{m}}_2)$, $F \cap \partial M_2 = \partial F$, but no component of $F$ a 2-sphere, or an admissible i-faced disc, $1 \le i \le 3$, in $(M_2,\underline{\underline{m}}_2)$.

Then any admissible map $f\colon (M_1,\underline{\underline{m}}_1) \to (M_2,\underline{\underline{m}}_2)$ is admissibly homotopic to a map, $g$, such that $g^{-1}F$ is an essential surface in $(M_1,\underline{\underline{m}}_1)$, and no component of $g^{-1}F$ is a 2-sphere or an admissible i-faced disc, $1 \le i \le 3$, in $(M_1,\underline{\underline{m}}_1)$.

Suppose in addition, that the complete boundary-pattern, $\bar{\underline{\underline{m}}}_1$, of $(M_1,\underline{\underline{m}}_1)$ is useful (but not necessarily the complete boundary-pattern of $(M_2,\underline{\underline{m}}_2)$), then we may choose $g$ such that $g^{-1}F$ is essential in $(M_1,\bar{\underline{\underline{m}}}_1)$ and no component of $g^{-1}F$ is a 2-sphere or an admissible i-faced disc, $1 \le i \le 3$, in $(M_1,\bar{\underline{\underline{m}}}_1)$.

Remark. If $f^{-1}F$ is already an admissible surface in $(M_1,\bar{\underline{\underline{m}}}_1)$, the following will be apparent from the proof:

1. $\beta_1(g^{-1}F) \le \beta_1(f^{-1}F)$, where $\beta_1$ denotes the first Betti number.
2. Let $(N_1,\underline{\underline{n}}_1)$ be a submanifold in $(M_1,\bar{\underline{\underline{m}}}_1)$ such that $(\partial N - \partial M)$ is an essential surface. Suppose $N \cap f^{-1}F = \emptyset$ and no component of $(\partial N - \partial M)^{-}$ is a 2-sphere or an admissible i-faced disc, $1 \le i \le 3$, in $(M_1,\bar{\underline{\underline{m}}})$. Then the homotopy of $f$ may be chosen constant on $N$.

Proof. We follow the proof of 1.1 of [Wa 3] whenever possible. First we may suppose $f$ is admissibly deformed into $g$ so that $G = g^{-1}F$ is an admissible surface in $(M_1,\bar{\underline{\underline{m}}}_1)$, with $G \cap \partial M_1 = \partial G$

(general position). We assume that one of the assertions of 4.4 is false and show that $g$ can be admissibly deformed so that $G$ becomes simpler.

Case 1. Suppose at least one component, $G_1$, of $G$ is either a 2-sphere or an admissible i-faced disc, $1 \leq i \leq 3$, in $(M_1, \underline{\underline{m}}_1)$ resp. $(M_1, \bar{\underline{\underline{m}}}_1)$.

Since $\underline{\underline{m}}_1$, resp. $\bar{\underline{\underline{m}}}_1$, is a useful boundary-pattern of $M_1$ and since $M_1$ is irreducible, $G_1$ separates a ball, $E$, from $M_1$. Let $G_1$ be chosen so that $E \cap G = G_1$. Furthermore we may suppose that $D = E \cap \partial M_1$ is a disc in $\partial M_1$ such that $J_1 \cap D$ is equal to the cone on $J_1 \cap \partial D$, where $J_j$ denotes the graph of $(M_j, \underline{\underline{m}}_j)$, $j = 1,2$. (For remark 2 note that $E \cap N = \emptyset$; apply [Wa 1, (1.4)].) We assert that $g$ can be admissibly deformed into $\hat{h}$ so that $\hat{h}^{-1}F = G - G_1$.

This is clear of $G_1$ is a 2-sphere, for $M_2$ is asperical and no component of $F$ is a 2-sphere (see [Wa 3, p. 508]).

Hence let $G_1$ be an admissible i-faced disc, $1 \leq i \leq 3$, in $(M_1, \underline{\underline{m}}_1)$, resp. $(M_1, \bar{\underline{\underline{m}}}_1)$.

(a) $\bar{\underline{\underline{m}}}_1$ is not a useful boundary-pattern of $M_1$.

In this case $G_1$ is an admissible i-faced disc, $1 \leq i \leq 3$, in $(M_1, \underline{\underline{m}}_1)$ and $g(\partial G_1) \subset \partial F$, since $g$ is an admissible map. Since no component of $F$ is an admissible i-faced disc, $1 \leq i \leq 3$, in $(M_2, \underline{\underline{m}}_2)$, the existence of the map $g|G_1$: $(G_1, \partial G_1) \to (F, \partial F)$ shows that $g|\partial G_1$ is contractible in $\partial F$. In particular, $G_1$ cannot be a 3-faced disc, and so $J_1 \cap D$ is either empty or an arc, $t$. We only deal with the latter case; the other one is analogous. $t$ splits $D$ into two discs, $D_1$, $D_2$. Define $t_j = D_j \cap \partial G_1$, and $H_j \in \underline{\underline{m}}_1$, $H'_j \in \underline{\underline{m}}_2$ as those surfaces which contain $D_j$, resp. $g(H_j)$, $j = 1,2$. Since $g|\partial G_1$ is contractible in $\partial F$, we may note that $g$ maps $\partial t$ into one point, $z$, of $J_2$, and that $g|t_j$, $j = 1,2$, can be contracted (rel $\partial t_j$) in $H'_j \cap \partial F$ into $z$.

Assume $g|t$: $t \to J_2$ cannot be contracted (rel $\partial t$) in $J_2$ into $z$. Then that component, $s$, of $J_2$ containing $g(t)$ must be a simple closed curve. The existence of the map $g|D_j: D_j \to H_j'$, $j = 1,2$, shows that a multiple of $s$ is contractible in $H_1'$ as well as in $H_2'$ (recall $g|t_j$ is contractible in $H_j'$). Hence $H_1'$ as well as $H_2'$ is a disc whose boundary is $s$. That means that $(M_2,\underline{\underline{m}}_2)$ is a ball ($M_2$ is irreducible) with precisely two sides. But this is impossible since $F$ is essential and no component of $F$ is an i-faced disc, $1 \leq i \leq 3$, in $(M_2,\underline{\underline{m}}_2)$.

This contradicts our assumption and so $g|t$ can be contracted (rel $\partial t$) in $J_2$ into $z$. Doing this contraction carefully, it can be extended to an admissible deformation of $g$ into $g_1$ so that $g_1^{-1}F = G \cup t$. Then $g_1(\partial D_j) \subset \partial F$, $j = 1,2$. Since $g|t_j$, $j = 1,2$, can be contracted (rel t) in $H_j' \cap \partial F$, then also $g_1|\partial D_j$ can be contracted (rel t) in $H_j' \cap \partial F$. Since $(M_2,\underline{\underline{m}}_2)$ cannot be a ball with one side, $H_j'$ is not a 2-sphere, and so $g_1|D_j$ can be deformed (rel $\partial D_j$) in $H_j'$ into $\partial F$. Doing this carefully, these homotopies of the $g_1|D_j$'s can be extended to an admissible deformation of $g_1$ into $g_2$ so that $g_2^{-1}F = G \cup D$. Then $g_2(\partial E) \subset F$. Since both $F$ and $M_2$ are aspherical, $g_2|E$ can be deformed (rel $\partial E$) into $F$. Doing this carefully, this homotopy can be extended to an admissible deformation of $g_2$ into $h$ such that $h^{-1}F = G \cup E$. Applying finally a small admissible deformation of $h$, which pulls $h|E$ in the right direction, we get a map $\hat{h}$ from $h$ with $\hat{h}^{-1}F = G - G_1$.

(b) $\underline{\underline{\bar{m}}}_1$ is a useful boundary-pattern of $M_1$.

If $G_1$ is an admissible i-faced disc, $1 \leq i \leq 3$, in $(M_1,\underline{\underline{m}}_1)$, we remove $G_1$ as in (a) above. Thus we suppose $G_1$ is not such a disc in $(M_1,\underline{\underline{m}}_1)$. $J_1$ splits $D$ into at most three discs, $D_1$, $D_2$, $D_3$, and, by our suposition, precisely one of them, say $D_1$, does lie in a free side of $(M_1,\underline{\underline{m}}_1)$. Define a contraction $H_t: (E,D_2,D_3) \to (E,D_2,D_3)$ $t \in I$, $H_0 = \mathrm{id}$, constant on $G_1$, which pulls $E$ into $G_1$. Since $D_1$ does not lie in a surface of $\underline{\underline{m}}_1$, $H_t$ can be extended to an admissible deformation $G_t: (M_1,\underline{\underline{m}}_1) \to (M_1,\underline{\underline{m}}_1)$. Doing this carefully, $g \circ G_t$ is an admissible homotopy of $g$ with $(g \circ G_1)^{-1}F = G \cup E$. As in the final part of (a), $g \circ G_1$ is admissibly homotopic to a map $\hat{h}$ with $\hat{h}^{-1}F = G - G_1$.

<u>Case 2. Suppose</u> G <u>is inessential in</u> $(M_1,\underline{\underline{m}}_1)$, <u>resp.</u> $(M_1,\bar{\underline{\underline{m}}}_1)$.

We apply 4.2. Let (D,d) be an admissible disc in $(M_1,\underline{\underline{m}}_1)$, resp. $(M_1,\bar{\underline{\underline{m}}}_1)$, as given in 4.2. Then $D \cap G$ is an essential curve, k, in G which is a side of $(D,\bar{\underline{\underline{d}}})$. Let U(D) be a regular neighborhood in the closure of $M_1$ - G, then U(D) can be identified with $D \times I$ such that $D = D \times 1/2$ (for remark 2 note that, by our suppositions on N, D, and so U(D), may be chosen disjoint to N).

(a) $\bar{\underline{\underline{m}}}_1$ <u>is not a useful boundary-pattern of</u> $M_1$.

In this case $(D,\underline{\underline{d}})$ is an admissible disc in $(M_1,\underline{\underline{m}}_1)$. Since $g: (M_1,\underline{\underline{m}}_1) \to (M_2,\underline{\underline{m}}_2)$ is an admissible map, $g|k$ is an admissible singular curve in F, and $g|D: (D,\underline{\underline{d}}) \to (M_2,\underline{\underline{m}}_2)$ is an admissible singular disc in $(M_2,\underline{\underline{m}}_2)$, which shows that $g|k$ is inessential in $(M_2,\underline{\underline{m}}_2)$. Hence $g|k$ must be inessential in F, since F is essential in $(M_2,\underline{\underline{m}}_2)$, and so there is an admissible map $\hat{g}: (D,\underline{\underline{d}}) \to (F,\underline{\underline{f}})$, with $\hat{g}|k = g|k$. Identifying $g|D$ with $\hat{g}$ at k, we either get a singular 2-sphere, or an admissible singular 1- or 2-faced disc in $(M_2,\underline{\underline{m}}_2)$. $\underline{\underline{m}}_2$ is a useful boundary-pattern of $M_2$ and $M_2$ is aspherical. Hence it follows that $g|D$ can be admissibly deformed (rel k) in $(M_2,\underline{\underline{m}}_2)$ into $\hat{g}$, i.e. into F. Doing this carefully, this homotopy can be extended to an admissible deformation of g into h so that $h^{-1}F = G \cup U(D)$. Applying then a small admissible homotopy to h which pulls $h|U(D)$ (rel $D \times \partial I$) in the right direction, we get a map $\hat{h}$ from h such that $\hat{h}^{-1}F = (G - U(D))^{-} \cup (D \times \partial I)$.

(b) $\bar{\underline{\underline{m}}}_1$ <u>is a useful boundary-pattern of</u> $M_1$.

If $(D,\underline{\underline{d}})$ is an admissible disc in $(M_1,\underline{\underline{m}}_1)$, we argue as in (a) above. Thus we suppose the contrary. Let $k_1$, $k_2$ (which may be equal) be the sides of $(D,\underline{\underline{d}})$. Then by our supposition, precisely one of them, say $k_1$, must lie in a free side of $(M_1,\underline{\underline{m}}_1)$. Define a contraction $H_t: (D,k_2) \to (D,k_2)$ $t \in I$, $H_0 = id$, constant on k, which pulls D into k. Since $k_1$ lies in a free side of $(M_1,\underline{\underline{m}}_1)$ $H_t \times id: (D \times I, k_2 \times I) \to (D \times I, k_2 \times I)$ can be extended to an admissible deformation $G_t$: $(M_1,\underline{m}_1) \to (M_1,\underline{m}_1)$. Doing this carefully, $g \cdot G_t$ is an admissible homotopy of $g$ with $(g \cdot G_1)^{-1}F = G \cup U(D)$. As in the final part of (a) above, $g \circ G_1$ is admissibly homotopic to a map $\hat{h}$ with $\hat{h}^{-1}F = (G - U(D))^{-} \cup (D \times \partial I)$. q.e.d.

Let $(B,\underline{b})$ be either a square, or an annulus (recall that the boundary-pattern of an annulus consists of its boundary curves), or a torus, and let $(A,\underline{a})$ be any surface such that $(A,\bar{\underline{a}})$ is neither a 2-sphere nor an i-faced disc, $1 \leq i \leq 3$. Then observe that the existence of any essential map $g$: $(A,\underline{a}) \to (B,\underline{b})$ implies that $(A,\bar{\underline{a}})$ is again either a square, annulus, or torus. Hence we obtain from 4.4 the following crucial property of essential maps.

*4.5 Corollary. Let $(M_i,\underline{m}_i)$, $i = 1,2$, be irreducible and aspherical 3-manifolds with useful boundary-patterns. Suppose the complete boundary-pattern, $\bar{\underline{m}}_1$, of $(M_1,\underline{m}_1)$ is also useful. Let $F$ be an essential surface in $(M_2,\underline{m}_2)$ such that every component of $F$ is a square, annulus, or torus in $(M_2,\underline{m}_2)$.*

*Then any essential map $f$: $(M_1,\underline{m}_1) \to (M_2,\underline{m}_2)$ is admissibly homotopic to a map $g$ such that $g^{-1}F$ is an essential surface in $(M_1,\bar{\underline{m}}_1)$ and such that every component of $g^{-1}F$ is a square, annulus, or torus in $(M_1,\bar{\underline{m}}_1)$.*

*Furthermore, the admissible homotopy of $f$ may be chosen constant on $N$, where $N$ is given as in remark 2 of 4.4.*

In the remainder of this paragraph we state and prove some facts about splitting 3-manifolds at admissible surfaces.

*4.6 Proposition. Let $(M,\underline{m})$ be a 3-manifold, and $F$ an admissible surface in $(M,\underline{m})$, with $F \cap \partial M = \partial F$. Suppose $(\tilde{M},\tilde{m})$ is the manifold, obtained by splitting $(M,\underline{m})$ at $F$. Let $G$ be any admissible surface in $(M,\underline{m})$, with $G \cap \partial M = \partial G$, and which is in general position with respect to $F$. Define $\tilde{G} = G \cap \tilde{M}$. Then the following holds:*

1. *Suppose $\underline{m}$ is a useful boundary-pattern of $M$ and $F$ is essential in $(M,\underline{m})$. Then $G$ is essential in $(M,\underline{m})$*

*if* $\tilde{G}$ *is essential in* $(\tilde{M},\tilde{m})$.

*Suppose, in addition,* $M$ *or* $\tilde{M}$ *is irreducible,* $\tilde{m}$ *is a useful boundary-pattern of* $\tilde{M}$, *and* $G$ *is admissibly isotoped so that the number of curves of* $G \cap F$ *is minimal. Then*

2. *every curve of* $G \cap F$ *is essential in* $G$ *(and therefore essential in* $F$, *provided* $G$ *is essential in* $(M,\underline{\underline{m}})$*).*
3. *If* $G$ *is chosen so that every component of* $G$ *is an admissible square, annulus, or torus, in* $(M,\bar{\underline{\underline{m}}})$, *which is essential in* $(M,\underline{\underline{m}})$, *then* $\tilde{G}$ *is such a surface in* $(\tilde{M},\tilde{\underline{\underline{m}}})$.
4. *Suppose every component of* $F$ *is an admissible square, annulus, or torus in* $(M,\bar{\underline{\underline{m}}})$ *which is essential in* $(M,\underline{\underline{m}})$. *If* $G$ *is an essential surface in* $(M,\underline{\underline{m}})$, *then* $\tilde{G}$ *is essential in* $(\tilde{M},\tilde{\underline{\underline{m}}})$.

*Remark*. In general, 4.6.4 is false for essential surfaces $F$ in $(M,\underline{\underline{m}})$ which are not squares, annuli, or tori.

*Proof of 1*. Assume $G$ is inessential in $(M,\underline{\underline{m}})$. Apply 4.1. Let $g\colon (D,\underline{\underline{d}}) \to (M,\underline{\underline{m}})$ be an admissible singular disc as given in 4.1. Then $g^{-1}G$ is equal to one side, $k$, of $(D,\bar{\underline{\underline{d}}})$, and $g|k$ is essential in $G$. Choose $g$ so that the number of intersection points of $g|k$ with $F$ is minimal. Then clearly every component of $k \cap g^{-1}\tilde{M}$ is mapped to an essential singular arc in $\tilde{G}$. Without loss of generality, $g^{-1}F$ is a system of curves. Since $F$ is essential, we may suppose that $g^{-1}F$ splits $D$ into a system of discs, $D_i$, $1 \leq i \leq n$, and that, moreover, every disc $D_i$ meets $k$. Then there is at least one disc, say $D_1$, which meets $g^{-1}F$ in precisely one arc. $g|D_1$ can be considered as an admissible singular disc in $(\tilde{M},\tilde{\underline{\underline{m}}})$ and this shows that the singular arc $g|D_1 \cap k$ in $\tilde{G}$ is inessential in $(\tilde{M},\tilde{\underline{\underline{m}}})$. Hence, it is inessential in $\tilde{G}$ since $\tilde{G}$ is essential in $(\tilde{M},\tilde{m})$. But this contradicts our choice of $g$.

*Proof of 2*. Assume there is a curve, $k$, of $F \cap G$ which is inessential in $G$. Then $k$ separates a disc, $D$, whose complete boundary-pattern has at most three sides, and we may suppose $k$ is chosen such that

$D \cap F = k$. $D$ can be considered as admissible i-faced disc, $1 \leq i \leq 3$, in $(\tilde{M},\tilde{\underline{\underline{m}}})$. Since $\tilde{\underline{\underline{m}}}$ is a useful boundary-pattern of $\tilde{M}$, $\partial D$ bounds a disc, $D^*$, in $\partial\tilde{M}$ such that $\tilde{J} \cap D^*$ is the cone on $\tilde{J} \cap \partial D^*$, where $\tilde{J}$ denotes the graph of $(\tilde{M},\tilde{\underline{\underline{m}}})$. $\tilde{J}$ splits $D^*$ into at most three discs and precisely one of them lies in $F$. Since $M$ or $\tilde{M}$ is irreducible, the 2-sphere $D \cup D^*$ bounds a ball in $\tilde{M}$. Hence the intersection $F \cap G$ can be diminished using an admissible isotopic deformation of $G$ (without enlarging the number of essential curves). But this contradicts our supposition on $G$.

Proof of 3. By 2 of 4.6, every curve of $F \cap G$ is essential in $G$. Since $G$ consists of admissible squares, annuli, or tori in $(M,\underline{\underline{m}})$ it follows that $\tilde{G}$ consists of admissible squares, annuli, or tori in $(\tilde{M},\tilde{\underline{\underline{m}}})$.

Assume $\tilde{G}$ is inessential in $(\tilde{M},\tilde{m})$. We may apply 4.2. Let $(D,\underline{\underline{d}})$ be an admissible disc in $(\tilde{M},\tilde{\underline{\underline{m}}})$ as given in 4.2. Then $D \cap \tilde{G}$ is an essential curve, $k$, in $\tilde{G}$ which is a side of $(D,\bar{\underline{\underline{d}}})$. Let $U(k)$ be a regular neighborhood in $G$ and $k_1$, $k_2$ the two copies of $k$ in $\partial U(k)$. Since $G$ is essential in $(M,\underline{\underline{m}})$, it follows that precisely one side of $(D,\bar{\underline{\underline{d}}})$ lies in $F$. Hence $k$ can be isotoped via $D$ into $F$, and so $G$ can be admissibly isotoped in $(M,\underline{\underline{m}})$ into $G'$ so that

$$G' \cap F = ((G \cap F) - U(k)) \cup k_1 \cup k_2.$$

Since $k$ is essential in $\tilde{G}$, and since every component of $G$ is a square, annulus, or torus, we find that (a) the number of essential curves of $G' \cap F$ is strictly less than that of $G \cap F$, and (b) at least one curve of $G' \cap F$ is inessential in $G'$. From the proof of 2 of 4.2, we see that the inessential curves of $G' \cap F$ can be removed without enlarging the number of essential curves. Thus, by (a), the number of curves of $G \cap F$ is not minimal. But this contradicts our supposition on $G$.

Proof of 4. Assume $\tilde{G}$ is inessential in $(\tilde{M},\tilde{m})$. Apply 4.1. Let $g\colon (D,\underline{\underline{d}}) \to (\tilde{M},\tilde{\underline{\underline{m}}})$ be an admissible singular disc as given in 4.1 Then $g^{-1}\tilde{G}$ is equal to a side, $k$, of $(D,\bar{\underline{\underline{d}}})$ and $g|k$ is essential in $\tilde{G}$.

Since $G$ is essential in $(M,\underline{\underline{m}})$, it follows that precisely one side, $k_1$ of $(D,\underline{\underline{d}})$ is mapped under $g$ into $F$. $G$ splits $F$ into connected surfaces, and one of them, say $F_1$, contains $g|k_1$ as an admissible singular arc.

Assume $g|k_1$ is inessential in $F_1$. By 2 of 4.6, every curve of $F \cap G$ is essential in $F$, as well as in $G$. In particular, by our choice of $F$, $F_1$ has to be a square or annulus, and so, without loss of generality, $g|k_1$ is an embedding. By our assumption, $g(k_1)$ separates an admissible disc from $F_1$ with at most two sides. This disc meets $F_1 \cap G$ in one arc, say $k_1'. k_1'$, together with $k$, defines an admissible singular arc in $G$ which either can be deformed into a point, or into a side of $(M,\underline{\underline{m}})$. Since $G$ is essential in $(M,\underline{\underline{m}})$, such deformations may be chosen within $G$. Since $F \cap G$ is essential in $G$, it follows that $g|k$ is inessential in $\tilde{G}$ which contradicts our choice of $g$.

Thus $g|k_1$ is essential in $F_1$. Let $(M^*,\underline{\underline{m}}^*)$ be the manifold obtained from $(M,\underline{\underline{m}})$ by splitting at $G$. The existence of $g$ shows that $F^* = F \cap M^*$ is inessential in $(M^*,\underline{\underline{m}}^*)$. $F$ consists of squares, annuli, and tori which are essential in $(M,\underline{\underline{m}})$. Hence, applying 3 of 4.6 to $F$ rather than to $G$, it follows that at least one curve of $G \cap F$ can be removed, using an admissible isotopic deformation of $F$. We extend this deformation to an admissible ambient isotopy, $\alpha_t$, $t \in I$, of $(M,\underline{\underline{m}})$, and see that $\alpha_t^{-1}|G$ is an admissible isotopy of $G$ which diminishes $G \cap F$. But this contradicts our suppositions on $G$. q.e.d.

<u>4.7 Proposition</u>. <u>Let</u> $(M,\underline{\underline{m}})$ <u>be a 3-manifold, and</u> $F$ <u>an admissible surface in</u> $(M,\underline{\underline{m}})$ <u>with</u> $F \cap \partial M = \partial F$. <u>Suppose</u> $(\tilde{M},\tilde{\underline{\underline{m}}})$ <u>is the manifold obtained by splitting</u> $(M,\underline{\underline{m}})$ <u>at</u> $F$. <u>Let</u> $f\colon G \to M$ <u>be any admissible singular surface in</u> $(M,\underline{\underline{m}})$, <u>with</u> $f^{-1}\partial M = \partial G$ <u>and which is in general position with respect to</u> $F$. <u>Define</u> $\tilde{f} = f|f^{-1}\tilde{M}$. <u>Then the following holds</u>:

1. <u>Suppose</u> $F$ <u>is essential in</u> $(M,\underline{\underline{m}})$. <u>Then</u> $f$ <u>is essential in</u> $(M,\underline{\underline{m}})$ <u>if</u> $\tilde{f}$ <u>is essential in</u> $(\tilde{M},\tilde{\underline{\underline{m}}})$.

<u>Suppose, in addition,</u> $M$ <u>or</u> $\tilde{M}$ <u>is aspherical,</u> $\tilde{\underline{\underline{m}}}$ <u>is a useful boundary-pattern of</u> $\tilde{M}$, <u>and</u> $f$ <u>is admissibly deformed in</u> $(M,\underline{\underline{m}})$ <u>so</u>

*that the number of curves of* $f^{-1}F$ *is minimal. Then:*

2. *Every curve of* $f^{-1}F$ *is essential in* G.
3. *If the restriction of* f *to every component of* G *is an admissible singular square, annulus, or torus in* $(M,\bar{\underline{\underline{m}}})$ *which is essential in* $(M,\underline{\underline{m}})$, *then the restriction of* f *to any component of* $f^{-1}\tilde{M}$ *is such a surface in* $(\tilde{M},\tilde{m})$.

*Remark.* We shall see in 4.10 that also 4.6.4 can be generalized to essential maps.

*Proof.* By 2.1, there is no problem to copy the proofs of 4.6.1-3.

*4.8 Proposition. Let* $(M,\underline{\underline{m}})$ *be an irreducible* 3-*manifold. Let* F *be an admissible surface in* $(M,\underline{\underline{m}})$, *with* $F \cap \partial M = \partial F$, *but no component of* F *a* 2-*sphere or an* i-*faced disc*, $1 \leq i \leq 3$. *Suppose* $(\tilde{M},\tilde{\underline{\underline{m}}})$ *is the manifold obtained by splitting* $(M,\underline{\underline{m}})$ *at* F. *Then the following holds:*

1. *If* $\tilde{\underline{\underline{m}}}$ *is a useful boundary-pattern of* $\tilde{M}$, *then* F *is essential in* $(M,\underline{\underline{m}})$ *and* $\underline{\underline{m}}$ *is a useful boundary-pattern of* M.

*Suppose in addition,* $\underline{\underline{m}}$ *is a useful boundary-pattern of* M, *then:*

2. F *is essential in* $(M,\underline{\underline{m}})$ *if and only if* $\tilde{\underline{\underline{m}}}$ *is a useful boundary-pattern of* $\tilde{M}$.

*Proof.* If F is inessential in $(M,\underline{\underline{m}})$ or $\underline{\underline{m}}$ not a useful boundary-pattern of M, it follows either from 4.1 or from 4.6.2, respectively, that $\tilde{\underline{\underline{m}}}$ is not a useful boundary-pattern of $\tilde{M}$. So it remains to show that $\tilde{\underline{\underline{m}}}$ is a useful boundary-pattern, provided F is essential. Since F is not an i-faced disc, $1 \leq i \leq 3$, this follows easily from the definitions of "essential" and "useful". q.e.d.

Finally let $(M,\underline{\underline{m}})$ be a 3-manifold whose boundary-pattern is useful. Let F be a surface in $(M,\bar{\underline{\underline{m}}})$ whose components are essential squares, annuli, or tori in $(M,\bar{\underline{\underline{m}}})$. Denote by $(\tilde{M},\tilde{\underline{\underline{m}}})$ the manifold obtained from $(M,\underline{\underline{m}})$ by splitting at $F$. Let $(G,\underline{\underline{g}})$ be a surface whose boundary-pattern is complete.

*4.9 Proposition.* *Suppose* $f\colon (G,\underline{\underline{g}}) \to (M,\underline{\underline{m}})$ *is an essential map such that* $f^{-1}F$ *consists of essential curves in* $(G,\underline{\underline{g}})$. *Denote by* $(\tilde{G},\tilde{\underline{\underline{g}}})$ *the surface obtained from* $(G,\underline{\underline{g}})$ *by splitting at* $f^{-1}F$, *and let* $(G_1,\underline{\underline{g}}_1)$ *be any component of* $(\tilde{G},\tilde{\underline{\underline{g}}})$.

*If* $f|G_1\colon (G_1,\underline{\underline{g}}_1) \to (\tilde{M},\tilde{\underline{\underline{m}}})$ *is inessential, then* $(G_1,\underline{\underline{g}}_1)$ *is either a square or an annulus.*

*Proof.* Suppose $f|G_1\colon (G_1,\underline{\underline{g}}_1) \to (\tilde{M},\tilde{\underline{\underline{m}}})$ is inessential. Then there is an essential singular curve $k$ in $(G_1,\underline{\underline{g}}_1)$, such that $f\circ k$ is inessential in $(\tilde{M},\tilde{\underline{\underline{m}}})$. At least one end-point of $k$ lies in a curve of $f^{-1}F$. For otherwise $k$ is an admissible singular curve in $(G,\underline{\underline{g}})$, and $f\circ k$ is inessential in $(M,\underline{\underline{m}})$. Hence $k$ is inessential in $(G,\underline{\underline{g}})$ since $f$ is essential. On the other hand, $k$ is essential in $(G_1,\underline{\underline{g}}_1)$. Thus we get a contradiction to the fact that $f^{-1}F$ consists of essential curves in $(G,\underline{\underline{g}})$.

*Case 1. Both the end-points of* $k$ *lie in curves of* $f^{-1}F$.

Let $k_1$, $k_2$ (possibly equal) the curves of $f^{-1}F$ which contain end-points of $k$.

(A) *Suppose* $k_1$, *say, is an arc.*

$f|k_1$ is an essential singular arc in a component, $F_1$, of $F$. Hence $F_1$ is either a square or an annulus, and $f|k_1$ joins opposite sides of $F_1$. Denote by $x_1$, $x_2$ the two end-points of $k_1$.

If $k_1 = k_2$ or if $k_2$ is a closed curve, it is easily seen that $k$ together with curves in $k_1 \cup k_2$ defines an essential singular arc $t$ in $(G,\underline{\underline{g}})$ whose end-points lie in $x_1$. Since $f\circ k$ is inessential in $(\tilde{M},\tilde{\underline{\underline{m}}})$, it follows that $f\circ t$ can be deformed (rel $x_1$) into $F_1$. Then $f\circ t$ is an admissible singular arc in the square or annulus $F_1$ whose end-points lie in the same side of $F_1$. Thus $f\circ t$ is inessential in $F_1$ and so in $(M,\underline{\underline{m}})$. Therefore $t$ is inessential in

$(G,\underline{\underline{g}})$ since $f$ is essential, and this contradicts our choice of $t$.

Thus $k_1 \neq k_2$ and $k_2$ must be an arc. Let $y_1$, $y_2$ be the two end-points of $k_2$. We may suppose the indices are chosen so that $f(y_1)$ lies in the same side of $F_1$ as $f(x_1)$. Hence $f(y_2)$ is mapped into the same side of $F_1$ as $f(x_2)$ since $F_1$ is either a square or an annulus and since $f|k_1$ as well as $f|k_2$ is an essential singular arc in $F_1$. $k$ together with arcs in $k_1 \cup k_2$ defines an admissible singular arc $t_i$, $i = 1$ or $2$, in $(G,\underline{\underline{g}})$ which joins $x_i$ with $y_i$. Since $f\cdot k$ is inessential in $(\tilde{M},\tilde{\underline{\underline{m}}})$, $f\cdot t_i$ can be deformed (rel $x_i \cup y_i$) into $F_1$ and so it is inessential in $(M,\underline{\underline{m}})$. Therefore $t_1$ as well as $t_2$ are inessential in $(G,\underline{\underline{g}})$ since $f$ is essential. Since the end-points of $t_i$, $i = 1,2$, are mapped into the same side of $(M,\underline{\underline{m}})$, it follows that moreover they lie in the same side of $(G,\underline{\underline{g}})$. Hence, altogether, an admissible homotopy in $(G,\underline{\underline{g}})$ can be defined which pulls $k_1$ via $G_1$ into $k_2$, and it follows that $G_1$ is a square.

(B) <u>Suppose</u> $k_1$ <u>as well as</u> $k_2$ <u>is a closed curve</u>.

Since $f\cdot k$ is inessential in $(\tilde{M}\ \tilde{\underline{\underline{m}}})$, $k_1$ and $k_2$ are both mapped under $f$ into the same component, $F_1$, of $F$. $F_1$ must be either an annulus or a torus since $k_1$ and $k_2$ are closed and since $f$ is essential. Let $x$ be one end-point of $k$ contained in $k_1$, and define $x$ and $f(x)$ to be the base points of $\pi_1 G$ and $\pi_1 M$. Notice that $f_*\colon \pi_1 G \to \pi_1 M$ is an injection since $f$ is essential. Consider the loops $k_1$ and $t$ defined by

$$t = \begin{cases} k * k_2 * k^{-1} & , \quad \text{if } k_1 \neq k_2 \\ k & , \quad \text{if } k_1 = k_2 \end{cases}$$

We show that (a multiple of) $t$ is homotopic (rel $x$) in $G$ to (a multiple of) $k_1$. Then, by our choice of $t$, this implies that $k_1 \neq k_2$ since $k$ is essential in $(G,\underline{\underline{g}})$, and hence that $(G,\underline{\underline{g}})$ is an annulus, and we are done (apply Nielsen's theorem).

Notice first that $f\cdot t$ can be deformed (rel $x$) into $F_1$ since $f\cdot k$ is inessential in $(\tilde{M},\tilde{\underline{\underline{m}}})$.

If $F_1$ is an annulus, this implies that (a multiple of) $f \cdot t$ is homotopic (rel x) to (a multiple of) $f \cdot k_1$. Since $f_*$ is an injection the above assertion follows.

If $F_1$ is a torus, this implies that $f \cdot (t * k_1 * t^{-1} * k_1^{-1})$ is nullhomotopic. Hence $t * k_1 * t^{-1} * k_1^{-1}$ is nullhomotopic in $G$ since $f_*$ is an injection. This means there is a map $g: S^1 \times S^1 \to G$ with $g|S^1 \times 0 = t$ and $g|0 \times S^1 = k_1$. By the transversality lemma [Wa 3], we may suppose $g$ is deformed, by a homotopy which is constant on $S^1 \times 0$ and $0 \times S^1$, so that $g^{-1}f^{-1}F$ consists of non-contractible, simple closed curves in $S^1 \times S^1$. Since $S^1 \times 0$ and $0 \times S^1$ are mapped under $g$ into $\tilde{G}$, it follows that $S^1 \times S^1$ is mapped under $g$ into $\tilde{G}$. $\partial\tilde{G} \neq \emptyset$, and so $g: S^1 \times S^1 \to \tilde{G}$ cannot be deformed into a covering map. Thus, applying Nielsen's theorem, $g_*: \pi_1(S^1 \times S^1) \to \pi_1\tilde{G}$ is not an injection. Since $\pi_1\tilde{G}$ has no element of finite order [Ep 1, Lemma (8.4)], it follows that there is a simple closed curve $l$ in $S^1 \times S^1$, such that $g|l$ is nullhomotopic in $\tilde{G}$. Hence $g|l$ can be extended to a map of a disc into $\tilde{G}$, and, since $\tilde{G}$ is aspherical, this implies that $g$ can be extended to a map of a tube into $\tilde{G}$. As a consequence we have that (a multiple of) $t = g|S^1 \times 0$ must be homotopic (rel x) to (a multiple of) $k_1 = g|0 \times S^1$.

Case 2. *Precisely one end-point of $k$ lies in a curve of $f^{-1}F$.*

We argue analogously as in Case 1. q.e.d.

4.10 Corollary. *Let $(M,\underline{\underline{m}})$, F, $(\tilde{M},\tilde{\underline{\underline{m}}})$, and $(G,\underline{g})$ be given as in 4.9. Suppose, in addition, $\tilde{M}$ is aspherical. Let $f: (G,\underline{g}) \to (M,\underline{\underline{m}})$ be an essential map. Suppose $f$ is admissibly deformed in $(M,\underline{\underline{m}})$ so that the number of curves of $f^{-1}F$ is as small as possible.*

*Then $f|\tilde{G}: (\tilde{G},\tilde{\underline{g}}) \to (\tilde{M},\tilde{\underline{\underline{m}}})$ is an essential map, where $(\tilde{G},\tilde{\underline{g}})$ is the surface obtained from $(G,\underline{g})$ by splitting at $f^{-1}F$.*

Proof. By 4.7.2, $f^{-1}F$ consists of essential curves in $(G,\underline{g})$, and so we may apply 4.9. Thus, if $(G_1,\underline{g}_1)$ is a component of $(\tilde{G},\tilde{\underline{g}})$ such that $f|G_1$ is inessential in $(\tilde{M},\tilde{\underline{\underline{m}}})$, then $(G_1,\underline{g}_1)$ must be either a square, or an annulus. Moreover, $f|G_1$ meets $F$. Notice that $\tilde{M}$ is aspherical, and that, by 4.8.2, $\tilde{\underline{\underline{m}}}$ is a useful boundary-pattern of $\tilde{M}$. Then it is easily seen that there is an admissible homotopy of $f$ in $(M,\underline{\underline{m}})$ constant outside of $G_1$ which pulls $f|G_1$ into $F$. After a small admissible general position deformation of $f$ in $(M,\underline{\underline{m}})$, the number of curves of $f^{-1}F$ is diminished which contradicts our minimaltiy condition on $f^{-1}F$. q.e.d.

## Chapter II: Essential singular surfaces in some special 3-manifolds.

For a large number of questions about Haken 3-manifolds (e.g. knot spaces), a good knowledge of such special 3-manifolds as I-bundles, Seifert fibre spaces, and Stallings fibrations is very helpful.

In this chapter we study singular essential squares, annuli, and tori in such special 3-manifolds. We shall see later that the study of such singular surfaces in general Haken 3-manifolds can always be completely reduced to these cases.

In other paragraphs (see §§ 25, 26, 31) we shall consider the mapping class group of I-bundles and Seifert fibre spaces, and we shall deduce certain relations between homeomorphisms (more general: homotopy equivalences) of I-bundles, and surfaces in their lids. The results will lead us to a description of exotic homotopy equivalences and will give us furthermore information about the mapping class group of Haken 3-manifolds.

### §5. I-bundles and Seifert fibre spaces

Let $(M,\underline{\underline{m}})$ be either an I-bundle over a connected surface or a Seifert fibre space (references for Seifert fibre spaces are [Se 1, Wa 1, Wa 3, OVZ 1, Or 1]). A _fibration_ of $(M,\underline{\underline{m}})$ with fibre projection $p: M \to F$ is called _admissible_ if there exists a boundary-pattern, $\underline{\underline{f}}$, of $F$ such that

$$\underline{\underline{m}} = \{G \mid G \text{ is either a component of } (\partial M - p^{-1}\partial F)^{-}, \text{ or } G = p^{-1}k, \text{ for some } k \in \underline{\underline{f}}\}.$$

Those sides of $(M,\underline{\underline{m}})$ which are components of $(\partial M - p^{-1}\partial F)^{-}$ are called _lids_ of $(M,\underline{\underline{m}})$ (Seifert fibre spaces have no lids).

_From now on it is to be understood that every I-bundle or Seifert fibre space, $(M,\underline{\underline{m}})$, admits an admissible fibration_. Note that the boundary-pattern of the base (or orbit surface) of any I-bundle (or Seifert fibre space) with fixed admissible fibration is uniquely determined by $\underline{\underline{m}}$ and vice versa.

In the following we often exclude some collection of the following exceptional cases:

5.1 Exceptional cases:

$(M,\underline{\underline{m}})$, together with a fixed admissible fibration is:

1. the I-bundle over an i-faced disc, $1 \leq i \leq 3$,
2. the $S^1$-bundle over an i-faced disc, $i = 2,3$ or a Seifert fibre space over a 1-faced disc with at most one exceptional fibre,
3. the I-bundle over the 2-sphere or projective plane,
4. a Seifert fibre space with the 2-sphere as orbit surface and at most three exceptional fibres,
5. a Seifert fibre space with the projective plane as orbit surface and at most one exceptional fibre, or
6. a Seifert fibre space with the 2-sphere (possibly with holes) as orbit surface and $\alpha$, $\alpha \geq 0$, boundary components and $\beta$, $\beta \geq 0$, exceptional fibres such that $\alpha + \beta \leq 3$.

It is well-known that an I-bundle or Seifert fibre space is irreducible provided it is not one of the exceptional cases 5.1.3-5.1.5 (for a proof note that the ball is irreducible, by the Schönflies theorem [Al 1] and apply (1.8) of [Wa 1] inductively).

The following lemma is a rather straightforward consequence of the definitions. Nevertheless we include the proof in order to get familiarity with the notations.

5.2 Lemma. Let $(M,\underline{\underline{m}})$ be an I-bundle or Seifert fibre space, with fixed admissible fibration, and neither 5.1.1 nor 5.1.2. Then $\underline{\underline{m}}$ is a useful boundary-pattern of M.

Proof. Let $(D,\underline{\underline{d}})$ be an admissible i-faced disc in $(M,\underline{\underline{m}})$, $1 \leq i \leq 3$. Then we have to show that $\partial D$ bounds a disc, $D'$, in $\partial M$ such that $J \cap D'$ is the cone on $J \cap \partial D'$, where J is the graph of $(M,\underline{\underline{m}})$. Denote by $p: M \to F$ the fibre projection, by A the surface $p^{-1}\partial F$, and let $\underline{\underline{f}}$ be the boundary-pattern of F induced by $\underline{\underline{m}}$.

Case 1. $\partial D$ lies entirely in $A$.

Assume $\partial D$ is not contractible in $A$. Then $\deg(p|\partial D: \partial D \to k) \neq 0$, where $k$ is that boundary curve of $F$ containing $p(\partial D)$. The existence of the map $p|D$ shows that $k$ is contractible in $F$. Hence $F$ is a disc. $p|D: (D,\underline{\underline{d}}) \to (F,\underline{\underline{f}})$ is an admissible singular i-faced disc, $1 \leq i \leq 3$, and so, by $\deg(p|\partial D) \neq 0$, $(F,\underline{\underline{f}})$ is also an i-faced disc, $1 \leq i \leq 3$. Thus, by our supposition on $(M,\underline{\underline{m}})$, $(M,\underline{\underline{m}})$ must be a Seifert fibre space. Hence $A$ is a torus. $\partial D$ is not contractible in $A$. So $(A - U(D))^-$ is an annulus, where $U(D)$ is a regular neighborhood in $M$. Define $\tilde{M} = (M - U(D))^-$, then $\partial\tilde{M}$ is a 2-sphere; a union of an annulus with two discs. This 2-sphere bounds a 3-ball in $M$ since $M$ is irreducible. So $\tilde{M}$ is a ball itself and $M$ a solid torus. Indeed, since $(D,\underline{\underline{d}})$ is an i-faced disc, $1 \leq i \leq 3$, it must be the exception 5.1.2 (recall that any Seifert fibration of the solid torus has at most one exceptional fibre; see [Wa 1, p. 205]) which was excluded.

Thus $\partial D$ is contractible in $A$, and so $\partial D$ bounds a disc, $D'$, in $A$. Define $J^* = (J \cap \overset{0}{A})^-$. $p$ is the fibre projection of an admissible fibration of $(M,\underline{\underline{m}})$, i.e. $p^{-1}pJ^* = J^*$. Hence $J^*$ is either a system of simple arcs, or a system of simple closed curves, with $J^* \cap \partial A = \partial J^*$. So every component of $J^* \cap D'$ is an arc. $J^* \cap \partial D'$ consists of at most three points since $(D,\underline{\underline{d}})$ is an admissible i-faced disc, $1 \leq i \leq 3$. Hence $J^* \cap D'$ is either empty or precisely one arc, and so $J \cap D'$ is the cone on $J \cap \partial D'$.

Case 2. $\partial D$ does not lie entirely in $A$.

In this case $(M,\underline{\underline{m}})$ must be an I-bundle and at least one side of $(D,\underline{\underline{d}})$ lies in a lid of $(M,\underline{\underline{m}})$. Since $(D,\underline{\underline{d}})$ has at most three sides, precisely one of them, say $k_1$, lies in a lid of $(M,\underline{\underline{m}})$, say $G_1$. $p|D: (D,\underline{\underline{d}}) \to (F,\underline{\underline{f}})$ is an admissible map. Lifting $p|D$ to $G_1$ we see that $k_1$ separates an admissible disc, $(D_1,\underline{\underline{d}}_1)$, from $G_1$ so that $(D_1,\bar{\underline{\underline{d}}}_1)$ is an i-faced disc, $1 \leq i \leq 3$ ($\bar{\underline{\underline{d}}}_1$ denotes the completed boundary-pattern of $(D_1,\underline{\underline{d}}_1)$).

If $k_1$ lies entirely in the interior of $G_1$, $D_1$ lies also entirely in $G_1^\circ$, i.e. $D_1 \cap J = \emptyset$. Define $D' = D_1$ and we are done.

If $k_1$ does not lie entirely in the interior of $G_1$, $\bar{D} = D \cup D_1$ is a disc. Pushing $\bar{D}$ a bit away from $G_1$ we get an admissible j-faced disc in $(M,\underline{\underline{m}})$, $j = 1,2$, with $\partial\bar{D} \subset A$. By Case 1, $\partial\bar{D}$ bounds a disc $D_2'$, in $A$ such that $J \cap D_2'$ is the cone on $J \cap \partial D_2'$. Hence $\partial D$ separates a disc, $D_2$, from $A$ such that $J^* \cap D_2$ is the cone on $J^* \cap \partial D_2$, where $J^* = (J \cap \overset{0}{A})^-$ again. Then $D' = D_1 \cup D_2$ is the required disc in $\partial M$. q.e.d.

The following notations were introduced in [Wa 1] for embeddings of surfaces in Seifert fibre spaces.

<u>5.3 Definition</u>. <u>Let</u> $(M.\underline{\underline{m}})$ <u>be an</u> I-<u>bundle or Seifert fibre space</u>, <u>with fixed admissible fibration</u>, <u>and</u> $p: M \to F$ <u>the fibre projection</u>. <u>Let</u> $G$ <u>be a surface</u>. <u>A map</u> $g: G \to M$ <u>is called</u> "<u>vertical</u>" (<u>with respect to</u> p), <u>if</u> $g(G) = p^{-1}pg(G)$ <u>and if</u> $g(G)$ <u>contains no exceptional fibres</u>. <u>A map</u> $g: G \to M$ <u>is called</u> "<u>horizontal</u>" (<u>with respect to</u> p), <u>if</u> $g^{-1}\partial M = \partial G$ <u>and if</u> $p \cdot g$ <u>is a branched covering map</u>; branching points only appear if $(M,\underline{\underline{m}})$ is a Seifert fibre space and then they lie over the exceptional points of the orbit surface.

<u>Furthermore</u>, <u>we call any</u> 3-<u>dim</u>. <u>submanifold</u>, X, <u>in</u> M "<u>vertical</u>", <u>if the surface</u> $(\partial X - \partial M)^-$ <u>is vertical</u> (note that the case that X contains exceptional fibres is not excluded).

The following lemma is an easy consequence of 4.8.2 and 5.2. (It is in general false for mappings of arcs and curves which are not embeddings.) As pointed out in [Wa 1], it gives us a useful tool which applies to I-bundles and a large class of Seifert fibre spaces, M. Indeed, with the help of 5.4, we are able to construct a "<u>vertical hierarchy</u>", i.e. a system of (pairwise disjoint) essential squares, annuli, or tori which splits M either into balls, or into solid tori. Using such a vertical hierarchy, problems about I-bundles and Seifert fibre spaces can often be reduced to local problems.

<u>5.4 Lemma.</u> <u>Let</u> $(M,\underline{\underline{m}})$ <u>be an</u> I-<u>bundle or Seifert fibre space</u>, <u>with fixed admissible fibration</u>, <u>and</u> $p: M \to F$ <u>the fibre projection</u>. <u>Let</u> $\underline{\underline{f}}$ <u>be that boundary-pattern of</u> F <u>induced by</u> $\underline{\underline{m}}$, <u>and</u> $x_1,\dots,x_n$, $n \geq 0$, <u>all the exceptional points of</u> F. <u>Define</u> $\tilde{F} = F - \overset{\circ}{U}(\cup x_i)$, <u>where</u> $U(\cup x_i)$ <u>is</u>

a regular neighborhood in $F$. Suppose $k$ is any essential arc or essential two-sided simple closed curve in $(F,\underline{\underline{f}})$ which is not parallel in $\tilde{F}$ to a curve of $\partial U(\cup x_i)$.

Then $p^{-1}k$ is an essential vertical square, annulus, or torus, resp. in $(M,\underline{\underline{m}})$.

$p^{-1}k$ is not boundary-parallel in $M$, if $k$ is not boundary-parallel in $\tilde{F}$.

5.5 Lemma. Every essential singular closed orientable surface in an I-bundle (twisted or not) can be deformed into the boundary.

Remark. Doubling the base and applying 4.7.1, we obtain from 5.5 a similar statement for essential maps of non-closed orientable surfaces whose boundaries do not meet the lids.

Proof. Let $M$ denote the given I-bundle and $f: G \to M$ the given essential singular surface. Let $p: \tilde{M} \to M$ be the covering map induced by the subgroup $f_*\pi_1 G$ in $\pi_1 M$. $f$ can be deformed, by Nielsen's theorem, into a covering map of a deformation retract of $M$; namely the section of the base. Hence $\tilde{M}$ is compact. The fibration of $M$ lifts to a fibration of $\tilde{M}$ as I-bundle. $\pi_1\tilde{M}$ is isomorphic to the fundamental group of a closed orientable surface, namely of $G$ since $f$ is essential. Hence $\tilde{M}$ must be a product I-bundle. Then a lift of $f$ can be deformed into $\partial\tilde{M}$ and so $f$ into $\partial M$. q.e.d.

The following result and 5.9 are already known for Seifert fibre spaces without boundary-patterns; see [Wa 1]. For convenience of the reader we reformulate the argument given in [Wa 1] in order to establish these results in the form they are needed later.

5.6 Proposition. Let $(M,\underline{\underline{m}})$ be an I-bundle or Seifert fibre space, with fixed admissible fibration, and $p: M \to F$ the fibre projection. Suppose $(M,\underline{\underline{m}})$ is not one of the exceptional cases 5.1.1-5.1.5. Let $G$ be any essential surface in $(M,\underline{\underline{m}})$ with $\partial G \subset \cup_{C\in\underline{\underline{m}}} C$, such that no component of $G$ is an i-faced disc, $1 \leq i \leq 3$, or a 2-sphere.

*Then* G *can be admissibly isotopic deformed in* (M,$\underline{\underline{m}}$) *so that either* 1 *or* 2 *holds*:

1. G *is vertical with respect to* p.
2. G *is horizontal with respect to* p.

*In addition*: *If* B *is any surface of* $\underline{\underline{m}}$ *which is not a lid of* (M,$\underline{\underline{m}}$), *then the admissible isotopy of* G *may be chosen constant on* B ∩ G, *provided* B ∩ G *is either vertical or horizontal*.

*Remark*. Using 4.10, this proposition can be generalized to essential maps f: G → M with G connected, by essentially the same proof.

*Proof*.

*Case* 1. F *is a disc with at most one exceptional point*.

Then M is either the ball or the solid torus. Consider a surface, B, of $\underline{\underline{m}}$ which is not a lid of (M,$\underline{\underline{m}}$). Since p is an admissible fibration of (M,$\underline{\underline{m}}$), B must be either a vertical square or a vertical annulus (it cannot be a torus since (M,$\underline{\underline{m}}$) is not 5.1.2). B ∩ G is a system of essential curves in B since G is essential and no component of G is an i-faced disc, $1 \leq i \leq 3$. Hence G can be admissibly isotoped so that B ∩ G is either vertical or horizontal. This holds for every B. Thus we may suppose G is admissibly isotoped so that G ∩ $p^{-1}\partial F$ is either vertical or horizontal.

Suppose M is a ball. Then (M,$\underline{\underline{m}}$), together with p is a product I-bundle, and G consists of discs [Wa 1,(1.4)]. If G ∩ $p^{-1}\partial F$ is vertical, we apply [Wa 4,3.4] and 5.6.1 follows. If G ∩ $p^{-1}\partial F$ is horizontal, we find easily a system of horizontal discs in M whose boundary is ∂G. But this system must be isotopic to G (relative boundary) since M is irreducible, and 5.6.2 follows.

Suppose M is a solid torus. Then (M,$\underline{\underline{m}}$), together with p, is a Seifert fibre space since F is a disc. By [Wa 1, (2.3)], G consists either of discs or of boundary-parallel annuli. G ∩ $p^{-1}\partial F$ is either vertical or horizontal. Thus we easily find a system which consists either of vertical annuli near ∂M or of horizontal meridian discs and which is isotopic to G, by an isotopy which is constant on the boundary (a solid torus is irreducible). Hence, in these cases, 5.6 follows too.

The additional remark follows by checking the arguments.

Case 2. F *is not a disc with at most one exceptional point.*

In this case we certainly may suppose that $\underline{\underline{m}}$ is a complete boundary-pattern of M. Then, applying 5.4, we find at least one essential vertical square, annulus, or torus, H, in $(M,\underline{\underline{m}})$ which is not boundary-parallel. Let $(\tilde{M},\tilde{\underline{\underline{m}}})$ be that manifold which we obtain by splitting $(M,\underline{\underline{m}})$ at H. Then $(\tilde{M},\tilde{\underline{\underline{m}}})$, together with $p|\tilde{M}$, consists of I-bundles of Seifert fibre spaces. H is an essential surface in $(M,\underline{\underline{m}})$ and $\underline{\underline{m}}$ is a useful boundary-pattern of M (see 5.2). Hence, by 4.8.2, $\tilde{\underline{\underline{m}}}$ is a useful boundary-pattern of $\tilde{M}$, and so, by 5.2, no component of $(\tilde{M},\tilde{\underline{\underline{m}}})$ is 5.1.1 or 5.1.2, hence it is clear that none of these is 5.1.1-5.1.5. Now isotope G admissibly in $(M,\underline{\underline{m}})$ so that the number of curves of $G \cap H$ is as small as possible (for the additional remark note that this isotopy may be chosen constant on $G \cap B$, if $G \cap B$ is vertical or horizontal, for $B \in \underline{\underline{m}}$). Hence, by 4.6.4, $\tilde{G} = G \cap \tilde{M}$ is an essential surface in $(\tilde{M},\tilde{\underline{\underline{m}}})$. Furthermore, no component of $\tilde{G}$ can be an i-faced disc, $1 \leq i \leq 3$, since the curves of $G \cap H$ are essential in H (see 4.6.2) and H is essential in $(M,\underline{\underline{m}})$. Thus we have an induction. q.e.d.

5.6 is a structure theorem for I bundles and Seifert fibre spaces which has a lot of applications. The remainder of this paragraph is devoted to describe some of them.

5.7 Corollary. *Let $(M,\underline{\underline{m}})$ be an I-bundle or Seifert fibre space with fixed admissible fibration. Suppose $(M,\underline{\underline{m}})$ is not one of the exceptional cases 5.1.1-5.1.5. Let T be any essential surface in $(M,\underline{\underline{m}})$. If each component of T is a square or annulus, one of the following holds:*

1. *There exists an admissible isotopic deformation of T in $(M,\underline{\underline{m}})$ which makes T vertical.*

2. *There exists an admissible fibration of* $(M,\underline{\underline{m}})$ *as I-bundle over the square, annulus, torus, Möbius band, or Klein bottle such that* $T$ *is vertical with respect to this fibration.*

*If* $T$ *consists of tori, then either* 1 *holds, or*

3. $M$ *is the I-bundle over the torus or Klein bottle.*
4. $M$ *is one of the closed 3-manifolds which can be obtained by glueing two I-bundles over the torus or Klein bottle together at their boundaries.*

*Proof*. We may apply 5.6. If 1 of 5.6 holds, we are done. So we may suppose $T$ can be admissibly isotoped into a horizontal surface. This isotopy extends to an admissible ambient isotopy $\alpha_t$, $t \in I$, of $(M,\underline{\underline{m}})$. Hence we may suppose the admissible fibration of $(M,\underline{\underline{m}})$ is admissibly isotoped, e.g. by $\alpha_t^{-1}$, so that $T$ is horizontal. Let $p\colon M \to F$ be the fibre projection.

Let $(\tilde{M},\tilde{\underline{\underline{m}}})$ be the manifold obtained from $(M,\underline{\underline{m}})$ by splitting at $T$. The fibration of $M$ induces a fibration of $(\tilde{M},\tilde{\underline{\underline{m}}})$ as a system of I-bundles. The copies of $T$ are the lids of $\tilde{M}$. So $M$ is as described in 5.7.3 and 5.7.4, if $T$ consists of tori. In the other cases there is also an admissible fibration of $(\tilde{M},\tilde{\underline{\underline{m}}})$ as a system of I-bundles such that the lids are contained in $\tilde{M} \cap \partial M$ and which corresponds via $T$. This fibration induces an admissible fibration of $(M,\underline{\underline{m}})$ as described in 5.7.2. q.e.d.

*5.8 Corollary.* *Let* $(N,\underline{\underline{n}})$ *be a connected, irreducible* 3-*manifold with useful boundary-pattern. Let* $(M,\underline{\underline{m}})$ *be an* I-*bundle such that* $\underline{\underline{m}}$ *is the complete boundary-pattern of* $M$. *Suppose* $(M,\underline{\underline{m}})$ *is neither* 5.1.1 *nor* 5.1.2. *Let* $p\colon (M,\underline{\underline{m}}) \to (N,\underline{\underline{n}})$ *be an essential map.*

*Then* $(N,\underline{\underline{n}})$ *is also an* I-*bundle.*

*In addition: If* $p$ *is a covering map, there are admissible fibrations of* $(M,\underline{\underline{m}})$ *and* $(N,\underline{\underline{n}})$, *as* I-*bundles, such that* $p$ *is fibre preserving.*

*Remark*. This extends results for manifolds without boundary-patterns

[Sc 1, Wa 4]. There is a similiar statement for Seifert fibre spaces, too, but to prove it requires more machinery; see 12.9.

Proof. By 5.2, $\underline{\underline{m}}$ is a useful boundary-pattern. Applying 3.4, we may suppose p is admissibly deformed into a covering map. Then we may suppose that N is not a ball since every covering map onto the ball is a homeomorphism, and we are done. $\partial N \neq \emptyset$ since $\partial M \neq \emptyset$. Hence, by 4.3, there exists an essential surface, F, in $(N,\underline{\underline{n}})$ with $F \cap \partial N = \partial F$ which is non-separating. It follows that $G = p^{-1}F$ is an essential surface in $(M,\underline{\underline{m}})$, $G \cap \partial M = \partial G$, such that each component of G is non-separating (apply [Wa 4, 5.3]). In particular, G cannot be admissibly isotoped in $(M,\underline{\underline{m}})$ into a horizontal surface, since $(M,\underline{\underline{m}})$ is an I-bundle. Therefore, by 5.6, G can be admissibly isotoped in $(M,\underline{\underline{m}})$ into a vertical surface. This isotopy extends to an admissible ambient isotopy, $\alpha_t$, $t \in I$, of $(M,\underline{\underline{m}})$. Then $\alpha_t^{-1}$ deforms admissibly the fibration of $(M,\underline{\underline{m}})$ so that G is vertical. Splitting $(N,\underline{\underline{n}})$ at F, and $(M,\underline{\underline{m}})$ at G, we have an induction.

The additional remark follows immediately. q.e.d.

Besides the forementioned applications, 5.6 is also a starting point for the study of homeomorphisms (or better, their isotopy classes), at least for homeomorphisms of I-bundles and Seifert fibre spaces. As a first consequence, we state a special version of Waldhausen's theorem as we need it (for more information the interested reader is referred to [Wa 1]). Later on we shall study homeomorphisms of I-bundles and Seifert fibre spaces more closely (see §§25 and 26), and we shall apply this to the study of the mapping class group of general Haken 3-manifolds.

5.9 Corollary. Let $(M_i,\underline{\underline{m}}_i)$, $i = 1,2$, be an I-bundle or Seifert fibre space with fixed admissible fibration. Suppose $(M_i,\underline{\underline{m}}_i)$, $i = 1,2$, is neither a solid torus with $\bar{\underline{\underline{m}}}_i = \{\partial M_i\}$, nor one of the exceptions described in 5.1.3-5.1.5 or 5.7.2-5.7.4.

Then every admissible homeomorphism $h: (M_1,\underline{\underline{m}}_1) \to (M_2,\underline{\underline{m}}_2)$ can be admissibly isotoped into a fibre preserving one.

In addition:

1. The conclusion holds, if $M_i$ is one of 5.7.2 and h as well as $h^{-1}$ map lids into lids.
2. If $M_1$ is an I-bundle (twisted or not) and $h: M_1 \to M_1$ is the identity on one lid, then the isotopy may be chosen to be constant on this lid.

Remark. 1. Observe that 5.9 is a way of expressing that the admissible fibrations of the I-bundles and Seifert fibre spaces in 5.9 are unique, up to admissible ambient isotopy.

2. One might also ask, whether or not an arbitrary essential map $f: (M_1, \underline{\underline{m}}_1) \to (M_2, \underline{\underline{m}}_2)$ can be admissibly deformed into a fibre preserving map. This question (besides others) will be considered in §28.

Proof. If $M_1$ is a ball or a solid torus, 5.9 follows by successive applications of Alexander's trick, or from [Wa 1, (5.1) and (5.4)]. In the other cases, we may suppose that $\underline{\underline{m}}_i$ is a complete boundary-pattern. Furthermore, applying 5.4, we find at least one essential vertical square, annulus, or torus, T, in $(M_1, \underline{\underline{m}}_1)$ which is not boundary-parallel. h(T) is an essential square, annulus, or torus in $(M_2, \underline{\underline{m}}_2)$ since h is an admissible homeomorphism. By our suppositions on $(M_2, \underline{\underline{m}}_2)$, we may apply 5.7. Hence it follows that $h|T$ can be admissibly isotoped in $(M_2, \underline{\underline{m}}_2)$ into a vertical map. Extending this isotopy to an admissible isotopy of h and splitting $(M_1, \underline{\underline{m}}_1)$ at T and $(M_2, \underline{\underline{m}}_2)$ at h(T), we have an induction.

A similar argument proves the first additional remark. The second additional remark is clear, if $M_1$ is a product I-bundle. In the other case it follows from a nice property of the Möbius band. To describe this, let B be a Möbius band, and let k, k' be two non-separating arcs in B. Denote by $x_1$ and $x_2$ the two end-points of k, and suppose that the end-points of k' are equal to $x_1$ and $x_2$. Isotop k (rel $\partial k$) so that $k \cap k'$ is minimal. Then the topology of the Möbius band implies that $k \cap k' = \partial k$. If k cannot be deformed (rel $\partial k$) into k' we isotop k (rel $x_1$) so that $x_2$ lies near $x_1$. Doing this in the right direction we see that $k$ is separating, which is a contradiction. Thus $k$ is isotopic to $k'$ (rel $\partial k$). This implies, by the Alexander trick, that a homeomorphism of the Möbius band is isotopic (rel $\partial B$) to the identity, provided the restriction to the boundary is the identity. Now let $M_1$ be a twisted I-bundle. By 5.4, we find a vertical non-separating square or annulus $A$ in $M_1$. By supposition, the sides $k_1$, $k_2$ of $A' = h(A)$, contained in the lid of $M_1$, are equal to sides of $A$. The projection of $A'$ to the base of $M_1$ is a map which maps $k_1$ and $k_2$ to the projection $t$ of $A$. By Nielsen's theorem, either this map can be admissibly deformed (rel $k_1$ and $k_2$) into $t$, or the base is a Möbius band or Klein bottle. In either case it is easily seen (lift the homotopy and recall the above property of the Möbius band) that $A'$ is admissibly isotopic to $A$ (rel $k_1 \cup k_2$). Hence again we have an induction. q.e.d.

5.7 can be considered as a classification of embedded essential squares, annuli and tori in I-bundles and Seifert fibre spaces. In the remainder of this paragraph we are going to classify also singular essential squares, annuli, or tori in these manifolds. Later on we apply these results to classify essential singular squares, annuli, or tori in Haken 3-manifolds (see §12).

<u>5.10 Proposition.</u> <u>Let</u> $(M,\underline{\underline{m}})$ <u>be an</u> I-<u>bundle or Seifert fibre space with fixed admissible fibration</u>. <u>Suppose</u> $(M,\underline{\underline{m}})$ <u>is neither the exception</u> 5.1.1 <u>nor</u> 5.1.2. <u>Let</u> $f: T \to M$ <u>be an essential singular square or annulus in</u> $(M,\underline{\underline{m}})$.

<u>Then either</u> 1 <u>or</u> 2 <u>holds</u> (<u>or both</u>):

1. <u>There exists an admissible deformation of</u> $f$ <u>in</u> $(M,\underline{\underline{m}})$ <u>into a vertical map</u>.
2. <u>There exists an admissible fibration of</u> $(M,\underline{\underline{m}})$ <u>as</u> I-<u>bundle over the square</u>, <u>annulus</u>, <u>Möbius band</u>, <u>torus</u>, <u>or Klein bottle</u>.

<u>In addition</u>: <u>If</u> $k$ <u>is any side of</u> $T$ <u>which is mapped by</u> $f$ <u>into a lid of</u> $(M,\underline{\underline{m}})$, <u>then</u> $f$ <u>can be admissibly deformed into a vertical map, using a homotopy which is constant on</u> $k$.

Proof. Let $p: M \to F$ be the fibre projection, and $\underline{\underline{f}}$ that boundary-pattern of $F$ induced by $\underline{\underline{m}}$.

Case 1. $T$ is a square and no side of $T$ is mapped by $f$ into a lid of $(M,\underline{\underline{m}})$.

We will show that in this case 2 of 5.10 holds. Since $f(\partial T) \subset p^{-1}\partial F$, $p \cdot f: T \to F$ is an admissible map into $(F,\underline{\underline{f}})$. The restriction of $f$ to each side of $T$ is an essential singular arc in some surface of $\underline{\underline{m}}$, since $f$ is essential and not an admissible singular i-faced disc, $1 \leq i \leq 3$. Thus $\deg(p \cdot f|\partial T: \partial T \to \partial F) \neq 0$, and so the existence of the admissible singular square $p \cdot f$ shows that $(F,\underline{\underline{f}})$ is either the 2- or 4-faced disc.

If $(M\ \underline{\underline{m}})$ is admissibly fibered as I-bundle, then $(F,\underline{\underline{f}})$ must be a square since $(M,\underline{\underline{m}})$ is not 5.1.1, and so 2 of 5.10 follows.

If $(M,\underline{\underline{m}})$ is admissibly fibered as Seifert fibre space, $\partial M$ consists of tori. $f|\partial T$ is not contractible in $\partial M$, since each side of $f$ is an essential singular arc in some surface of $\underline{\underline{m}}$. Hence the existence of $f$ shows that $M$ is boundary-reducible, and so it follows that $M$ must be a solid torus ($M$ is irreducible). Since $(F,\underline{\underline{f}})$ is a 2- or 4-faced disc, $\underline{\underline{m}}$ is a complete boundary-pattern of $M$ which consists of two or four annuli. Consequently $(M,\underline{\underline{m}})$ can be admissibly fibered also as I-bundle over the annulus or Möbius band ($(M,\underline{\underline{m}})$ is not 5.1.2) and so 2 of 5.10 follows again.

Case 2. $T$ is an annulus and no side of $T$ is mapped by $f$ into a lid of $(M,\underline{\underline{m}})$.

If $(M,\underline{\underline{m}})$ is admissibly fibered as I-bundle, then, projecting the essential singular annulus $f$ onto the base and applying Nielsen's theorem, we see that the base must be an annulus or Möbius band; hence 2 of 5.10 follows.

If $(M,\underline{\underline{m}})$ is a solid torus, the boundary curves of $T$ are mapped into different annuli of $\underline{\underline{m}}$, since $f$ is essential, and then it is easily seen that $f$ can be deformed (rel $\partial T$) into $\partial M$, and that this implies 1 of 5.10.

Thus we may suppose that $(M,\underline{\underline{m}})$ is admissibly fibered as Seifert fibre space, and that $M$ is not a solid torus. Then we may assume that $\underline{\underline{m}}$ is a complete boundary-pattern of $M$, and, applying 5.4, we find at least one essential vertical annulus, $H$, in $(M,\underline{\underline{m}})$ which is not boundary-parallel. Let $(\tilde{M},\tilde{\underline{\underline{m}}})$ be that manifold which we obtain by splitting $(M,\underline{\underline{m}})$ at $H$. Then $(\tilde{M},\tilde{\underline{\underline{m}}})$, together with $p|\tilde{M}$, consists of Seifert fibre spaces, again. $H$ is an essential surface in $(M,\underline{\underline{m}})$ and $\underline{\underline{m}}$ is a useful boundary-pattern of $M$ (see 5.2). Hence by 4.8.2, $\tilde{m}$ is a useful boundary-pattern of $\tilde{M}$, and so, by 5.2, no component of $(\tilde{M},\tilde{\underline{\underline{m}}})$ is 5.1.2. Now deform $f$ admissibly in $(M,\underline{\underline{m}})$ so that the number of curves of $f^{-1}H$ is as small as possible. Hence, by 4.7.3, the restriction of $f$ to any component of $f^{-1}\tilde{M}$ is an essential singular square, or annulus in $(\tilde{M},\tilde{\underline{\underline{m}}})$ (recall $T$ is an annulus).

1. Suppose $f^{-1}\tilde{M}$ is a system of squares. No side of these squares is mapped by $f$ into a lid of $(\tilde{M},\tilde{m})$ Hence, by Case 1, each component of $(\tilde{M},\tilde{m})$ can be admissibly fibered as I-bundle over the annulus or Möbius band. Moreover, these fibrations can be chosen so that neither $H_1$ nor $H_2$ is a lid, where $H_1$, $H_2$ are the copies of $H$ in $(\tilde{M},\tilde{\underline{\underline{m}}})$. Hence they induce an admissible fibration of $(M,\underline{\underline{m}})$ as I-bundle, as described in 2 of 5.10.

2. Suppose $f^{-1}\tilde{M}$ is a system of annuli. If every component of $\tilde{M}$ is a solid torus, $f|f^{-1}\tilde{M}$ can be admissibly deformed in $(\tilde{M},\tilde{\underline{\underline{m}}})$ into a vertical map. If not, applying 5.4 again, we find at least one essential vertical annulus, $H'$, in $(\tilde{M},\tilde{\underline{\underline{m}}})$ such that at least one boundary curve of $H'$ lies in $H_1$, say. Using the above argument, $f|f^{-1}\tilde{M}$ can be admissibly deformed in $(\tilde{M},\tilde{\underline{\underline{m}}})$ so that $f^{-1}H'$ splits $f^{-1}\tilde{M}$ into a system either of squares, or of annuli. But, by our choice of $H'$, it follows that $f^{-1}\tilde{M}$ must be a system of annuli. Thus we have an induction, and 1 of 5.10 follows.

<u>Case 3</u>. <u>At least one side</u>, $k$, <u>of</u> $T$ <u>is mapped by</u> $f$ <u>into a lid of</u> $(M,\underline{\underline{m}})$.

Assume only the side $k$ is mapped into a lid of $(M,\underline{\underline{m}})$. Then let $t$ be any essential arc in $T$ which meets $k$. Considering $p^{-1}p(t)$, it follows that $f|t$ is inessential in $(M,\underline{\underline{m}})$. But this contradicts the fact that $f$ is essential in $(M,\underline{\underline{m}})$. Thus that side of $T$ opposite to $k$ is also mapped into a lid of $(M,\underline{\underline{m}})$. Notice that for the square or annulus, $k \times I$, there is a contraction $\alpha_t\colon k \times I \to k \times I$, $t \in I$, defined by $\alpha_t(x,s) = (x,(1-t)s)$. Since that side of $T$ opposite to $k$ is also mapped into a lid, there is a boundary-pattern of $T$ such that $p\circ f\circ\alpha_t\colon T \to F$ is an admissible homotopy of $p\circ f$ in $(F,\underline{\underline{f}})$. $(M,\underline{\underline{m}})$ is an I-bundle. Lift the homotopy $p\circ f\circ\alpha_t$ to a "level preserving" admissible homotopy of $f$ in $(M,\underline{\underline{m}})$. This is clear if $(M,\underline{\underline{m}})$ is a product I-bundle. If $(M,\underline{\underline{m}})$ is a twisted I-bundle, let $p\colon (\bar{M},\bar{\underline{\underline{m}}}) \to (M,\underline{\underline{m}})$ be the canonical admissible 2-sheeted covering such that $(\tilde{M},\tilde{\underline{\underline{m}}})$ is a product I-bundle, and let $\tilde{f}\colon T \to \tilde{M}$ some lifting of $f$ (to see that this lifting exists note that $T$ is an annulus). Then first lift $p\circ f\circ\alpha_t$ to a level preserving admissible homotopy of $\tilde{f}$ in $(\tilde{M},\tilde{\underline{\underline{m}}})$, and project this finally down to $(M,\underline{\underline{m}})$. The homotopy of $f$ defined in this way clearly pushes $f$ into a vertical map and is constant on $k$. Thus 1 of 5.10 and the additional remark follow.

q.e.d.

In the following two corollaries we describe what happens if the sides of an essential singular annulus satisfy certain conditions, especially if they are disjoint or non-singular.

<u>5.11 Corollary</u>. <u>Let</u> $(M,\underline{\underline{m}})$ <u>be a Seifert fibre space, but not</u> 5.1.2. <u>Let</u> $f\colon T \to M$ <u>be an essential singular annulus in</u> $(M,\underline{\underline{m}})$, <u>and</u> $k_1$, $k_2$ <u>both the boundary curves of</u> $T$. <u>Suppose</u> $f(k_1) \cap f(k_2) = \emptyset$.

<u>Then there exists an admissible fibration of</u> $(M,\underline{\underline{m}})$ <u>as Seifert fibre space and an admissible deformation of</u> $f$ <u>in</u> $(M,\underline{\underline{m}})$ <u>into a vertical map with respect to this fibration</u>.

<u>Proof</u>. Fix, for a moment, an admissible fibration of $(M,\underline{\underline{m}})$ as Seifert fibre space and suppose $f$ cannot be admissibly deformed into a vertical map with respect to this fibration. Then $M$ cannot

be a solid torus since then, by $f(k_1) \cap f(k_2) = \emptyset$, it is easily seen that $f$ can be admissibly deformed into a vertical map (push $f$ near $\partial M$). Hence, by 5.10, $(M,\underline{\underline{m}})$ can be admissibly fibered as I-bundle over the torus of Klein bottle. Fix such an admissible fibration and let $p: M \to F$ be the fibre projection. $\partial M$ consists of tori and so we may suppose $f|k_i$, $i = 1,2$, is a multiple of a simple closed curve, $t_i$.

We will show that $f$ can be admissibly deformed into a non-singular annulus or Möbius band. It suffices to prove that $pf|k_1$ can be deformed into a (multiple of a) non-singular curve in $F$ (one- or two-sided), since then we find the required homotopy of $f$ by the lifting-argument in Case 3 of 5.10. The existence of such a deformation of $pf|k_1$ is clear if $F$ is the torus and so we suppose $(M,\underline{\underline{m}})$, together with $p$, is the I-bundle over the Klein bottle. Then $\partial M$ is connected, $p|\partial M: \partial M \to F$ is a 2-sheeted covering map, and the non-trivial covering translation $d: \partial M \to \partial M$ is given by the reflections in the fibres of $(M,\underline{\underline{m}})$. By the additional remark of 5.10, $f$ can be admissibly deformed (rel $k_1$) into a vertical map. Hence $df|k_1 \simeq (f|k_2)^{\pm 1}$. Furthermore, $f(k_1) \cap f(k_2) = \emptyset$. Thus $t_1 = f(k_1)$ can be deformed in $\partial M$ into $t_1'$ so that $t_1' \cap dt_1 = \emptyset$. In this case we find discs, $D$, in $F$ with $\partial D = D \cap (t_1 \cup dt_1)$, and using these discs it is not difficult to define an equivariant homotopy of $t_1$ into $t_1'$ such that $t_1' \cap dt_1' = \emptyset$ or $t_1' = dt_1'$. Thus, in any case, $pf|k_1$ can be deformed into (a multiple of) a non-singular curve, $t$, in $F$.

Denote $B = p^{-1}(t)$. A regular neighborhood $U(B)$ in $M$ is a solid torus, since $B$ is either an annulus or a Möbius band. Furthermore, $t$ splits $F$ either into one annulus, or into one or two Möbius bands, since $F$ is either a torus or a Klein bottle. Thus $(M - U(B))^-$ consists of solid tori as well. Hence, denoting $(\tilde{M},\tilde{\underline{\underline{m}}})$ as that manifold obtained from $(M,\underline{\underline{m}})$ by splitting at $(\partial U(B) - \partial M)^-$, $(\tilde{M},\tilde{\underline{\underline{m}}})$ consists of solid tori. These can be admissibly fibered as Seifert fibre spaces and these fibrations induce an admissible fibration of $(M,\underline{\underline{m}})$ as required (see [Wa 1,(5.1)]). q.e.d.

With a similiar argument as used in 5.11 we can prove the

following fact about singular squares or annuli in I-bundles.

5.12 Corollary. *Let $(M,\underline{\underline{m}})$ be an I-bundle, but not 5.1.1. Let $f: T \to M$ be an essential singular square or annulus in $(M,\underline{\underline{m}})$. Let $k_1,k_2$ be two sides of $T$ mapped under $f$ into the lids of $M$. Suppose that $f(k_1)$ is a non-singular curve and that $f(k_1) \cap f(k_2) = \emptyset$. Then $f$ can be admissibly deformed into a (non-singular) vertical, essential square, annulus, or Möbius band in $(M,\underline{\underline{m}})$.*

We now finally come to the classification of essential singular tori in certain I-bundles and Seifert fibre spaces.

5.13 Proposition. *Let $(M,\underline{\underline{m}})$ be an I-bundle or Seifert fibre space with fixed admissible fibration. Suppose that $(M,\underline{\underline{m}})$ is not one of the exceptional cases 5.1.1-5.1.5, and not as described in 5.7.4. Let $f: T \to M$ be an essential singular torus.*

*Then there eixsts an admissible fibration of $(M,\underline{\underline{m}})$ as Seifert fibre space and an admissible deformation of $f$ into a vertical map with respect to the latter Seifert fibration.*

Remark. In 7.1 we shall see that it is not really necessary to assume that $M$ is not as described in 5.7.4 (compare the proofs).

Proof. If $(M,\underline{\underline{m}})$ is admissibly fibered as I-bundle, then projecting the essential singular torus $f$ onto the base and applying Nielsen's theorem, it follows that the base must be a torus or a Klein bottle. Hence $(M,\underline{\underline{m}})$ can be admissibly fibered as Seifert fibre space. Furthermore, by 5.5, $f$ can be admissibly deformed near $\partial M$, and so we are done.

Thus we may suppose $(M,\underline{\underline{m}})$ is admissibly fibered as Seifert fibre space. Let $p: M \to F$ be the fibre projection. $M$ is not the solid torus since $f$ is essential. Hence we may suppose that $\underline{\underline{m}}$ is a complete boundary-pattern of $M$. Furthermore, $M$ is not 5.1.3-5.1.5. Therefore, applying 5.4, we find at least one essential vertical annulus or torus, $H$, in $(M,\underline{\underline{m}})$ which is not boundary-parallel. Let $(\tilde{M},\tilde{\underline{\underline{m}}})$ be that manifold which we obtain by splitting $(M,\underline{\underline{m}})$ at $H$.

Then, $(\tilde{M},\tilde{\underline{\underline{m}}})$, together with $p|\tilde{M}$, again consists of Seifert fibre spaces. H is an essential surface in $(M,\underline{\underline{m}})$ and $\underline{\underline{m}}$ is a useful boundary-pattern of M (see 5.2). Hence, by 4.8.2, $\tilde{\underline{\underline{m}}}$ is a useful boundary-pattern of $\tilde{M}$, and so, by 5.2, no component of $(\tilde{M},\tilde{\underline{\underline{m}}})$ is 5.1.2. Moreover, clearly no component of $(\tilde{M},\tilde{\underline{\underline{m}}})$ is 5.1.1 or 5.1.3-5.1.5. Now deform f admissibly in $(M,\underline{\underline{m}})$ so that the number of curves of $f^{-1}H$ is as small as possible. Hence, by 4.7.3, the restriction of f to any component of $f^{-1}\tilde{M}$ is an essential singular annulus or torus in $(\tilde{M},\tilde{\underline{\underline{m}}})$ (T is a torus). Without loss of generality $f^{-1}\tilde{M}$ is not a torus since otherwise we apply the above construction to $(\tilde{M},\tilde{\underline{\underline{m}}})$ and so on.

If no component of $(\tilde{M},\tilde{\underline{\underline{m}}})$ admits an admissible fibration as I-bundle over the torus of Klein bottle, then, by 5.10, it follows that the restriction of f to any component of $f^{-1}\tilde{M}$ can be admissibly deformed in $(\tilde{M},\tilde{\underline{\underline{m}}})$ into a vertical map. Then 5.12 follows immediately.

Assume at least one component of $(\tilde{M},\tilde{\underline{\underline{m}}})$ admits an admissible fibration as I-bundle over the torus. Then at least one copy of H is a boundary component of this product I-bundle. Hence H is non-separating since H is not boundary-parallel. But then it follows that M is one of those closed manifolds which can be obtained by glueing two I-bundles over the torus together at their boundaries, which was excluded.

Finally we assume at least one component, $(\tilde{M}_1,\tilde{\underline{\underline{m}}}_1)$, of $(\tilde{M},\tilde{\underline{\underline{m}}})$ admits an admissible fibration as I-bundle over the Klein bottle. Then again at least one copy of H is a boundary component of $\tilde{M}_1$. Hence H must be separating (H is two-sided in M). The other component, $(\tilde{M}_2,\tilde{\underline{\underline{m}}}_2)$, of $(\tilde{M},\tilde{\underline{\underline{m}}})$ can be neither an I-bundle over the torus nor an I-bundle over the Klein bottle since H is not boundary-parallel and since M cannot be obtained by glueing two I-bundles over the Klein bottle together at their boundaries. Hence, by 5.10, the restriction of f to any component of $f^{-1}\tilde{M}_2$ can be admissibly deformed in $(\tilde{M}_2,\tilde{\underline{\underline{m}}}_2)$ into a vertical map with respect to $p|\tilde{M}_2$. Now let, A, be any annulus of $f^{-1}\tilde{M}_1$ and $k_1$, $k_2$ both its boundary curves. Then, by the fact that $f|f^{-1}\tilde{M}_2$ is vertical up to admissible homotopy, it follows that $f|A$ can be admissibly deformed in $(\tilde{M}_1,\tilde{\underline{\underline{m}}}_1)$ so that then $f(k_1) \cap f(k_2) = \emptyset$. Therefore, by 5.11, there exists an admissible fibration of $(\tilde{M}_1, \tilde{\underline{\underline{m}}}_1)$ as Seifert fibre space and an admissible deformation of $f|A$ in $(\tilde{M}_1, \tilde{\underline{\underline{m}}}_1)$ into a vertical map with respect to this fibration. In this way every annulus of $f^{-1}\tilde{M}_1$ induces an admissible fibration of $(\tilde{M}_1, \tilde{\underline{\underline{m}}}_1)$ as Seifert fibre space. By 5.9 (set h = id), all these fibrations are equal, up to admissible ambient isotopy of $(\tilde{M}_1, \tilde{\underline{\underline{m}}}_1)$. Moreover, by our choice of these fibrations, they can be admissibly isotoped in $(\tilde{M}_1, \tilde{\underline{\underline{m}}}_1)$ so that they coincide on H with the fibration induced by $p|\tilde{M}_2$ (apply [Wa 1, (5.2)]. Thus the conclusion of 5.12 follows. q.e.d.

## §6. Stallings manifolds

A 3-manifold is called a <u>Stallings manifold</u> if there is a connected essential surface $F$ in $(M,\underline{\underline{m}})$, $F \cap \partial M = \partial F$, such that the manifold $(\tilde{M},\tilde{\underline{\underline{m}}})$ obtained from $(M,\underline{\underline{m}})$ by splitting at $F$ consists of I-bundles (twisted or not) whose lids may be chosen as copies of $F$. Two cases arise.

1. $\tilde{M}$ is connected. Then $(\tilde{M},\tilde{\underline{\underline{m}}})$ is a product I-bundle and $(M,\underline{\underline{m}})$ admits a structure of a Stallings fibration.
2. $\tilde{M}$ is disconnected. Then each component of $(\tilde{M},\tilde{\underline{\underline{m}}})$ is a twisted I-bundle.

If we denote by $F_1$, $F_2$ the two sides of $\tilde{M}$ which are copies of $F$, then we reobtain $M$ from $\tilde{M}$ by attaching $F_1$ and $F_2$ via a canonic homeomorphism $\varphi: F_2 \to F_1$. $\varphi$ gives rise to the definition of a homeomorphism $F \to F$, respectively a pair of such homeomorphisms, depending on whether or not $\tilde{M}$ is connected. To construct them observe that the reflections in the I-fibres of $(\tilde{M},\tilde{\underline{\underline{m}}})$ define a homeomorphism $g: F_1 \to F_2$ (respectively, homeomorphisms $g_i: F_i \to F_i$, $i = 1,2$). We define $f = \varphi \cdot g$ if $\tilde{M}$ is connected, or $f_1 = g_1$ and $f_2 = \varphi \cdot g_2 \cdot \varphi^{-1}$ otherwise. These homeomorphisms will be called the <u>autohomeomorphisms of $F$ induced by</u> $(\tilde{M},\tilde{\underline{\underline{m}}})$.

We only deal with the case that $(M,\underline{\underline{m}})$ is irreducible and not the solid torus. Then $F$ is neither the disc, nor the 2-sphere. Furthermore observe that $\tilde{\underline{\underline{m}}}$ and $\underline{\underline{m}}$ are useful boundary-patterns of $\tilde{M}$ resp. $M$ (5.2 and 4.8), and that $M$ is aspherical (sphere theorem).

<u>6.1 Proposition.</u> <u>Let $(M,\underline{\underline{m}})$ be an irreducible Stallings manifold. Suppose there exists an essential singular annulus in $(M,\underline{\underline{m}})$. Then there exists a non-singular essential annulus in $(M,\underline{\underline{m}})$.</u>

<u>6.2 Proposition.</u> <u>Let $(M,\underline{\underline{m}})$ be an irreducible Stallings manifold. Suppose there exists an essential singular torus in $(M,\underline{\underline{m}})$ which cannot be deformed into $\partial M$. Then at least one of the following assertions is true:</u>

1. <u>There exists a non-singular essential torus in $(M,\underline{\underline{m}})$</u>

*which is not boundary-parallel.*

2. *(M,$\underline{\underline{m}}$) is a Seifert fibre space.*

The idea is to reduce 6.1 and 6.2 to 2-manifold problems and then to use arguments of Nielsen on the structure of surface-homeomorphisms.

As a first reduction step we state and prove the following lemma. For this let (M,$\underline{\underline{m}}$) be an irreducible Stallings manifold, and define F and $f_1$, $f_2$ as in the beginning of this paragraph (if M is connected, set $f_2$ = id).

6.3 Lemma.

1. *Suppose* $\chi(F) < 0$ *and there exists an essential singular annulus in* (M,$\underline{\underline{m}}$). *Then there exists an essential singular arc* $k: I \to F$ *and an integer* $n \geq 1$ *such that*

$$(f_2 f_1)^n k \text{ is admissibly homotopic in } F \text{ to } k.$$

2. *Suppose* $\chi(F) < 0$ *and there exists an essential singular torus in* (M,$\underline{\underline{m}}$). *Then there exists an essential singular closed curve* $k: S^1 \to F$ *and an integer* $n \geq 1$ *such that*

$$(f_2 f_1)^n k \simeq k.$$

*If, in addition, the singular torus cannot be deformed into* $\partial M$, k *may be chosen so that it cannot be deformed in* F *into* $\partial F$.

Remark. The lemma is also true when the condition $\chi(F) < 0$ is replaced by the condition that the given singular annulus or torus cannot be admissibly deformed into ($\tilde{M}$,$\tilde{\underline{\underline{m}}}$).

Proof. Let $f: T \to M$ be a singular essential annulus or torus, respectively. Deform f admissibly so that the number of curves of $f^{-1}F$ is as small as possible. If $f^{-1}F = \emptyset$, f(T) lies entirely in one I-bundle, ($\tilde{M}_1$,$\tilde{\underline{\underline{m}}}_1$), of ($\tilde{M}$,$\tilde{\underline{\underline{m}}}$), and does not meet any lid.

Projecting $f$ onto the base of $(\widetilde{M}_1, \underline{\widetilde{m}}_1)$, and applying Nielsen's theorem we see that the Euler characteristic of the base is greater than or equal to zero ($f$ is essential). Hence $\chi(F) \geq 0$ which was excluded. Therefore $f^{-1}F \neq \emptyset$, and it follows, by 4.7.3, that $f|B$ is an essential singular square or annulus in $(\widetilde{M}_i, \underline{\widetilde{m}}_i)$, $i = 1$ or $2$, for every component, $B$, of $f^{-1}\widetilde{M}_i$. By 5.10, $f|B$ can be admissibly deformed in $(\widetilde{M}_i, \underline{\widetilde{m}}_i)$ into a vertical map. Thus we conclude that, for any two curves, $k$, $k'$, of $f^{-1}F$, there exists an integer $i$ such that, with $g = (f_2 f_1)^i$ or $= f_1(f_2 f_1)^i$, we have: $gf|k$ is admissibly homotopic in $F$ to $f|k'$. Furthermore, given any curve $k$, of $f^{-1}F$, there exists an integer $n \geq 1$ (e.g. $n$ equal to the number of components of $T - f^{-1}F$) such that $(f_2 f_1)^n f|k$ is admissibly homotopic in $F$ to $f|k$. Since, by 4.7.2, $k$ is essential in $T$, and since $f$ is essential, $f|k$ is essential in $F$.

We still have to show the addition to 6.3.2. Assume $f|k$ can be deformed in $F$ into $\partial F$. Then, for every curve, $k'$, of $f^{-1}F$, $f|k'$ can be deformed in $F$ into $\partial F$, since there is a homeomorphism $g: F \to F$ such that $gf|k \simeq f|k'$ in $F$. Every component, $B$, of $f^{-1}\widetilde{M}_i$, $i = 1,2$, is an annulus and we may suppose that $f$ is deformed so that $f(\partial B)$ lies in $\partial M$. Moreover, if we denote by $p_i: \widetilde{M}_i \to G_i$ the fibre projection, we have $p_i f(\partial B) \subset \partial G_i$. $G_i$ is neither an annulus nor a Möbius band since $\chi(F) < 0$, and so, by Nielsen's theorem, $pf|B$ can be deformed in $G_i$ (rel $\partial B$) into $\partial G_i$. Lifting this deformation ($\widetilde{M}_i$ is an I-bundle) to a homotopy of $f$ in $\widetilde{M}_i$ which is constant on $\partial B$, we see that $f|B$ can be deformed (rel $\partial B$) into $\partial M$. Thus, altogether, $f$ can be deformed into $\partial M$. q.e.d.

<u>Proof of 6.1 in the Stallings fibration case.</u>

<u>6.4 Lemma</u>. <u>Let</u> $F$ <u>be an orientable, connected surface. Suppose</u> $\chi(F) < 0$. <u>Let</u> $f: F \to F$ <u>be an orientation preserving autohomeomorphism. Suppose there exists an essential singular arc</u>, $k: I \to F$, <u>and an integer</u>, $n \geq 1$, <u>such that</u> $f^n k$ <u>is admissibly homotopic in</u> $F$ <u>to</u> $k$.

<u>Then there exists a system</u>, $\mathfrak{S}$, <u>of non-singular (pairwise disjoint) essential arcs in</u> $F$ <u>and an admissible ambient isotopy of</u> $F$ <u>which</u>

deforms $f\mathfrak{S}$ into $\mathfrak{S}$.

This lemma implies 6.1 in the Stallings fibration case. To see this recall that in this case there exists a connected surface, F, in $(M,\underline{\underline{m}})$ which splits $(M,\underline{\underline{m}})$ into a product I-bundle, $(\tilde{M},\tilde{\underline{\underline{m}}})$. We may fix an actual product structure of $(\tilde{M},\tilde{\underline{\underline{m}}})$ such that the lids are copies, $F_1$, $F_2$, of F. Let $p: \tilde{M} \to G$ be the fibre projection, $g: F_1 \to F_2$ the reflection in the fibres, $\varphi: F_2 \to F_1$ the attaching homeomorphism, and $f = \varphi \circ g$ (i.e. f is the autohomeomorphism induced by $(\tilde{M},\tilde{\underline{\underline{m}}})$). F is neither a disc, nor a torus since $(M,\underline{\underline{m}})$ contains an essential singular annulus. If F is an annulus we are done. Hence we may suppose $\chi(F) < 0$. Then, by 6.3, we may apply 6.4, and we get a system, $\mathfrak{S}$, of non-singular arcs in $F_1$ such that $f\mathfrak{S}$ is admissibly isotopic in $F_1$ to $\mathfrak{S}$. $\tilde{C} = p^{-1}p(\mathfrak{S})$ is a system of essential squares in $(\tilde{M},\tilde{m})$ (see (5.4) such that $\mathfrak{S} \cup g\mathfrak{S}$ is the set of all those sides of $\tilde{C}$ which lie in the lids of $(\tilde{M},\tilde{\underline{\underline{m}}})$. Since $(\tilde{M},\tilde{\underline{\underline{m}}})$ is a product I-bundle any admissible isotopic deformation of $g\mathfrak{S}$ in $F_2$ can be extended to an admissible isotopy of $\tilde{C}$. Now $g\mathfrak{S}$ can be admissibly isotoped such that $\varphi g\mathfrak{S} = f\mathfrak{S} = \mathfrak{S}$. This means that $\tilde{C}$ fits together to a system, C, of non-singular annuli or Möbius bands. Hence either a component of C or of $(\partial U(C) - \partial M)^-$ is an annulus which is essential in $(M,\underline{\underline{m}})$ (see (4.6.3), where $U(C)$ is a regular neighborhood in M.

Proof of 6.4. Without loss of generality, the boundary-pattern of F consists of all boundary components of F.

The proof of 6.4 uses the machinery of Nielsen [Ni 4]. But observe that it involves constructions which are not piecewise linear.

Recall that one can identify the hyperbolic plane to the interior of the complex unit disc, D, in such a way that the hyperbolic straight lines are the circular arcs perpendicular to the boundary and the hyperbolic translations are given by certain fractional linear transformations.

If F is any (compact) connected, orientable surface, closed or not, with $\chi(F) < 0$, then it is possible to identify the universal cover, $\tilde{F}$, of F to a part of the hyperbolic plane such that the covering translation group $\pi_1 F$ is a discontinuous group of hyperbolic translations. In particular, a covering translation, d, with $d \neq 1$, has precisely two different fixed points, U(d), V(d) in $\partial D$. The hyperbolic straight line joining these two points is called <u>axis</u> of d. The axis of d is preserved by the action of d, and, in particular, it covers a (not necessarily simple) closed curve in F.

The closure $\tilde{F}^- \subset D$ of the universal cover $\tilde{F}$ in D is equal to the closure of the smallest subset of the disc D which contains the axis of any $d \in \pi_1 F - 1$ and is convex in the non-euclidean sense. The action of $\pi_1 F$ extends to an action on $\tilde{F}^-$.

Two cases arise:

1. $\tilde{F}^- = D$ and
2. $\tilde{F}^- \neq D$.

In the second case $\partial D - \tilde{F}^-$ consists of countably many intervals, and $\tilde{F} - \tilde{F}^0$ consists of axes, covering $\partial F$.

Let $f: F \to F$ be the given orientation preserving homeomorphism and $\tilde{f}: \tilde{F} \to \tilde{F}$ some lifting of f. It is an important fact that $\tilde{f}$ extends continuously to the closure of $\tilde{F}$.

Let $\tilde{k}$ be a lifting of k.

<u>Case 1. Suppose every curve</u> $t\tilde{k}$ <u>with</u> $t = d\tilde{f}^j$, $d \in \pi_1 F$ <u>and</u> $0 \leq j \leq n-1$, <u>can be deformed so that</u> $\tilde{k}$ <u>and</u> $t\tilde{k}$ <u>are disjoint</u>.

Fix a hyperbolic structure of F such that $\tilde{F} \to F$ is hyperbolic. For every $f^j k$, $0 \leq j \leq n-1$, choose that geodesic curve, $k_j$, admissibly homotopic in F to $f^j k$ (note that $k_j$ is not pl). Let $\tilde{k}_j$ be a lifting of $k_j$. The curves $\tilde{k}_j$ are hyperbolic straight lines and the covering translations are hyperbolic translations. Hence, by our supposition, every curve $d\tilde{k}_i$, $0 \leq i \leq n-1$ and $d \in \pi_1 F$, which meets $\tilde{k}_j$, $0 \leq j \leq n-1$, is equal to $\tilde{k}_j$. Therefore the curves $k_j$, $0 \leq j \leq n-1$, define a system, $\mathfrak{S}'$, of pairwise disjoint simple curves in F. $k_j$ is admissibly homotopic to $f^j k$, and so $fk_j$ is admissibly homotopic to $k_{j+1}$, $0 \leq j \leq n-1$ (indices mod n). Thus every curve of $f\mathfrak{S}'$ is admissibly homotopic to a curve of $\mathfrak{S}'$. Make $\mathfrak{S}'$pl, using a small ambient isotopy of F.

Let $\mathfrak{S}$ be any maximal subsystem of curves of $\mathfrak{S}'$ which are pairwise not admissibly homotopic. Suppose $f\mathfrak{S}$ is deformed into a system, $\mathfrak{S}^*$, using an admissible ambient isotopy of F, so that as many curves of $\mathfrak{S}^*$ lie in $\mathfrak{S}$ as possible. Let $\mathfrak{T}$ be that subsystem of curves of $\mathfrak{S}$ which lie in $\mathfrak{S}^*$.

If $\mathfrak{T} = \mathfrak{S}$, then clearly $\mathfrak{S}$ satisfies the conclusion of 6.4. So we assume the contrary. Then there exists a curve, $\ell$, of $\mathfrak{S}^*$ which does not lie in $\mathfrak{S}$. But, by our definition of $\mathfrak{S}$ and the properties of $\mathfrak{S}'$, there is an admissible homotopy $g: \ell \times I \to F$ with $g|\ell \times 0 = \ell$ and such that $g|\ell \times 1$ is a curve in $\mathfrak{S}$. Since $\mathfrak{T}$ consists of essential curves we may suppose that g is deformed so that $g^{-1}\mathfrak{T}$ also consists of essential curves. Then $g^{-1}\mathfrak{T} = \emptyset$ since the curves of $\mathfrak{S}$ are pairwise not admissible homotopic (see above). Hence, defining $F^*$ as that surface obtained from F by splitting at $\mathfrak{T}$, $\ell$ is admissibly homotopic and so admissibly isotopic in $F^*$ to a curve of $\mathfrak{S}$. That means that there is an admissible ambient isotopy of F constant on $\mathfrak{T}$ which deforms $\ell$ into a curve of $\mathfrak{S}$. But this contradicts our maximality condition on $\mathfrak{T}$.

<u>Case 2</u>. <u>Suppose Case</u> 1 <u>does not hold</u>.

Then there exists at least one $t(=d\tilde{f}^j)$ such that $t\tilde{k}$ cannot be deformed so that $\tilde{k}$ is disjoint to $t\tilde{k}$. Let $A_1$, $A_2$ and $tA_1$, $tA_2$ be the boundary components of $\tilde{F}$ joined by $\tilde{k}$ and $t\tilde{k}$ resp. Then $A_1$, $A_2$, $tA_1$, $tA_2$ are pairwise disjoint, for we are in Case 2.

<u>6.5 Assertion</u>. <u>There is a covering translation</u>, $d_0$, <u>such that</u> $d_0\tilde{f}^n(A_i) = A_i$ <u>and</u> $d_0\tilde{f}^n(tA_i) = tA_i$, $i = 1,2$.

Since $f^n k$ is admissibly homotopic to k, we have $\tilde{f}^n\tilde{k} \simeq d_1\tilde{k}$, for some $d_1 \in \pi_1 F$, and therefore $d_1^{-1}\tilde{f}^n(A_i) = A_i$, $i = 1,2$.

Define $\varphi = d_1^{-1}\tilde{f}^n$. If $\varphi(tA_i) = tA_i$, $i = 1,2$, we set $d_0 = d_1^{-1}$ and we are done. So we assume $\varphi(tA_1) \neq tA_1$, say. Since $\varphi(tA_1) \neq A_1, A_2$ and since $\varphi$ is orientation preserving, $\varphi(tA_1)$ must lie between $tA_1$ and $A_2$ or between $A_1$ and $tA_1$. Without loss of generality we assume it lies between $tA_1$ and $A_2$. We join $A_1$, $tA_1$ and $\varphi(tA_1)$, $A_2$ by two disjoint arcs, w, w'. As $\varphi(w) \cap \varphi(w') = \emptyset$, $\varphi^2(tA_1)$ lies between $\varphi(tA_1)$ and $A_2$. Thus we get for each $p \geq 1$ at least p intersection points of arcs $\varphi^j t\tilde{k}$, $1 \leq j \leq p$, with $\tilde{k}$ which cannot be pulled away using a homotopy. Hence we conclude that $\tilde{k}$ intersects properly an infinite number of translates of $t\tilde{k}$. But this is impossible and so we have proved the assertion.

To continue the proof of 6.4, let, for any two different boundary components, $C_1$, $C_2$, of $\tilde{F}$, $\delta(C_1,C_2)$ be defined as the non-euclidean length of the shortest hyperbolic straight arc joining $C_1$ and $C_2$. We have $\delta(C_1,C_2) = \delta(dC_1,dC_2)$, for every $d \in \pi_1 F$, but in general $\delta(C_1,C_2) \neq \delta(t'C_1,t'C_2)$, for $t' = d\tilde{f}^j$. If $d\tilde{f}^n(C_i) = C_i$, $i = 1,2$, for some $d \in \pi_1 F$, then we may define, following [Ni 4], the <u>medium distance</u> of $C_1, C_2$ to be

$$\delta^*(C_1,C_2) = \frac{1}{n}(\Sigma_{0\leq i\leq n}\delta(\tilde{f}^i(C_1),\tilde{f}^i(C_2))).$$

Then even $\delta^*(C_1,C_2) = \delta^*(t'C_1,t'C_2)$, for every $t' = d\tilde{f}^j$.

By 6.5, the medium distance is defined for any two of $\{A_1,A_2,tA_1,tA_2\}$. Clearly

$$\delta^*(tA_1,A_2) + \delta^*(A_2,tA_2) + \delta^*(tA_1,A_1) + \delta^*(A_1,tA_2)$$

is strictly smaller than

$$2(\delta^*(tA_1,tA_2) + \delta^*(A_1,A_2)) = 4\delta^*(A_1,A_2).$$

Hence there is a pair of components of $\partial\tilde{F}$, invariant under $d_0\tilde{f}^n$, let us say $(A_1,tA_1)$ without loss of generality, with smaller medium distance than $A_1$ and $A_2$. We replace $\tilde{k}$ by any straight arc joining this pair (this arc is not pl).

But the medium distances of components of $\partial\tilde{F}$ are bounded below and nowhere dense, since F is compact and has a hyperbolic structure so that $\tilde{F} \to F$ is hyperbolic. So the above procedure must terminate after finitely many steps. Thus we find a straight arc $\tilde{\ell}$ such that $t\tilde{\ell}$ can be deformed in such a way that $t\tilde{\ell}$ is disjoint to $\tilde{\ell}$, for every $t = d\tilde{f}^j$. Since that pair of components of $\partial\tilde{F}$ joined by $\tilde{\ell}$ is invariant under $\varphi$, $\tilde{\ell}$ covers an essential arc, $\ell$, in F such that $f^n\ell$ is admissibly homotopic in F to $\ell$. Make $\ell$

pl and then we are in Case 1. q.e.d.

Proof of 6.2 in the Stallings fibration case. As above we also reduce 6.2 in the Stallings fibration case to a 2-manifold problem:

6.6 Lemma. *Let* F, f *be as in* 6.4. *Suppose there exists an essential singular closed curve* $k: S^1 \to F$ *which cannot be deformed into* $\partial F$ *and an integer* $n \geq 1$ *such that* $f^n k \simeq k$.

*Then at least one of the following assertions is true*:

1. *There exists a system*, $\mathfrak{S}$, *of essential (pairwise disjoint), simple closed curves in* F *which cannot be isotoped into* $\partial F$ *and an admissible ambient isotopy of* F *which deforms* $f\mathfrak{S}$ *into* $\mathfrak{S}$.
2. *The homeomorphism* f *is isotopic to a homeomorphism of finite order*.

To show that 6.6 implies 6.2 in the Stallings fibration case, we refer to the argument given after 6.4 and so we only have to prove that 6.6.2 implies 6.2.2. But the latter follows precisely as in [Wa 3, pp. 514].

Proof of 6.6. If there exists some $i \geq 1$ such that $f^i \simeq \mathrm{id}$, then, by Nielsen [Ni 4] (see also [Fn 1][Ma 1], [Zim 1]) f is homotopic, and so isotopic, to a homeomorphism of finite order. Thus 6.6.2 follows, and so we suppose $f^i$ is not homotopic to the identity, for any $i \geq 1$.

In the following we shall use the notation described in the beginning of 6.4. We fix a hyperbolic structure of F such that $\tilde{F} \to F$ is hyperbolic, and we suppose k is deformed so that it is (a multiple of) a geodesic curve (this again is not pl). In particular k has at most finitely many self-intersections. Furthermore we may suppose that f is deformed so that, for every $1 \leq j \leq n-1$, $f^j k$ meets k in at most finitely many points. Let $\tilde{k}$ be a covering of k. If every curve $t\tilde{k}$ with $t = d\tilde{f}^j$, $d \in \pi_1 F$ and $0 \leq j \leq n-1$, can be deformed so that $\tilde{k}$ and $t\tilde{k}$ are disjoint, then

6.6.1 follows as in Case 1 of the proof of 6.4. So we assume the contrary.

Then $\tilde{k}$ meets properly $t\tilde{k}$. In particular $\tilde{k}$ and $t\tilde{k}$ join pairwise different points in $\partial D$, say $P_1$, $P_2$ and $tP_1$, $tP_2$.

<u>6.7 Assertion</u>. <u>There is an integer</u> $r \geq 1$ <u>and a covering translation</u> $d_r$ <u>such that</u> $d_r(\tilde{f}^n)^r(P_i) = P_i$ <u>and</u> $d_r(\tilde{f}^n)^r(tP_i) = tP_i$, $i = 1,2$.

Since $f^n k \simeq k$, there is a $d_0 \in \pi_1 F$ with $\tilde{f}^n P_i = d_0 P_i$, $i = 1,2$. Define $\varphi_0 = d_0^{-1}\tilde{f}^n$. If $\varphi_0(tP_i) = tP_i$, $i = 1,2$, then we are done. So we suppose the contrary. Then $\varphi_0(tP_1) \neq tP_1$, say, and we get, as in the proof of 6.5, a sequence of pairwise different lines $\varphi_0^j \circ t\tilde{k}$ which properly intersect $k$. By our definition of $\varphi_0$, we have $\varphi_0^j t\tilde{k} \simeq d_j t\tilde{k}$, for some $d_j \in \pi_1 F$. Since $k$ has only finitely many self-intersection points and meets $f^j k$, $1 \leq j \leq n-1$, in only finitely many points, there is an integer $r \geq 1$ and a covering translation $d_r^*$ such that $\varphi_0^r t\tilde{k} \simeq d_r^* t\tilde{k}$ and $d_r^* \tilde{k} \simeq \tilde{k}$. But $d_r^{*-1}\varphi_0^r$ can be written as $d_r(\tilde{f}^n)^r$, with $d_r \in \pi_1 F$, and so the assertion follows.

To continue the proof of 6.6 we call, following [Ni 2], the <u>kernel region</u> of a homeomorphism $\varphi\colon \tilde{F} \to \tilde{F}$ the smallest convex subset of $D$ that contains every axis belonging to a covering translation, $d$, with $\varphi d = d\varphi$. Defining $\varphi_1 = d_r(\tilde{f}^n)^r$, we have just verified that the kernel region of $\varphi_1$ contains at least two different axes. In this case, by [Ni 2, Satz 8] (cf. [Ni 5] if $F$ is bounded), the kernel region covers a subsurface, $G$, of $F$. Since we supposed $\tilde{f}^i$ is not homotopic to id, for all $i \geq 1$, this surface cannot be all of $F$.

That means that there exists at least one boundary curve, $\ell$, of $G$ which cannot be deformed into $\partial F$. $\ell$ is a geodesic curve in $F$ and a covering, $\tilde{\ell}$, of $\ell$ is an axis in $\tilde{F}$ joining two points, $Q_1, Q_2$, in $\partial D$. Let $d$ be a covering translation with axis $\tilde{\ell}$. Then $\varphi_1 Q_i = \varphi_1 d Q_i = d\varphi_1 Q_i$, $i = 1,2$, by definition of "kernel region", i.e. $\varphi_1 Q_1$ and $\varphi_1 Q_2$ are fixed points of $d$. But $d$ has only two fixed points, and so $\varphi_1^2\tilde{\ell} \simeq \tilde{\ell}$. Hence there is an integer $m \geq 1$ such that $f^m \ell \simeq \ell$.

If there is another curve $t\tilde{\ell}$, with $t = d\tilde{f}^j$, such that $t\tilde{\ell}$

cannot be deformed so that $\tilde{\ell}$ and $t\tilde{\ell}$ are disjoint, then we get, using the preceding argument, a covering $\varphi_2$ of some $f^p$, $p \geq 1$, such that the kernel region of $\varphi_2$ covers a surface which is larger than G. Iteration of this process must eventually stop since F is compact. Hence we find a line, $\tilde{\ell}$, such that every line $t\tilde{\ell}$ can be deformed so that $\tilde{\ell}$ and $t\tilde{\ell}$ are disjoint. Then we again get the system, $\mathfrak{S}$, as in Case 1 of the proof of 6.4. q.e.d.

<u>Proof of 6.1 and 6.2 in the case of twisted Stallings manifolds.</u> In this case there exists a connected surface, F, in $(M,\underline{\underline{m}})$ which splits $(M,\underline{\underline{m}})$ into two twisted I-bundles, $(\tilde{M}_1,\tilde{\underline{\underline{m}}}_1)$, $(\tilde{M}_2,\tilde{\underline{\underline{m}}}_2)$, which admit admissible fibrations such that the lids are copies of F. There is a canonical two sheeted covering $p\colon \bar{M} \to M$ with the property that each of $\tilde{M}_1$ and $\tilde{M}_2$ is covered by a product I-bundle. Therefore $\bar{M}$ is a Stallings fibration. Denote by d the covering translation. Let $f\colon T \to M$ be either an essential singular annulus in $(M,\underline{\underline{m}})$ or an essential singular torus which cannot be deformed into $\partial M$. It follows from 6.3 that there exists also an essential singular annulus or torus $\bar{f}$ resp. in $\bar{M}$ which cannot be deformed into $\partial\bar{M}$. To be precise there is a commuting diagram

$$\begin{array}{ccc} \bar{T} & \xrightarrow{\ \bar{f}\ } & \bar{M} \\ {\scriptstyle q}\downarrow & & \downarrow{\scriptstyle p} \\ T & \xrightarrow{\ f\ } & M \end{array}$$

where q is a covering of index 1 or 2. Hence, by 6.1 and 6.2 in the Stallings fibration case, at least one of the following must be true:

1. There exists a non-singular essential annulus in $\bar{M}$, if T is an annulus, and a non-singular essential torus in $\bar{M}$ which is not boundary-parallel, if T is a torus.
2. $\bar{M}$ is a Seifert fibre space (this case must be considered only if T is a torus).

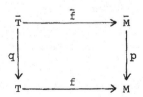

A) Suppose 1 holds.

Let $B$ be a non-singular annulus or torus, resp., in $\bar{M}$ as given in 1. We may assume that $B \cap dB$ is a system of simple arcs or simple closed curves. Furthermore, by an exchange-of-discs argument [Wa 4, p. 70], we may assume that no curve of $B \cap dB$ is inessential. It follows that a regular neighborhood $U(pB)$ of the image of $pB$ in $(M,\underline{\underline{m}})$ is either an I- or $S^1$-bundle with boundary-pattern such that the inclusion is an admissible map. Fix an actual admissible fibration of $U(pB)$. Since $p|B\colon B \to U(pB)$ is vertical in $U(pB)$ and since $p|B\colon B \to M$ is essential in $(M,\underline{\underline{m}})$, the fibres of $U(pB)$ are essential in $(M,\underline{\underline{m}})$. Hence, adding trivial components of $M - \mathring{U}(pB)$ if necessary, we see that $p(B)$ is contained in an essential I-bundle or Seifert fibre space $N$ in $(M,\underline{\underline{m}})$. By 5.4, $N$ contains a vertical annulus or torus, resp., which is essential in $N$ and so also in $M$ (see 4.6). If $M$ is not a Seifert fibre space and if $B$ is not boundary-parallel, then not every essential, vertical torus in $N$ is boundary-parallel in $M$. This proves 6.1 and 6.2, respectively.

B) Suppose 2 holds.

$\bar{M}$ is a Seifert fibre space and we want to prove that either $M$ contains an essential torus which is not boundary-parallel, or that $M$ itself is a Seifert fibre space. For this we still have to consider the following cases:

(i) $\partial\bar{M} \neq \emptyset$.

(ii) $\partial\bar{M} = \emptyset$, and $\bar{M}$ does not contain any incompressible torus.

Case (i). $\bar{M}$ contains an essential annulus (see 5.4). Hence, by A) above, so does $M$. Consider a regular neighborhood, $U$, of the union of the essential annulus in $M$ with the adjacent boundary component(s) of $M$. If any of the tori in $M^0 \cap \partial U$ are inessential or boundary-parallel, then $M$ must be a Seifert fibre space.

Case (ii). Since $\partial\bar{M} = \emptyset$, and $\bar{M}$ is sufficiently large but does not contain an incompressible torus, $\bar{M}$ must be a Seifert fibre space over the 2-sphere with exactly three exceptional fibres (see 5.4). Furthermore, by construction of $\bar{M}$, the covering translation acts non-trivially on the center of $\pi_1\bar{M}$. But this case is not possible in view of the following lemma due to F. Waldhausen:

6.8 Lemma. Let N be a Seifert fibre space over the 2-sphere with precisely three exceptional fibres and infinite fundamental group. Then there does not exist any torsion free centerless extension of $\pi_1 N$ by $\mathbb{Z}_2$.

Proof. $\pi_1 N$ has a presentation $\{a_1,a_2,a_3,z \mid a_1a_2a_3z^\gamma, [a_i,z], a_i^{\alpha_i}z^{\beta_i}\}$, where $0 < \beta_i < \alpha_i$ and g.c.d.$(\alpha_i,\beta_i) = 1$, $1 \leq i \leq 3$. The center, C, of $\pi_1 N$ is free cyclic, generated by z.

Let t be an automorphism of $\pi_1 N$, well defined up to composition with an inner automorphism, which represents the given action of $\mathbb{Z}_2$ on the set of conjugacy classes of $\pi_1 N$. We have $t(z) = z^{-1}$, because we want the extension to be centerless. Let $\bar{t}\colon \pi_1 N/C \to \pi_1 N/C$ be the automorphism induced by t. By V.11 of [ZVC 1], $\bar{t}$ must satisfy the equation

$$\bar{t}(\bar{a}_i) = \bar{w}_i\bar{a}_{\sigma(i)}^{\varepsilon_i}\bar{w}_i^{-1}, \qquad i = 1,2,3,$$

where $\varepsilon_i = \pm 1$, and $\sigma$ is a permutation $\begin{pmatrix}1 & 2 & 3\\ \sigma(1) & \sigma(2) & \sigma(3)\end{pmatrix}$. $\bar{t}^2$ is an inner automorphism and no two of $\bar{a}_1,\bar{a}_2,\bar{a}_3$ are conjugate to each other. Hence $\sigma$ must be an involution. Now 3 is an odd number, therefore $\sigma$ must leave fixed at least one of 1,2,3. Without loss of generality we assume $\sigma(1) = 1$. Returning to $\pi_1 N$, from what has been proved so far we must have

$$t(a_1) = z^{\delta_1}w_1a_1^{\varepsilon_1}w_1^{-1}, \quad \text{for some integer } \delta_1.$$

Composing t with an inner automorphism if necessary, we may assume

$$t(a_1) = z^{\delta_1}a_1^{\varepsilon_1}.$$

Applying this to the relator $a_1^{\alpha_1} z^{\beta_1} = 1$ we obtain (recall that $t(z) = z^{-1}$)

$$1 = (z^{\delta_1} a_1^{\varepsilon_1})^{\alpha_1} z^{-\beta_1} = a_1^{\varepsilon_1\alpha_1} z^{\delta_1\alpha_1-\beta_1} = z^{-\varepsilon_1\beta_1+\delta_1\alpha_1-\beta_1}$$

hence

$$\delta_1\alpha_1 = \beta_1(1 + \varepsilon_1). \qquad (*)$$

This implies

$$\delta_1 = 0, \quad \text{if} \quad \varepsilon_1 = -1, \text{ resp.}$$

$$\delta_1 = 1, \quad \text{if} \quad \varepsilon_1 = +1, \text{ since } 0 < \beta_1 < \alpha_1.$$

Hence

$$t(a_1) = \begin{cases} za_1, & \text{if} \quad \varepsilon_1 = +1 \\ a_1^{-1}, & \text{if} \quad \varepsilon_1 = -1 \end{cases}$$

and so, in either case, $t^2(a_1) = a_1$.

Let the inner automorphism $t^2$ be induced by $b \in \pi_1 N$. Then $ba_1b^{-1} = t^2(a_1) = a_1$. But, by IV.12 c) of [ZVC 1], any element of $\pi_1 N/C$ of finite order, such as $\bar{a}_1$, does not commute with any other element of $\pi_1 N/C$ unless they are both multiples of some other element.

As $\bar{a}_1$ is not non-trivially a multiple of any other element of $\pi_1 N/C$, $\bar{b}$ must be a multiple of $\bar{a}_1$. Hence, multiplying $b$ by an element of $C$ if necessary we may assume that

$$b = a_1^{\rho}.$$

We now distinguish two cases, $b \in C$, and $b \notin C$. In either case we will refer to the obstruction theory for group extensions as described in [ML 1].

Case 1. $b \in C$.

The assumption of this case means that $t$ is an involution on $\pi_1 N$. There certainly exists an extension of $\pi_1 N$ by $\mathbb{Z}_2$ inducing $t$, namely the split extension. On the other hand, the extensions of $\pi_1 N$ by $\mathbb{Z}_2$, inducing the automorphism class of $t$, are classified, up to equivalence, by $H^2(\mathbb{Z}_2, C)$, cf. [ML 1, IV theorem 7.1] which is the zero group. So there is only one equivalence class of extensions inducing $t$, and $t$ is not induced by any torsion free extension.

Case 2. $b \notin C$.

Recall that the cohomology of a group $\Pi$ can be computed as the cohomology of a specific chain complex, the bar resolution $B(\mathbb{Z}\Pi)$. In the case at hand, $\Pi = \mathbb{Z}_2$, the degree $n$ part $B_n(\mathbb{Z}\mathbb{Z}_2)$ of the bar resolution is just the integral group ring $\mathbb{Z}\mathbb{Z}_2$. Hence the cohomology group $H^n(\Pi, A)$ with coefficients in a $\mathbb{Z}_2$-module $A$ is the n-th homology group of a chain complex which in degree $n$ is

$$\mathrm{Hom}_{\mathbb{Z}_2}(B_n(\mathbb{Z},\mathbb{Z}_2), A) \cong A.$$

We will be interested below in $H^3(\mathbb{Z}_2, C)$, where as above $C$ = center of $\pi_1 N$, and the action is the non-trivial one. One has $H^3(\mathbb{Z}_2, C) \cong \mathbb{Z}_2$, cf. [ML 1, IV theorem 7.1], so, by what was said above, the non-zero element of $H^3(\mathbb{Z}_2, C)$ is represented by odd multiples of the generator of $C$.

Theorem IV 8.7 of [ML 1] says that the given $\mathbb{Z}_2$-action is induced by an extension if and only if a certain obstruction in $H^3(\mathbb{Z}_2, C)$ is zero. A formula for 3-cocycles representing the obstruction is given in [ML 1, p. 116]. In the case at hand the formula simply says that $t(b)\cdot b^{-1}$ belongs to $C$ and represents the obstruction.

We have two cases to consider.

In case $\varepsilon_1 = +1$, cf. above, we have, by (*), $\alpha_1 = 2\beta_1$ and hence $\alpha_1 = 2$, $\beta_1 = 1$ since $\alpha_1$, $\beta_1$ are coprime. Since $b = a_1^{\rho} \notin C$, by assumption, we must have $|\rho| = 1$. Hence $t(b)\cdot b^{-1} = z^{\rho} = z^{\pm 1}$.

In case $\varepsilon_1 = -1$, we have, cf. above, $t(a_1) = a_1^{-1}$, so $t(b)\cdot b^{-1} = a_1^{-2\rho}$. Since the latter is an element of C, but $b = a_1^{\rho}$ is not, we must have $2\rho = \alpha_1 \cdot \varphi$, where $\varphi$ is odd. Hence $\alpha_1$ is even, so $\beta_1$ is odd since $\alpha_1$, $\beta_1$ are coprime. So $t(b)\cdot b^{-1} = a_1^{-2\rho}$ $= (a_1^{-\alpha_1})^{\varphi} = (z^{\beta_1})^{\varphi}$, is an odd multiple of z.

In either case $t(b)\cdot b^{-1}$ is an odd multiple of the generator of C. So it represents the non-trivial element of $H^3(\mathbb{Z}_2, C)$, and hence the automorphism t is not induced by any extension at all.

q.e.d.

P. Scott has pointed out that the preceding proof can be short cut if one assumes the fact that $\mathbb{Z}$ has no torsion free centerless extension by a finite cyclic group (for in the course of the proof it was pointed out that C and a suitable lifting of t generate just such a group). Conversely, the obstruction theory part of this proof can be used to prove the latter fact.

## §7. Generalized Seifert fibre spaces.

In this section we consider 3-manifolds, $(M,\underline{\underline{m}})$, with the property that in $(M,\underline{\underline{m}})$ there exists an essential torus, G, (separating or not) which splits $(M,\underline{\underline{m}})$ into Seifert fibre spaces. These manifolds are called generalized Seifert fibre spaces (they are special cases of the "Graphenmannigfaltigkeiten" considered in [Wa 1]). M is irreducible and aspherical and $\underline{\underline{m}}$ is a useful boundary-pattern of M.

7.1 Proposition. *Let $(M,\underline{\underline{m}})$ be a generalized Seifert fibre space and G an essential torus in $(M,\underline{\underline{m}})$ which splits $(M,\underline{\underline{m}})$ into Seifert fibre spaces. Let $f\colon T \to M$ be an essential singular annulus or torus in $(M,\underline{\underline{m}})$.*

*Then either $(M\ \underline{\underline{m}})$ is a Seifert fibre space or f can be admissibly deformed in $(M,\underline{\underline{m}})$ so that $f^{-1}G = \emptyset$.*

Proof. Deform f admissibly in $(M,\underline{\underline{m}})$ so that the number of curves of $f^{-1}G$ is as small as possible. If $f^{-1}G = \emptyset$, we are done. So we suppose $f^{-1}G \neq \emptyset$ and we have to show that $(M,\underline{\underline{m}})$ admits an admissible fibration as Seifert fibre space.

Let $(\tilde{M},\tilde{\underline{\underline{m}}})$ be the manifold obtained from $(M,\underline{\underline{m}})$ by splitting at G, and $A_1$, $A_2$ two neighboring components of $f^{-1}\tilde{M}$. By 4.7.3, $f|A_i$, $i = 1,2$, is an essential singular annulus in $(\tilde{M},\tilde{\underline{\underline{m}}})$. $(\tilde{M},\tilde{\underline{\underline{m}}})$ is connected or not. But in either case it would suffice to show that there exists an admissible fibration of the components of $(\tilde{M},\tilde{\underline{\underline{m}}})$ as Seifert fibre space and an admissible deformation of $f|A_i$, $i = 1,2$, such that $f|A_i$ is vertical with respect to this fibration (apply [Wa 1, (5.2)]). For this let $(\tilde{M}_1,\tilde{\underline{\underline{m}}}_1)$, $(\tilde{M}_2,\tilde{\underline{\underline{m}}}_2)$ (possibly equal) be the components of $(\tilde{M},\tilde{\underline{\underline{m}}})$. By supposition, $(\tilde{M}_i,\tilde{\underline{\underline{m}}}_i)$, $i = 1,2$, are Seifert fibre spaces. Suppose $(\tilde{M}_1,\tilde{\underline{\underline{m}}}_1)$ admits no admissible fibration as an I-bundle over the torus or Klein bottle. Then, by 5.10, $f|f^{-1}\tilde{M}_1$ can be admissibly deformed in $(\tilde{M}_1,\tilde{\underline{\underline{m}}}_1)$ into a vertical map. If G is non-separating, we are done. If not, suppose $A_2$ is a component of $f^{-1}\tilde{M}_2$ and let $k_1$, $k_2$ be the two boundary curves of $A_2$.

Since $f|f^{-1}\tilde{M}_1$ can be admissibly deformed into a vertical map, it follows that $f|A_2$ can be admissibly deformed in $(\tilde{M}_2,\tilde{\underline{\underline{m}}}_2)$ so that $f(k_1) \cap f(k_2) = \emptyset$. Hence, we may apply 5.11, and again we are done.

Thus we suppose every component of $(\tilde{M},\tilde{\underline{\underline{m}}})$ admits an admissible fibration as I-bundle over the torus or Klein bottle. Then $(M,\underline{\underline{m}})$ is clearly a Seifert fibre space if $G$ is boundary-parallel. Hence we may suppose $(M,\underline{\underline{m}})$ is a Stallings manifold (without boundary). Let $g_2,g_1\colon G \to G$ be the autohomeomorphism of $G$ induced by $(\tilde{M},\tilde{\underline{\underline{m}}})$ (for definition see the beginning of paragraph 6; set $g_1 = \mathrm{id}$ if $\tilde{M}$ is connected). Since $\partial M = \emptyset$, $T$ must be a torus. By the remark of 6.3, there is an essential singular curve $k\colon S^1 \to G$ and an integer $n \geq 1$, such that $(g_2g_1)^n k \simeq k$ in $G$. Since $G$ is a torus, we may suppose $k$ is deformed in $G$ so that it is a multiple of some simple closed curve.

<u>Case 1</u>. $(g_2g_1)k$ <u>is not homotopic in</u> $G$ <u>to</u> $k$.

In this case two essential curves in $G$, which are not homotopic, are invariant under $(g_2g_1)^n$ (up to homotopy), namely $k$ and $g_2g_1(k)$. $k$ and $g_2g_1(k)$ induce two linear independent elements $\alpha_1,\alpha_2$ of $\pi_1G$, generating $\pi_1G$ since $\pi_1G \approx \mathbb{Z} \oplus \mathbb{Z}$. $(g_2g_1)^n_*(\alpha_i) = \pm\alpha_i$, $i = 1,2$. Therefore $(g_2g_1)^{2n}_* = 1$ since $\alpha_1,\alpha_2$ generate $\pi_1G$. Therefore $(g_2g_1)^{2n} \simeq 1$, and so $g_2g_1$ is homotopic, [Ni 4], and so isotopic, to a homeomorphism of finite order. Let $p\colon \bar{M} \to M$ be the canonical two sheeted covering with the property that each of $\tilde{M}_1$ and $\tilde{M}_2$ is covered by a product I-bundle, $\bar{M}_i$, $i = 1,2$. Since $g_2g_1$ is of finite order (up to isotopy) the product fibrations of $\bar{M}_i$ fit together to a fibration of $\bar{M}$ as Seifert fibre space (cf. [Wa 3, p. 514]). Fix such a fibration of $\bar{M}$, then $p^{-1}G$ is horizontal. By 6.8, it follows that $\bar{M}$ is not the exceptional case 5.1.4. Hence, by 5.4, there is a vertical essential torus, $\bar{B}$, in $\bar{M}$. As in A) of the proof of 6.1, and 6.2 in the case of twisted Stallings manifolds, we find either that $M$ is a Seifert fibre space or that $M$ contains a non-singular essential torus, $B$, in $M$ which cannot be isotoped into $\tilde{M}$ ($\bar{B}$ cannot be deformed into $\bar{M} - p^{-1}G$). In the latter case isotop

B so that the number of curves of $B \cap G$ is as small as possible. Then $B \cap \tilde{M}$ consists of essential annuli in $(\tilde{M},\underline{\underline{\tilde{m}}})$ (recall B cannot be isotoped into $\tilde{M}$). Since these annuli are non-singular, their boundary curves are disjoint. Hence, by 5.11, it follows that there are admissible fibrations of $(\tilde{M}_i,\underline{\underline{\tilde{m}}}_i)$, $i = 1,2$, as Seifert fibre spaces such that $B \cap \tilde{M}$ is vertical. As usual these fibrations can be admissibly isotoped so that they define a fibration of M as Seifert fibre space [Wa 1, (5.2)].

<u>Case 2</u>. $(g_2g_1)k$ <u>is homotopic in</u> G <u>to</u> k, <u>but not</u> $g_2k$.

In this case $\tilde{M}$ is disconnected. $g_2k \simeq g_1k$ since $g_2g_1k \simeq k$ and since $g_2$ is an involution. Therefore $g_2k \simeq (g_2g_1)g_2k$ since $g_1$ is an involution. Hence two essential curves in G, which are not homotopic, are invariant under $g_2g_1$ (up to homotopy), namely k and $g_2(k)$. Then we conclude, as in Case 1, that M is a Seifert fibre space.

<u>Case 3</u>. $(g_2g_1)k$ <u>as well as</u> $g_2k$ <u>is homotopic in</u> G <u>to</u> k.

By our definition of $g_1,g_2$, the curves k, $g_1k$ and $g_1k,(g_2g_1)k$ can be considered as boundary curves of vertical essential singular annuli $\tilde{f}_1,\tilde{f}_2$ in $(\tilde{M}_1,\underline{\underline{\tilde{m}}}_1)$, $(\tilde{M}_2,\underline{\underline{\tilde{m}}}_2)$ resp. Since we are in Case 3, $\tilde{f}_i$, $i = 1,2$, can be admissibly deformed in $(\tilde{M}_i,\underline{\underline{\tilde{m}}}_i)$ so that the boundary curves are disjoint. Hence, by 5.11, there exists an admissible fibration of $(\tilde{M}_i,\underline{\underline{\tilde{m}}}_i)$ as Seifert fibre space and an asmissible deformation of $\tilde{f}_i$ into a vertical map with respect to this fibration. Then, as usual, these fibrations can be admissibly isotoped so that they define a fibration of M as Seifert fibre space.

q.e.d.

## *Chapter III.* *Characteristic submanifolds.*

In this chapter we give rigorous definitions of a characteristic submanifold, and show its existence. In §10 we establish a useful property of such submanifolds, which we call "complete." It turns out e.g. that all full (see 8.2), essential F-manifolds which are complete are already characteristic submanifolds. Furthermore, it will be the completeness of characteristic submanifolds which enables us to prove special cases of the enclosing theorems, namely their non-singular versions (see 10.7 and 10.8). An immediate corollary of this is then the uniqueness of the characteristic submanifolds.

### §8. Definition of a characteristic submanifold

We begin by defining some notations.

Let $(A,\underline{\underline{a}})$ be an essential surface in the surface $(F,\underline{\underline{f}})$. Suppose that, in addition, $(A,\underline{\underline{a}})$ is admissible in $(F,\bar{\underline{\underline{f}}})$. Here $\bar{\underline{\underline{f}}}$ and $\bar{\underline{\underline{a}}}$ denote the complete boundary-patterns of $(F,\underline{\underline{f}})$ resp. $(A,\underline{\underline{a}})$. We say that $(A,\underline{\underline{a}})$ is an *inner square in* $(F,\underline{\underline{f}})$, if $(A,\bar{\underline{\underline{a}}})$ is a square, and furthermore we say that $(A,\underline{\underline{a}})$ is an *inner annulus* or *Möbius band in* $(F,\underline{\underline{f}})$, if $A$ is homeomorphic to an annulus or Möbius band, respectively, and if $\bar{\underline{\underline{a}}}$ consists of all the boundary curves of $A$. If in addition, $F$ is a surface of a boundary-pattern, $\underline{\underline{m}}$, of some 3-manifold $M$ (i.e. a side of $(M,\underline{\underline{m}})$) and if $\underline{\underline{f}}$ is the boundary-pattern of $F$ induced by $\underline{\underline{m}}$, then we also say that $A$ is an inner square, annulus, or Möbius band, respectively, in $\underline{\underline{m}}$.

Let $(M,\underline{\underline{m}})$ be a 3-manifold. An *admissible* I-*bundle* or *Seifert fibre space* $(X,\underline{\underline{x}})$, in $(M,\underline{\underline{m}})$ is an I-bundle or Seifert fibre space (see paragraph 5) such that the inclusion defines admissible maps $(X,\underline{\underline{x}}) \rightarrow (M,\underline{\underline{m}})$ and $(X,\underline{\underline{x}}) \rightarrow (M,\bar{\underline{\underline{m}}})$. An admissible I-bundle or Seifert fibre space $(X,\underline{\underline{x}})$ in $(M,\underline{\underline{m}})$ is called *essential in* $(M,\underline{\underline{m}})$ if $(\partial X - \partial M)^-$ is an essential surface. It is easily seen that then the inclusion $(X,\underline{\underline{x}}) \subset (M,\underline{\underline{m}})$ defines an essential map.

A submanifold, $W$, of a 3-manifold $(M,\underline{\underline{m}})$ is called an *admissible* (*essential*) F-*manifold in* $(M,\underline{\underline{m}})$ if each component of $W$ is either an admissible (essential) I-bundle or Seifert fibre space in $(M,\underline{\underline{m}})$. If $W$ is an essential F-manifold in $(M,\underline{\underline{m}})$, it follows that $(\partial W - \partial M)^-$ is an essential surface in $(M,\underline{\underline{m}})$ which consists of squares, annuli, or tori, and that, for every $G \in \underline{\underline{m}}$, $W \cap G$ is an essential surface in $G$, where $G$ carries the boundary-pattern induced by $\underline{\underline{m}}$ (for the latter fact note that every point of the graph of $(M,\underline{\underline{m}})$ has order at most three). Furthermore, every side of $W$ which is not a lid of an I-bundle is either a torus, or an inner square, or an inner annulus in $\underline{\underline{m}}$.

<u>8.1 Definitions of complexities</u>. Let $F$ be a surface with boundary-pattern $\underline{\underline{f}}$. Then the <u>complexity of</u> $(F,\underline{\underline{f}})$ is defined to be

$$c(F,\underline{\underline{f}}) = 2\beta_1(F) + \mathrm{card}(\underline{\underline{f}}),$$

where $\beta_1(F)$ is the first Betti number of $F$. Note that $c(F,\underline{\underline{f}})$ is always a non-negative integer since $F$ is compact and $\underline{\underline{f}}$ is finite.

Let $(X,\underline{\underline{x}})$ be an I-bundle or Seifert fibre space with fixed admissible fibration and fibre projection $p\colon X \to F$. Let $\underline{\underline{f}}$ be the boundary-pattern of $F$ induced by $\underline{\underline{x}}$. Let $x_1,\ldots,x_n$, $n \geq 0$, be all the exceptional points of $F$. Denote by $\tilde{F}$ the surface $F - \dot{U}(\cup\, x_i)$, where $U(\cup\, x_i)$ is a regular neighborhood in $F$. Then the <u>complexity of</u> $(X,\underline{\underline{x}},p)$ is defined to be

$$c(X,\underline{\underline{x}},p) = c(\tilde{F},\underline{\underline{f}}).$$

The complexity of an I-bundle or Seifert fibre space without fixed admissible fibration is defined to be

$$c(X,\underline{\underline{x}}) = \min c(X,\underline{\underline{x}},p),$$

where the minimum is taken over all admissible fibrations, $p$, of $(X,\underline{\underline{x}})$. In order to simplify the notation we sometimes write $c(X)$ for $c(X,\underline{\underline{x}})$. Note that for almost all I-bundles or Seifert fibre spaces the admissible fibration is unique up to admissible ambient isotopy (set $h = \mathrm{id}$ in 5.9).

Let $(M,\underline{\underline{m}})$ be a 3-manifold and $V$ an essential F-manifold in $(M,\underline{\underline{m}})$. Let $\alpha_V\colon \mathbb{N} \to \mathbb{N}$ be the map defined by

$\alpha_V(i)$ = the number of components of $V$ which have complexity $i$.

Then the *complexity of* $V$ is defined to be

$$c(V) = (\ldots,\alpha_V(2),\alpha_V(1),\alpha_V(0)).$$

*8.2 Definition. Let* $V$ *be an essential* F-*manifold in a* 3-*manifold* $(M,\underline{\underline{m}})$. $V$ *is called a "characteristic submanifold in* $(M,\underline{\underline{m}})$*" if the following holds:*

1. $V$ *is "full" that is, for any non-empty submanifold,* $W$, *of* $M$ *whose components are components of* $(M - V)^-$, $V \cup W$ *is not an essential* F-*manifold in* $(M,\underline{\underline{m}})$, *and*
2. *there is no essential* F-*manifold in* $(M,\underline{\underline{m}})$ *with property* 1 *and larger complexity than* $V$ *(with respect to the lexicographical order of the complexities).*

*Remark*. For equivalent, but more geometric, definitions of a characteristic submanifold see 10.10.

*8.3 Corollary. Any essential* F-*manifold,* $V$, *is contained in a full essential* F-*manifold.*

*Proof*. In the notation above suppose $V \cup W$ is an admissible F-manifold. Then $(\partial(V \cup W) - \partial M)^-$ is a subset of $(\partial V - \partial M)^-$. Hence $(\partial(V \cup W) - \partial M)^-$ is essential since $(\partial V - \partial M)^-$ is, and so, by definition, $V \cup W$ is essential F-manifold. The corollary follows now by downward induction on the number of components of $W$. q.e.d.

## §9. Existence of a characteristic submanifold

Let $T_1$, $T_2$ be two disjoint, connected, codimension one submanifolds in a 2- or 3-manifold $(X,\underline{\underline{x}})$. $T_1$ and $T_2$ are called *admissibly parallel in* $(X,\underline{\underline{x}})$ if at least one component of $(\tilde{X},\tilde{\underline{\underline{x}}})$ is a product I-bundle whose lids are copies of $T_1$ *and* $T_2$ (or a square or annulus with sides near $T_1$ and $T_2$ ), where $(\tilde{X},\tilde{\underline{\underline{x}}})$ is the manifold obtained from $(X,\underline{\underline{x}})$ by splitting at $T_1 \cup T_2$. Analogously, we say $T_1$ is *admissibly parallel in* $(X,\underline{\underline{x}})$ *to a side* Z, *of* $(X,\underline{\underline{x}})$ if, in the notation above, one component of $(\tilde{X},\tilde{\underline{\underline{x}}})$ is a product I-bundle (resp. a square or annulus) such that one lid is Z.

The following lemma is a translation of the Kneser-Haken finiteness-theorem into our language. It is crucial for the existence of the characteristic submanifold.

*9.1 Lemma. Let* $(M,\underline{\underline{m}})$ *be an irreducible* 3-*manifold with useful boundary-pattern. Then there exists an integer* $n(M,\underline{\underline{m}})$ *such that the following holds:*

*For every essential surface in* $(M,\underline{\underline{m}})$ *with no component an i-faced disc,* i = 1,2, *or a* 2-*sphere, and with more than* $n(M,\underline{\underline{m}})$ *components, at least two components are admissibly parallel in* $(M,\underline{\underline{m}})$.

*Proof*. We show how the proof of the finiteness theorem [Ha 3] (see also [Kn 1]) can be modified so as to yield a proof of 9.1. Fix a triangulation, $\Delta$, of M in such a way that the graph, J, of $(M,\underline{\underline{m}})$ is a subcomplex. Then, by the proof of Lemma 4 of [Ha 3] it suffices to show that any essential surface, F, in $(M,\underline{\underline{m}})$ with no component an i-faced disc, i = 1,2, or a 2-sphere can be admissibly isotoped in $(M,\underline{\underline{m}})$ such that it is in "nice" position with respect to the triangulation (for definition see [Ha 3, p. 48]). For this we have to check that the steps (i) - (v) of the proof of Lemma 3 of [Ha 3] can be realized by an admissible isotopic deformation of F in $(M,\underline{\underline{m}})$. F is essential in $(M,\underline{\underline{m}})$, no component of F is a 1-faced disc or a 2-sphere, and M is irreducible. Hence a simple inspection of the steps (i), (ii), and (iv) shows that they can be realized by an admissible isotopy of F in $(M,\underline{\underline{m}})$. Now consider step (iii). Let $E^2$ be a 2-simplex of $\Delta$ such that at least one component of $F \cap E^2$ is an arc, k, with both end points in the same edge, $E^1$, of $\Delta$. In [Ha 3, p. 58] there are three cases distinguished:

Case A: Both $E^2$ and $E^1$ lie in $M^0$.

Case B: Both $E^2$ and $E^1$ lie in $\partial M$.

Case C: $E^2 \subset M^0$ and $E^1 \subset \partial M$.

Since $F$ is essential in $(M,\underline{\underline{m}})$ and no component of $F$ is a 2-faced disc, it follows that in any case above the end points of $k$ cannot lie in the graph $J$. Hence the small isotopy described in [Ha 3] for Case A and B is an admissible isotopy of $F$ in $(M,\underline{\underline{m}})$. $F$ is essential, $M$ is irreducible and $\underline{\underline{m}}$ is a useful boundary-pattern. Therefore it finally follows that the operation described in Case C in [Ha 3] can be realized by an admissible isotopy of $F$ in $(M,\underline{\underline{m}})$. In the same way we see that step (v) can be realized by an admissible isotopy of $F$ in $(M,\underline{\underline{m}})$. q.e.d.

9.2 Lemma. Let $(F,\underline{\underline{f}})$ be a surface. Then there exists an integer $n(F,\underline{\underline{f}})$ such that the following holds:

For every system of non-singular essential curves in $(F,\underline{\underline{f}})$, closed or not, with more than $n(F,\underline{\underline{f}})$ curves, at least two curves are admissibly parallel in $(F,\underline{\underline{f}})$.

Proof. Apply Kneser's idea [Kn 1] [Ha 3] to 2-manifolds, or apply 9.1 to the product I-bundle over $(F,\underline{\underline{f}})$. q.e.d.

For later use we mention the following:

9.3 Corollary. Let $(F,\underline{\underline{f}})$ be a surface. Let $(G_i,g_i)_{i\in N}$ be a sequence of essential surfaces in $(F,\underline{\underline{f}})$. Denote by $\bar{\underline{\underline{g}}}_i$ the complete boundary-pattern of $(G_i,\underline{\underline{g}}_i)$. Suppose $(G_{i+1},\underline{\underline{g}}_{i+1})$ is admissible in $(G_i,\bar{\underline{\underline{g}}}_i)$, for all $i \in N$.

Then there is an integer $n \in N$ and an admissible ambient isotopy of $(F,\underline{\underline{f}})$, constant outside a regular neighborhood of $G_n$ in $F$, which contracts $G_n$ into $G_{n+1}$.

Remark. We do not assert that the inclusions $G_n \subset F$ and $G_{n+1} \subset F$ are admissibly isotopic in $(F,\underline{\underline{f}})$. Indeed, there is in general no integer $n \in N$ with such a property.

Proof. Since $(G_{i+1},\underline{\underline{g}}_{i+1})$ is admissible in $(G_i,\bar{g}_i)$, we have that $(\partial G_{i+1} - \partial F)^- \cap (\partial G_i - \partial F)^- = \emptyset$ and $G_{i+1} \subset G_i$, for all $i \geq 1$. Thus $\cup_{1\leq i\leq m}(\partial G_i - \partial F)^-$ is a system of non-singular essential curves in $(F,\underline{\underline{f}})$, for all $m \geq 1$. Hence, by 9.2, there is an integer, s, $s \geq 1$, such that every curve of $(\partial G_t - \partial F)^-$, for all $t > s$, is admissibly isotopic in $(F,\underline{\underline{f}})$ to some curve of $\cup_{1\leq i\leq s}(\partial G_i - \partial F)^-$. Since $G_{i+1} \subset G_i$, for all $i \geq 1$, this implies that for each component, B, of $G_t$ one of the following holds:

1. $B \cap G_{t+1}$ is a system of inner squares or annuli in $(F,\underline{\underline{f}})$ (entirely contained in a neighborhood of $(\partial B - \partial F)^-$).
2. B can be admissibly isotoped into $B \cap G_{t+1}$, using an admissible ambient isotopy of $(F,\underline{\underline{f}})$ which is constant outside a regular neighborhood of B in F.

Hence the existence of the required integer n is obvious. q.e.d.

As a consequence of 9.1 we finally prove the existence of the characteristic submanifold.

9.4 Proposition. Let $(M,\underline{\underline{m}})$ be an irreducible 3-manifold with useful boundary-pattern. Then there exists a characteristic submanifold in $(M,\underline{\underline{m}})$.

Proof. By the very definition of characteristic submanifolds, it suffices to show that the integers $n(W)$: = number of components of W, and $c(X,\underline{\underline{x}})$ have an upper bound, where W resp. $(X,\underline{\underline{x}})$ are taken from the set of all full essential F-manifolds, respectively all essential I-bundles or Seifert fibre spaces in $(M,\underline{\underline{m}})$. This follows from 9.1: for the integers $n(W)$ use the fact that no three components of $(\partial W - \partial M)^-$ can be pairwise admissibly parallel in $(M,\underline{\underline{m}})$ since W is a full, essential F-manifold. To see the claim for the integers $c(X,\underline{\underline{x}})$, recall that $c(X,\underline{\underline{x}}) = \min c(X,\underline{\underline{x}},p) = 2\beta_1(\tilde{F}) + \text{card}(\underline{\underline{f}})$, where $(\tilde{F},\underline{\underline{f}})$ denotes an appropriate orbit surface of $(X,\underline{\underline{x}})$ minus the exceptional points. Now the number of pairwise not admissibly parallel, essential squares or annuli in $(M,\underline{\underline{m}})$ cannot be larger than a certain integer, say m. Hence it follows from 5.4, 4.6.1, and our definition of $c(X,\underline{\underline{x}})$ that the number of pairwise not admissibly parallel, essential arcs in $(\tilde{F},\underline{\underline{f}})$ cannot be larger than m. In particular, card $(\underline{\underline{f}})$ cannot be arbitrarily large. By the same argument, the integers $\beta_1(\tilde{F})$ have an upper bound. q.e.d.

Remark. Using [Ha 1], one can also give a constructive proof of 9.4 (cf. §29).

## §10. Uniqueness of the characteristic submanifold

Let $W$ be an admissible F-manifold in a 3-manifold $(M,\underline{\underline{m}})$. Define

$M' = (M - W)^-$, and let $\underline{\underline{m}}'$ be the proper boundary-pattern of $M'$ induced by $W$.

For convenience we call $(M',\underline{\underline{m}}')$ the <u>manifold obtained from</u> $(M,\underline{\underline{m}})$ <u>by splitting at</u> $W$ (cf. §1). Using 4.8, it follows that $\underline{\underline{m}}'$ is a useful boundary-pattern of $M'$, provided $W$ is an essential F-manifold and $\underline{\underline{m}}$ is useful.

$W$ is called <u>complete</u> (with respect to essential squares, annuli, or tori in $(M',\underline{\underline{m}}')$), if, for every essential square, annulus, or torus, $T$, in $(M',\underline{\underline{m}}')$, either 1 or 2 holds:

1. $T \cap W \neq \emptyset$, and the component of $(M',\underline{\underline{m}}')$ which contains $T$ admits an admissible fibration as a product I- or $S^1$-bundle over the square or annulus.
2. $T \cap W = \emptyset$, and $T$ is admissibly parallel in $(M',\underline{\underline{m}}')$ to a side of $(M',\underline{\underline{m}}')$ which is contained in $(\partial W - \partial M)^-$.

We shall show that every characteristic submanifold in an irreducible 3-manifold $(M,\underline{\underline{m}})$ with useful boundary-pattern is complete (see 10.4). This needs some preparation.

<u>10.1 Lemma.</u> <u>Let $(M,\underline{\underline{m}})$ be an irreducible 3-manifold with useful boundary-pattern. Suppose $T$ is an admissible square or annulus, inessential in $(M,\underline{\underline{m}})$, but containing a curve essential in $(M,\underline{\underline{m}})$. Let $(\tilde{M},\tilde{\underline{\underline{m}}})$ be obtained from $(M,\underline{\underline{m}})$ by splitting at the surface $T$.</u>

<u>Then at least one component of $(\tilde{M},\tilde{\underline{\underline{m}}})$ can be admissibly fibered as I- or $S^1$-bundle over an i-faced disc</u>, $1 \leq i \leq 3$.

<u>Proof.</u> We may apply 4.2. Let $(D,\underline{\underline{d}})$ be an admissible disc as given in 4.2. Denote by $U(D)$ a regular neighborhood of $D$ in $M$, and let $D_1, D_2$ be the two copies of $D$ in $\partial U(D)$. Then of course, each component of $\tilde{T} = (T - U(D)) \cup D_1 \cup D_2$ is an admissible i-faced disc,

$1 \leq i \leq 3$, in $(M,\underline{\underline{m}})$ and 10.1 follows from the definition of "useful" and the irreducibility of $M$. q.e.d.

10.2 Lemma. Let $(F,\underline{\underline{f}})$ be a surface with non-empty boundary. Let $(G,\underline{\underline{g}})$ be a connected, essential surface in $(F,\underline{\underline{f}})$. Suppose that, for no component, $A$, of $(F - G)^{-}$, $(\partial A \cap \partial F)^{-}$ is contained in $\cup_{k \in \underline{\underline{f}}} k$. Then card $(\underline{\underline{f}}) \geq$ card$(\underline{\underline{g}})$, if $c(G,\underline{\underline{g}}) \geq c(F,\underline{\underline{f}})$.

Proof. Assume card$(\underline{\underline{f}}) <$ card$(\underline{\underline{g}})$ and $c(G,\underline{\underline{g}}) \geq c(F,\underline{\underline{f}})$. Then defining $m =$ card$(\underline{\underline{g}})$ - card$(\underline{\underline{f}})$, we have $m \geq 1$. Hence

$$\beta_1(F) = \frac{1}{2}(c(F,\underline{\underline{f}}) - \mathrm{card}(\underline{\underline{f}})) \leq \frac{1}{2}(c(G,\underline{\underline{g}}) + m - \mathrm{card}(\underline{\underline{g}}))$$

$$= \frac{1}{2}(\beta_1(G) + m) < \beta_1(G) + m, \text{ since } m \geq 1.$$

$F$ and $G$ are surfaces with non-empty boundary and so their fundamental groups are free. Recall that, for a free group, any two generating sets of independent elements have the same cardinality. Hence the rank of a free group is well-defined. Furthermore, $\beta_1(F) = \mathrm{rank}\ \pi_1 F$ and $\beta_1(G) = \mathrm{rank}\ \pi_1 G$. Therefore

$$\mathrm{rank}\ \pi_1 F < m + \mathrm{rank}\ \pi_1 G.$$

On the other hand, by our definition of $m$, there are at least $m$ components, $b_1,\ldots,b_m$, of $(\partial F - G)^{-}$ which are arcs contained entirely in the interior of sides of $(F,\underline{\underline{f}})$. Let $U(b_i)$ be a regular neighborhood of $b_i$ in $F$, and define $G^{+} = G \cup \cup U(b_i)$. Then clearly rank $\pi_1 G^{+} = m + \mathrm{rank}\ \pi_1 G$ since $G$ is essential. Now $G^{+}$ has $n, n \geq 0$, boundary curves which do not meet $\partial F$. For each of them fix one point (lying in the specified curve) and remove from $F$ a regular neighborhood of this point. In this way we obtain from $F$ a surface $F^*$ with rank $\pi_1 F^* = n + \mathrm{rank}\ \pi_1 F$. Let $G^* = G^{+} \cap F^*$, and denote by $H_i^*$ the components of $(F^* - G^*)^{-}$. Then the components of $G^* \cap H_i^*$ are simply connected. In particular, by Seifert-van Kampen's theorem [ZVC 1], rank $\pi_1 F^* \geq \mathrm{rank}\ \pi_1 G^* + \Sigma_i \mathrm{rank}\ \pi_1 H_i^*$. Now observe that, by our suppositions on the components of $(F - G)^-$, at least one boundary curve of each $H_i^*$ meets $\partial F$. Hence $\Sigma_i$ rank $\pi_1 H_i^* \geq n$, and therefore

$$\text{rank } \pi_1 F = \text{rank } \pi_1 F^* - n \geq \text{rank } \pi_1 G^* = \text{rank } \pi_1 G^+ = m + \text{rank } \pi_1 G.$$

This is the required contradiction. q.e.d.

<u>10.3 Lemma</u>. <u>Let</u> $(F,\underline{\underline{f}})$ <u>and</u> $(G,\underline{\underline{g}})$ <u>be given as in</u> 10.2. <u>Furthermore suppose</u> $c(G,\underline{\underline{g}}) \geq c(F,\underline{\underline{f}})$.

<u>Then each component of</u> $(F - G)^-$ <u>is a regular neighborhood of a free side of</u> $(F,\underline{\underline{f}})$ (in particular, an inner square or annulus in $(F,\bar{\underline{\underline{f}}})$).

<u>Proof</u>. Consider one component, A, of $(F - G)^-$. By supposition, $c(G,\underline{\underline{g}}) \geq c(F,\underline{\underline{f}})$ and, by 10.2., $\text{card}(\underline{\underline{f}}) \geq \text{card}(\underline{\underline{g}})$. Thus $\beta_1(F) \leq \beta_1(G)$. On the other hand, $\beta_1(F) \geq \beta_1(G)$ since G is essential in F. Hence $\beta_1(F) = \beta_1(G)$ and $\text{card}(\underline{\underline{f}}) = \text{card}(\underline{\underline{g}})$.

We assert that $(\partial A - \partial F)^- = A \cap G$ is connected. Assume the contrary. Let $r_1, r_2$ be two different components of $(\partial A - \partial F)^-$ and $x_i$ be a point in $r_i$, $i = 1,2$. Both A and G are connected. Hence there are arcs, $k_1$, $k_2$, in A, G, resp., joining $x_1$ with $x_2$. Then $k = k_1 \cup k_2$ is a non-contractible simple closed curve in F which cannot be deformed into G, i.e. $\beta_1(F) = \text{rank } \pi_1 F > \text{rank } \pi_1 G = \beta_1(G)$ which is a contradiction.

Every closed curve in A can be deformed into G, i.e. into $(F - A)^-$, since $\text{rank } \pi_1 F = \beta_1(F) = \beta_1(G) = \text{rank } \pi_1 G$. Since $(\partial A - \partial F)^-$ is connected, it follows that A must be either homeomorphic to a disc or an annulus.

Suppose A is homeomorphic to a disc. G is essential in F, by supposition. Hence $(\partial A - \partial F)^-$ is an essential arc in $(F,\underline{\underline{f}})$. Furthermore, $\text{card}(\underline{\underline{f}}) = \text{card}(\underline{\underline{g}})$. Therefore $\underline{\underline{f}}$ induces a boundary-pattern of A with precisely two elements and these are disjoint. This means that A is an inner square in $(F,\bar{\underline{\underline{f}}})$ such that one side does not lie in an element of $\underline{\underline{f}}$. Analogously, if A is homeomorphic to an annulus. q.e.d.

10.4 Proposition. Let $(M,\underline{\underline{m}})$ be an irreducible 3-manifold with useful boundary-pattern. Then each characteristic submanifold, $V$, in $(M,\underline{\underline{m}})$ is complete with respect to essential squares, annuli, or tori in $(M',\underline{\underline{m}}')$, where $(M',\underline{\underline{m}})$ is the manifold obtained from $(M,\underline{\underline{m}})$ by splitting at $V$.

Proof. Let $T$ be an essential square, annulus, or torus in $(M',\underline{\underline{m}}')$. Let $U(V)$ be a regular neighborhood in $M$, and let $R_1$, $R_2$ be all the components (possibly empty or equal) of $(\partial U(V) - \partial M)^-$ with $R_i \cap T \neq \emptyset$, $i = 1,2$. Denote by

$$N = U(R_1 \cup (T - \dot{U}(V)) \cup R_2)$$

a regular neighborhood in $M'$. Then $N$ is an admissible I-bundle or Seifert fibre space in $(M,\underline{\underline{m}})$ with $N \cap V = \emptyset$.

Suppose there is a component, $B$, of $(\partial N - \partial M)^-$ which is inessential in $(M,\underline{\underline{m}})$. Then $B$ separates $M$ into two components, say $W$ and $W'$. At least one of them, say $W$, is an I- or $S^1$-bundle (in $(M,\underline{\underline{m}})$) over an i-faced disc, $1 \leq i \leq 3$. This is immediate from 10.1 if $B$ is a square or annulus. If $B$ is a torus use the following slightly more refined argument. First recall that $V$ is non-empty, and let $W'$ contain some component of $V$. Indeed, by our choice of $B$, we may specify an essential annulus $A$ in $(W' - V)^-$ which joins a component of $V$ with $B$. Furthermore, there is a disc $D$ in $M$ with $D \cap B = \partial D \subset B$ so that $\partial D$ is not contractible in $B$. This disc cannot lie in $W'$: otherwise either $A \cap \partial D \neq \emptyset$ or $\partial D$ can be moved via $B \cup A$ into $(\partial V - \partial M)^-$ which is both impossible since $(\partial V - \partial M)^-$ is essential. If we denote by $U(D)$ a regular neighborhood of $D$, then $(B - U(D))^-$, together with two copies of $D$ gives us a 2-sphere which in turn bounds a 3-ball since $M$ is irreducible. Since $D \subset W$, this means that either $(W - U(D))^-$ or $W' \cup U(D)$ is a 3-ball. The latter is impossible since $W'$ contains a component of $V$, and so $W$ is a solid torus, which proves our claim.

In particular, no component of $(\partial V - \partial M)^-$ lies in $W$. It follows that $W \subset (M - (N \cup V))^-$. Moreover, it is easily seen that $N \cup W$ is again an admissible I-bundle or Seifert fibre space in $(M,\underline{\underline{m}})$

with $(N \cup W) \cap V = \emptyset$, and we attach $W$ to $N$. After finitely many such steps all inessential components of $(\partial N - \partial M)^-$ are eliminated, and we finally get an essential I-bundle or Seifert fibre space. By 8.3, the union of this I-bundle or Seifert fibre space with $V$ is contained in a full essential F-manifold, $V^*$, in $(M,\underline{\underline{m}})$.

Now let $X^*$ be the component of $V^*$ containing $N$, and denote by $\underline{\underline{x}}^*$ the boundary-pattern of $X^*$ induced by $\underline{\underline{m}}$. $(X^*,\underline{\underline{x}}^*)$ is an essential I-bundle or Seifert fibre space in $(M,\underline{\underline{m}})$. Fix an actual admissible fibration of $(X^*,\underline{\underline{x}}^*)$ with fibre projection $p: X^* \to F^*$ such that $c(X^*,\underline{\underline{x}}^*) = c(X^*,\underline{\underline{x}}^*,p)$ (see definition of complexity).

There is at least one component of $V$ which is contained in $X^*$, for otherwise $V \cup X^*$ is a full essential F-manifold in $(M,\underline{\underline{m}})$ (note that $V$ is a full essential F-manifold) with more components than $V$. But then $V \cup X^*$ has, by definition of complexity, a larger complexity than $V$, which contradicts the fact that $V$ is a characteristic submanifold in $(M,\underline{\underline{m}})$.

Moreover, we may suppose $V \cap X^*$ is in fact vertical in $X^*$ with respect to $p$. To see this consider the surface $G = (\partial V - \partial X^*) \cap X^*$. $G$ is a surface such that each component is an essential square, annulus, or torus in $(M,\underline{\underline{m}})$, and so in $(X^*,\underline{\underline{x}}^*)$ (see 4.6.3). Since $V$ is a full F-manifold in $(M,\underline{\underline{m}})$, $X^*$ cannot be a component of $M$. Hence it follows from 5.6, that $G$ can be admissibly isotoped in $(X^*,\underline{\underline{x}}^*)$ into a vertical surface. Extend this isotopy to an admissible ambient isotopy, $\alpha_t$, $t \in I$, of $(X^*,\underline{\underline{x}}^*)$. Then $\alpha_t^{-1}$ deforms the fibration of $(X^*,\underline{\underline{x}}^*)$ admissibly so that afterwards $p^{-1}p(V \cap X^*) = V \cap X^*$.

Let $x_1,\ldots,x_n$, $n \geq 0$, be all the exceptional points of $F^*$. Choose a regular neighborhood $U(\cup x_i)$ in $F^*$, and define $\tilde{F} = F^* - \dot{U}(\cup x_i)$. $\underline{\underline{x}}^*$ induces uniquely a boundary-pattern, $\tilde{\underline{\underline{f}}}$, of $\tilde{F}$. By our choice of $p$ and by the definitions of the complexities, $c(X^*,\underline{\underline{x}}^*) = c(\tilde{F},\tilde{\underline{\underline{f}}})$ holds.

Let $Y = (X^* - V)^-$, and let $Z$ be a component of $V \cap X^*$ with the property that the complexity of any component of $V \cap X^*$ is less than or equal to $c(Z)$. Define $\tilde{F}_0 = \tilde{F} \cap p(Y)$, $\tilde{G} = \tilde{F} \cap p(Z)$, and $\tilde{H} = (\tilde{F} - \tilde{G})^-$ (recall $Z$ and $Y$ are vertical). Then denote by $\tilde{\underline{\underline{g}}}$ and $\tilde{\underline{\underline{h}}}$ the boundary-patterns of $\tilde{G}$ and $\tilde{H}$ resp. induced by $\tilde{\underline{\underline{f}}}$.

10.5 Assertion.

1. $\tilde{F}_0$ and $\tilde{G}$ are connected.
2. $\tilde{F}_0$ and $\tilde{G}$ are essential in $(\tilde{F},\tilde{\underline{\underline{f}}})$.
3. For no component, A, of $(\tilde{F} - \tilde{G})^-$, $A \cap \partial\tilde{F}$ is contained in $\cup_{k\in\tilde{\underline{\underline{f}}}} k$,
4. $c(\tilde{F},\tilde{\underline{\underline{f}}}) \leq c(\tilde{G},\tilde{\underline{\underline{g}}})$.

ad 1. $\tilde{F}_0$ is connected, for otherwise there is a component of Y which does not meet N, and so this component is a component of $(M - V)^-$, which is impossible since V is a full F-manifold. $\tilde{G}$ is connected, since Z is.

ad 2. $(\partial V - \partial M)^-$ is essential in $(M,\underline{\underline{m}})$.

ad 3. V is a full F-manifold in $(M,\underline{\underline{m}})$.

ad 4. By definition, $\alpha_{V*}(i)$, $\alpha_V(i)$ are the numbers of all those components of V*, V, respectively, which have complexity i. X* is a component of V*, and Z a component of V with $Z \subset X^*$. Moreover, 1. $V - X^* = V^* - X^*$ since V is a full F-manifold in $(M,\underline{\underline{m}})$, 2. the complexities of all the components of $V \cap X^*$ are less than or equal to c(Z), by our choice of Z. Hence it follows

$$\alpha_{V*}(i) \geq \alpha_V(i), \quad \text{for all} \quad i > c(Z).$$

On the other hand V is a characteristic submanifold in $(M,\underline{\underline{m}})$ and V* is a full essential F-manifold in $(M,\underline{\underline{m}})$. Consequently, $c(V^*) \leq c(V)$, by 8.2. Therefore, by the definition of complexity,

$$\alpha_{V*}(i) = \alpha_V(i), \quad \text{for all} \quad i > c(Z),$$

which implies, $c(X^*) \leq c(Z)$. Now, $c(\tilde{F},\tilde{\underline{\underline{f}}}) = c(X^*,\underline{\underline{x}}^*)$, by our choice of p, and $c(Z) \leq c(\tilde{G},\tilde{\underline{\underline{g}}})$, by the definition of complexity. Thus, altogether, $c(\tilde{F},\tilde{\underline{\underline{f}}}) \leq c(\tilde{G},\tilde{\underline{\underline{g}}})$. This proves 10.5.

By 10.5, we may apply 10.3. Hence each component of $\tilde{H} = (\tilde{F} - \tilde{G})^-$ is an inner square or annulus in $(\tilde{F},\tilde{\underline{\underline{f}}})$ (where $\bar{\underline{\underline{f}}}$ is the complete boundary-pattern of $(\tilde{F},\tilde{\underline{\underline{f}}})$) such that one side does not lie in an element of $\underline{\underline{f}}$. Now $\tilde{F}_0$ is contained in $\tilde{H}$. Moreover, $\tilde{F}_0$ is connected (10.5.1), and essential in $(\tilde{F},\tilde{\bar{f}})$ (10.5.2). Since $V$ is full, these facts imply that $\tilde{F}_0$ is a component of $\tilde{H}$. Hence $\tilde{F}_0 \cap U(\cup x_i) = \emptyset$ (recall $x_i$, $1 \leq i \leq n$, are the exceptional points of $F^*$). This is clear, if $\tilde{F}_0$ is an inner square. If $\tilde{F}_0$ is an inner annulus, it follows since $(\partial V - \partial M)^-$ is an essential surface in $(M,\underline{\underline{m}})$. Hence $Y$ is an I- or $S^1$-bundle over the square or annulus, and $(\partial Y - \partial M)^-$ has precisely two components. Finally, by construction of $X^*$, $Y$ contains $N$ and so $T$ is nearly contained in $Y$.

If $T \cap V \neq \emptyset$, then $T$ must join the two components of $(\partial Y - \partial M)^-$ since $T$ is essential in $(M',\underline{\underline{m}}')$. But then $Y$ is nearly a component of $(M - V)^- = M'$, by the defintion of $N$ and $X^*$, and so $T$ lies in a component of $(M',\underline{\underline{m}}')$ which is an I- or $S^1$-bundle over the square or annulus.

If $T \cap V = \emptyset$, then it follows that $T$ splits $Y$ into two different components which are I- or $S^1$-bundles over the square or annulus. By the definition of $Y$, at least one of them must meet $V$ since $X^*$ contains at least one component of $V$ (see above). Hence $T$ is admissibly parallel in $(M',\underline{\underline{m}}')$ to a side of $(M',\underline{\underline{m}}')$ which is contained in $(\partial V - \partial M)^-$.

Hence by definition, $V$ is complete. q.e.d.

In the following proposition we list some useful properties of complete F-manifolds:

<u>10.6 Proposition.</u> <u>Let</u> $(M,\underline{\underline{m}})$ <u>be an irreducible</u> 3-<u>manifold with useful boundary-pattern.</u> <u>Suppose</u> $V$ <u>is a non-empty, complete, essential</u> F-<u>manifold in</u> $(M,\underline{\underline{m}})$. <u>Then the following holds:</u>

1. <u>If</u> $W$ <u>is an essential</u> F-<u>manifold in</u> $(M,\underline{\underline{m}})$ <u>entirely contained in</u> $M - V$, <u>then</u> $W$ <u>is a regular neighborhood of an essential surface in</u> $(M,\underline{\underline{m}})$ <u>whose components are admissibly parallel to components of</u> $(\partial V - \partial M)^-$.

<u>Suppose, in addition,</u> $V$ <u>is a full</u> F-<u>manifold. Then:</u>

2. *If* G *is a bound side of* $(M,\underline{\underline{m}})$ *which is either a square, annulus, or torus (with respect to the boundary-pattern induced by* $\underline{\underline{m}}$*), then* G *is entirely contained in* V.

3. *If* W *is an essential* F-*manifold in* $(M,\underline{\underline{m}})$ *containing* V, *then there is an admissible isotopic deformation of* W *in* $(M,\underline{\underline{m}})$ *into* W' *such that each component of* V *is a component of* W'.

   *In addition: If* W *is a full* F-*manifold in* $(M,\underline{\underline{m}})$, *then* W' = V.

*Remark*. 10.6 holds, by 10.4, if V is a characteristic submanifold.

*Proof*. Without loss of generality, W is connected, i.e. contained in a component $(Z,\underline{\underline{z}})$ of $(M - V)^{-}$. If $(Z,\underline{\underline{z}})$ is an essential I-bundle or Seifert fibre space in $(M,\underline{\underline{m}})$, then, by 5.2 and 5.4, we find an essential square or annulus in $(Z,\bar{\underline{\underline{z}}})$ which meets V, and we are done since V is complete.

In the other case, 1 of 10.6 follows from the fact that each component of $(\partial W - \partial M)^{-}$ is essential and hence admissibly parallel to a component of $(\partial V - \partial M)^{-}$ since V is complete.

*Proof to 2*. Assume the contrary. Let $(M',\underline{\underline{m}}')$ be the manifold obtained from $(M,\underline{\underline{m}})$ by splitting at V. Let G* be an admissible surface in $(M,\underline{\underline{m}})$ near G. Then G* is an essential square, annulus, or torus in $(M,\underline{\underline{m}})$ since $\underline{\underline{m}}$ is a useful boundary-pattern of M. Since V is a full F manifold, G cannot be a side of a component of $(M',\underline{\underline{m}}')$, which is the product I- or $S^1$-bundle over the square, annulus, or torus. Hence, since V is complete, G*, and so G, must meet V. Since V is an essential F-manifold, $V \cap G$ is a system of inner squares or annuli in G. This implies that the components of V can be admissibly fibered so that they do not meet G in a lid. Since V is a full F-manifold, there is no component, W, of $(M',\underline{\underline{m}}')$ such that $V \cup W$ is an essential F manifold in $(M,\underline{\underline{m}})$. Hence we conclude that W cannot be a product I- or $S^1$-bundle over the square or annulus which meets G. On the other hand each component of $G' = G^* \cap M'$ is an admissible square or annulus in $(M',\underline{\underline{m}}')$ which meets V. Therefore, since V is complete, the components of G' are inessential squares or annuli in $(M',\underline{\underline{m}}')$. But $\underline{\underline{m}}'$ is a useful boundary-pattern of M' (see 4.8.2). Thus, by 10.1 and our choice of G', it follows that $(\partial V - \partial M)^-$ is inessential in $(M,\underline{\underline{m}})$. This contradicts the fact that V is an essential F-manifold in $(M,\underline{\underline{m}})$.

Proof to 3. Without loss of generality we may suppose that $V \cap (\partial W - \partial M)^- = \emptyset$. Now let X be any component of V. Then X is contained in some component, Y, of W. $(Y - V)^-$ is not empty since $V \cap (\partial W - \partial M)^- = \emptyset$. By 5.7, we may fix an admissible fibration, p, of Y such that $(\partial X - \partial M)^- \cap Y$ is vertical with respect to this fibration. Then $(Y - V)^-$ consists of I-bundles or Seifert fibre spaces. Let $Y_1$ be one of them which meets X, and define $Y_1' = (Y_1 - U(V))^-$, where U(V) is a regular neighborhood in $(M,\underline{\underline{m}})$. Since $(\partial Y - \partial M)^- \cap (\partial V - \partial M)^- = \emptyset$, it follows that $Y_1'$ is an essential I-bundle or Seifert fibre space in $(M,\underline{\underline{m}})$. Hence $V \cup Y_1'$ is an essential F-manifold in $(M,\underline{\underline{m}})$. Therefore, applying 1 above, it follows that $Y_1'$, and so $Y_1$ must be the product I- or $S^1$-bundle over the square or annulus. This implies that $Y_1$ cannot be a component of $(M - V)^-$, since otherwise $V \cup Y_1$ is an essential F-manifold (note that p induces an admissible fibration of $(V \cup Y_1) \cap Y$ and that $V \cap (\partial Y - \partial M)^- = \emptyset$) which contradicts the fact that V is a full F-manifold. That means that Y is a regular neighborhood of X in $(M,\underline{\underline{m}})$. Certainly there is then an admissible isotopic deformation of W in $(M,\underline{\underline{m}})$ which contracts Y to X and which is constant on W - Y. Thus we have an induction.

The additional remark follows from 1 above, since V is a full F-manifold. q.e.d.

The following two propositions are special cases of the enclosing theorems (see §§12 and 13 for the general versions).

10.7 Proposition. *Let* $(M,\underline{\underline{m}})$ *be an irreducible 3-manifold with useful boundary-pattern. Let* $V$ *be a full, essential F-manifold in* $(M,\underline{\underline{m}})$ *which is complete. Let* $T$ *be a surface such that each component of* $T$ *is an essential square, annulus, or torus in* $(M,\underline{\underline{m}})$. *Then there exists an admissible isotopic deformation of* $T$ *in* $(M,\underline{\underline{m}})$ *which pulls* $T$ *into* $V$.

Remark. 10.7 holds, by 10.4, if $V$ is a characteristic submanifold.

Proof. Suppose $T$ is admissibly isotoped in $(M,\underline{\underline{m}})$ so that it meets the surface $(\partial V - \partial M)^-$ in a minimal number of curves. Since $V$ is an essential F-manifold in $(M,\underline{\underline{m}})$, each component of $(\partial V - \partial M)^-$ is an essential square, annulus, or torus in $(M,\underline{\underline{m}})$. Let $(\tilde{M},\tilde{\underline{\underline{m}}})$ be the manifold obtained from $(M,\underline{\underline{m}})$ by splitting at $(\partial V - \partial M)^-$. Then, by 4.8, $\tilde{\underline{\underline{m}}}$ is a useful boundary-pattern of $\tilde{M}$, and so, by 4.6, every component of $T \cap \tilde{M}$ is an essential square, annulus, or torus in $(\tilde{M},\tilde{\underline{\underline{m}}})$.

Assume there is a component, $A$, of $(T - V)^-$ with $A \cap V \neq \emptyset$. Let $A_1$, $A_2$ be the components of $T \cap V$ neighborhing $A$ (possibly $A_2$ empty or equal to $A_1$). Then $A_i$, $i = 1,2$, is an essential square or annulus in an I-bundle or Seifert fibre space, $X_i$, of $V$. Applying 5.7, $X_i$ admits an admissible fibration such that $A_i$ is vertical with respect to this fibration. Moreover (see 5.7), this fibration may be chosen so that no component of $A_i \cap A$ lies in a lid of $X_i$. On the other hand, since $V$ is complete, $A$ lies in a component, $W$, of $(M - V)^-$ which is a product I- or $S^1$-bundle over the square or annulus. Hence it follows that $V \cup W$ is an essential F-manifold in $(M,\underline{\underline{m}})$ which contradicts the fact that $V$ is a full F-manifold in $(M,\underline{\underline{m}})$.

Thus each component of $T$ is either contained in $V$ or in $M - V$, and 10.7 follows immediately since $V$ is complete. q.e.d.

10.8 Proposition. *Let* $(M,\underline{\underline{m}})$ *be an irreducible 3-manifold with useful boundary-pattern. Let* $V$ *be a full, essential F-manifold in* $(M,\underline{\underline{m}})$ *which is complete. Let* $W$ *be any essential F-manifold in* $(M,\underline{\underline{m}})$.

*Then there is an admissible isotopic deformation of* $W$ *in* $(M,\underline{\underline{m}})$ *which pulls* $W$ *into* $V$.

*Remark.* 10.8 holds, by 10.4, if $V$ is a characteristic submanifold.

*Proof.* Since $W$ is an essential F-manifold in $(M,\underline{\underline{m}})$, $(\partial W - \partial M)^-$ is a surface whose components are essential squares, annuli, or tori in $(M,\underline{\underline{m}})$. Hence, by 10.7, we may suppose that $(\partial W - \partial M)^- \subset V$ and that, in addition, $(\partial W - \partial M)^- \cap (\partial V - \partial M)^- = \emptyset$. Furthermore, we suppose $W$ is admissibly isotoped in $(M,\underline{\underline{m}})$ so that it has the preceding properties and that, in addition, the number of components of $(W - V)^-$ is as small as possible.

We assert $(W - V)^- = \emptyset$, and then we are done.

Assume the converse. Then there is at least one I-bundle or Seifert fibre space, $X$, of $W$ with $(X - V)^- \neq \emptyset$. By 5.7, we may fix an admissible fibration of $X$, with fibre projection $p: X \to F$, such that $(\partial V - \partial M)^- \cap X$ is vertical with respect to this fibration. Hence $(X - V)^-$ consists of I-bundles or Seifert fibre spaces. Since $(\partial X - \partial M)^- \subset V$ (see above), each component of $(X - V)^-$ is a component of $(M - V)^-$. Therefore, by 10.6, $(X - V)^-$, together with $p|(X - V)^-$, consists of product I- or $S^1$-bundles over the square or annulus.

Let $\underline{\underline{f}}$ be the boundary-pattern of $F$ induced by $\underline{\underline{m}}$, and $\bar{\underline{\underline{f}}}$ as usual the complete boundary-pattern of $(F,\underline{\underline{f}})$. Denote by $x_1,\ldots,x_n$, $n \geq 0$, all the exceptional points in $F$ and define $\tilde{F} = F - \mathring{U}(\cup x_i)$, where $U(\cup x_i)$ is a regular neighborhood in $F$. Let $\tilde{G} = \tilde{F} \cap p(X \cap V)$ and $\tilde{H} = \tilde{F} \cap p(X - V)^- = (\tilde{F} - \tilde{G})^-$. Then, by what we have proved so far, $\tilde{H}$ consists of inner squares or annuli in $(\tilde{F},\underline{\underline{f}})$. Let $\tilde{H}_1$ be one of them. Then $(\partial\tilde{H}_1 - \partial\tilde{F})^-$ is disconnected since $V$ is a full F-manifold in $(M,\underline{\underline{m}})$. Let $k_1$, $k_2$ be the two components of $(\partial\tilde{H}_1 - \partial\tilde{F})^-$. Then $k_i$, $i = 1,2$, lies in a component, $\tilde{G}_i$, of $\tilde{G}$ (possibly $\tilde{G}_1 = \tilde{G}_2$).

The number of components of $(W - V)^-$ is minimal. Hence there is no admissible isotopy of $(\tilde{F},\underline{\underline{f}})$ which contracts $\tilde{F}$ into $\tilde{F} - (\tilde{H}_1 \cup \tilde{G}_1)$ since such an isotopy can be lifted to an admissible isotopy of $X$, and, moreover, of $W$, which is constant on $W - X$. This implies that, if $\tilde{G}_i$, $i = 1,2$, is an inner square or annulus in $(\tilde{F},\bar{\underline{\underline{f}}})$, then it is also one in $(F,\underline{\underline{f}})$. From this it is easily seen that there is at least one essential arc, $t_i$, in $(\tilde{G}_i,\bar{\underline{\underline{g}}})$ such that one boundary point of $t_i$ lies in $k_i$ and the other one either in $k_1 \cup k_2$ or in an element of $\underline{\underline{f}}$.

Let $Z_i$ be that component of $W$ containing $p^{-1}\tilde{G}_i$ and $\underline{\underline{z}}_i$ the boundary-pattern of $Z_i$ induced by $\bar{\underline{\underline{m}}}$. Then, by 5.2, it follows that $\bar{\underline{\underline{z}}}_i$ is a useful boundary-pattern of $Z_i$, and so, by 4.6, it follows that $(\partial X - \partial M)^- \cap Z_i$ is an essential surface in $(Z_i,\underline{\underline{z}}_i)$. Let $(\tilde{Z}_i,\tilde{\underline{\underline{z}}}_i)$ be the manifold obtained from $(Z_i,\bar{\underline{\underline{z}}}_i)$ by splitting at $(\partial X - \partial M)^- \cap Z_i$. Then $T_i = p^{-1}t_i$ is an admissible square or annulus in $(Z_i,\bar{\underline{\underline{z}}}_i)$ as well as in $(\tilde{Z}_i,\tilde{\underline{\underline{z}}}_i)$. By 5.4 and our definition of $T_i$, $T_i$ is essential in $(\tilde{Z}_i,\tilde{\underline{\underline{z}}}_i)$, and so, by 4.6, it is also essential in $(Z_i,\bar{\underline{\underline{z}}}_i)$. Hence, by 5.7, there is an admissible fibration of $(Z_i,\bar{\underline{\underline{z}}}_i)$, such that $T_i$ is vertical with respect to this fibration. Moreover (see 5.7), this fibration may be chosen so that $p^{-1}k_i$ does not lie in a lid of $(Z_i,\bar{\underline{\underline{z}}}_i)$. But this implies that $V \cup p^{-1}\tilde{H}_1$ is an essential F-manifold in $(M,\underline{\underline{m}})$ (recall $\tilde{H}_1$ is an inner square or annulus in $(\tilde{F},\underline{\underline{f}})$) which contradicts the fact that $V$ is a full F-manifold in $(M,\underline{\underline{m}})$.

q.e.d.

As an immediate consequence of 10.8 and 10.6 we obtain the uniqueness of the characteristic submanifold (up to admissible ambient isotopy).

<u>10.9 Corollary</u>. <u>Let</u> $(M,\underline{\underline{m}})$ <u>be an irreducible</u> 3-<u>manifold with useful boundary-pattern</u>. <u>Let</u> V, V' <u>be two characteristic submanifolds in</u> $(M,\underline{\underline{m}})$.

<u>Then</u> V <u>can be admissibly isotoped in</u> $(M,\underline{\underline{m}})$ <u>so that</u> V = V'.

<u>Remark</u>. The isotopy of V can be extended to an admissible ambient isotopy of $(M,\underline{\underline{m}})$.

We are now finally in the position to give equivalent but more geometric descriptions of characteristic submanifolds. It will be these descriptions which we shall use in the following.

*10.10 Corollary (characterization of characteristic submanifolds). Let* $(M,\underline{\underline{m}})$ *be an irreducible 3-manifold with useful boundary-pattern. Let* V *be a full, essential F-manifold in* $(M,\underline{\underline{m}})$.

*Then the following statements are equivalent:*

1. V *is a characteristic submanifold.*
2. V *is complete.*
3. *Every essential F-manifold in* $(M,\underline{\underline{m}})$ *can be admissibly isotoped in* $(M,\underline{\underline{m}})$ *into* V.

*Proof.*

1 implies 2: See 10.4.

2 implies 3: See 10.8.

3 implies 1: By 9.4, there exists a characteristic submanifold, W, in $(M,\underline{\underline{m}})$. This is, in particular, an essential F-manifold in $(M,\underline{\underline{m}})$. Hence, by supposition, W can be admissibly isotoped in $(M,\underline{\underline{m}})$ into V. Therefore, by the additional remark of 10.6, V can be admissibly isotoped in $(M,\underline{\underline{m}})$ so that V = W, and so 1 of 10.10 follows.

q.e.d.

*Remark 1.* Let W be a full, essential F-manifold in $(M,\underline{\underline{m}})$. Suppose W has the property that every essential square, annulus, or torus in $(M,\underline{\underline{m}})$ can be admissibly isotoped in $(M,\underline{\underline{m}})$ into W. Then simple examples show that W is in general *not* a characteristic submanifold in $(M,\underline{\underline{m}})$.

*Remark 2.* Observe that, by 10.9, every admissible homeomorphism $h: (M,\underline{\underline{m}}) \to (M,\underline{\underline{m}})$ can be admissibly isotoped so that afterwards $h(V) = V$ This point of view will be applied in §27 to the study of the mapping class group of $(M,\underline{\underline{m}})$. But note that 10.9 does not assert that, for any given admissible homeomorphism h, V can be admissibly isotoped so that afterwards $h(V) = V$. At this point, the question arises naturally, whether or not there is any reasonable equivariant theory of characteristic submanifolds. By "reasonable" we mean a theory which extends at least some of the results in this paper to almost sufficiently large 3-manifolds. This looks plausible for $\mathbb{Z}_2$-actions, and should be one way to extend the theory at hand to non-orientable 3-manifolds (for some work in this direction, see [Bo 1]).

## Part II. THE ENCLOSING THEOREM

The enclosing theorem says that the characteristic submanifolds contain all essential maps of squares, annuli, or tori into Haken 3-manifolds, up to admissible homotopy. This will be proved in §12. In §13 we extend this to essential maps of I-bundles and Seifert fibre spaces. These properties make the characteristic submanifold a useful object to work with, both in the study of homotopy equivalences and 3-manifold groups. We will come to these points later.

## Chapter IV: Singular surfaces and characteristic submanifolds.

### §11. A lemma on essential intersections

This paragraph is devoted to the introduction of the "essential intersection", or, dually, the "essential union". Later (see §§26, 30, and 31) we shall have to study essential intersections in connection with homeomorphisms and homotopy equivalences. Here, however, we restrict ourselves to the study of essential intersections with respect to essential singular squares, annuli, or tori in I-bundles, and we prove a technical result which is needed in the induction-argument of the enclosing theorem.

To begin with let $(F,\underline{\underline{f}})$ be any orientable surface, and let $F_1$, $F_2$ be two (not necessarily disjoint) essential surfaces in $(F,\underline{\underline{f}})$. We say that $F_1$ is in a good position to $F_2$, if

1. $(\partial F_1 - \partial F)^-$ is transversal to $(\partial F_2 - \partial F)^-$, and
2. the number of points of $(\partial F_1 - \partial F)^- \cap (\partial F_2 - \partial F)^-$ cannot be diminished, using an admissible isotopic deformation of $F_1$ in $(F,\underline{\underline{f}})$.

Furthermore, we say that $F_1$ is in a very good position to $F_2$, if, in addition,

3. there is no admissible isotopic deformation of $F_1$ in $(F,\underline{\underline{f}})$ which at the same time enlarges the number of components of both $F_1 \cap F_2$ and $(\partial F_1 - \partial F)^-$ contained in $F_2$.

Observe that $F_1$ is in a good position to $F_2$ if and only if

$(F - F_1)^-$ is in a good position to $F_2$.

The essential intersection (essential union) of two essential surfaces $F_1$ and $F_2$ in $(F,\underline{\underline{f}})$ in good position is defined to be the largest (smallest) essential surface contained in (containing) $F_1 \cap F_2$ $(F_1 \cup F_2)$. Observe that the essential union of $F_1$, $F_2$ is the complement of the essential intersection of $(F - F_1)^-$, $(F - F_2)^-$.

The next lemma is an easy consequence of the definitions and the transversality lemma [Wa 3].

11.1 Lemma. Let $F_1$, $F_2$ be two essential surfaces in $(F,\underline{\underline{f}})$ such that $(\partial F_1 - \partial F)^-$ is transversal to $(\partial F_2 - \partial F)^-$. Then the following holds:

1. $F_1$ and $F_2$ are in good position if and only if no component of $(F - U)^-$ is a 1- or 2-faced disc, where $U$ is a regular neighborhood of $(\partial F_1 - \partial F)^- \cup (\partial F_2 - \partial F)^-$.

Suppose that $F_1$ and $F_2$ are in good position. Then

2. $F_2$ is in a very good position to $F_1$ if and only if there is no inner square or annulus $A$ in $(F,\underline{\underline{f}})$ such that one component of $(\partial A - \partial F)^-$ lies in $(\partial F_1 - \partial F)^-$ and the other one in $(\partial F_2 - \partial F)^-$, and $A \subset (F - F_1 \cup F_2)^-$.
3. If $G$ denotes the essential intersection (resp. union) of $F_1$ and $F_2$, then $(F_1 \cap F_2 - G)^-$ resp. $(G - F_1 \cup F_2)^-$) consists of 3-faced discs (with respect to the completed boundary-patters).

We shall see that this lemma is a convenient criterion for very good positions of surfaces.

As mentioned above, 11.1 is obvious in the case when $F_2$ is independent from $F_1$. But for later use we need an extension. Indeed, in the applications below we shall often encounter the following situation: $F_2$ is defined to be $d(F_1)$, where $F_1$ is an essential surface in $(F,\underline{\underline{f}})$, and where $d:(F,\underline{\underline{f}}) \to (F,\underline{\underline{f}})$ is an admissible, fixpoint-free, orientation reversing involution. This means that $F_2$ depends on $F_1$ in the sense that any isotopy of $F_1$ enforces equally well an isotopy of $F_2$. But we shall see that also in this case the criterion of 11.1 applies, i.e. also when $F_1 = F_1$ and $F_2 = dF_1$. This is based on the following observation:

11.2 Lemma. Suppose that $k$ is an essential curve in $(F,\underline{\underline{f}})$ with $dU(k) = U(k)$, where $U(k)$ is a regular neighborhood of $k$ in $F$. Then (in the notation above) $d$ interchanges the components of $(\partial U(k) - \partial F)^-$.

Proof. Since d is orientation reversing, it follows that the restriction $(d|U(k)$ is orientation reversing. Hence, if d does not interchange the components of $(\partial U(k) - \partial F)^-$, then $d|r: r \to r$ has to be orientation reversing, for each component r of $(\partial U(k) - \partial F)^-$. But this contradicts the fact that d is fixpoint-free.

q.e.d.

11.3 Lemma. Using the notations above, $F_1$ is in a very good position to $F_2 = dF_1$, if and only if the criterion of 11.1 is satisfied.

Proof. First of all it is easily checked that $F_1$ can be admissibly isotoped so that afterwards $(\partial F_1 - \partial F)^-$ is transverse with respect to $(\partial dF_1 - \partial F)^-$, i.e. the hypothesis of 11.1 is satisfied.

Now, one direction of the criterion of 11.1 is obvious. For the other one we first suppose that there is a 2- or 3-faced disc as described in 11.1.2. This means that there is an embedding $i: D \to F$ (not admissible embedding), such that $\underline{\underline{d}}$ is the disjoint union of $i^{-1}(\partial F_1 - \partial F)^-$, $i^{-1}(\partial dF_1 - \partial F)^-$, and $i^{-1}k$, for some $k \in \underline{\underline{f}}$. We write $D = i(D)$. Then $D^\circ \cap dD^\circ = \emptyset$, for d is fixpoint free. If $D \cap dD = \emptyset$, clearly the number of points of $(\partial F_1 - \partial F)^- \cap (\partial dF_1 - \partial F)^-$ can be diminished, using an admissible isotopic deformation of $F_1$. This is also true if $D \cap dD \neq \emptyset$. Because in this case $D \cap dD$ must consist of precisely two points (d is fixpoint-free), i.e. a regular neighborhood A of $D \cup dD$ is an annulus. Without loss of generality, $dA = A$ and our claim follows from 11.2. Thus $F_1$ and $F_2$ are not in a good position.

So we suppose that $F_1$ and $dF_1$ are in a good position and that there is an inner square or annulus as described in 11.1.2. Then either $A \cap dA = \emptyset$ or $A = dA$. In both cases if follows from 11.2 again that $F_1$ and $dF_1$ cannot be in a very good position.

q.e.d.

Having established our criterion for very good positions, we are now going to prove the first interesting property of such a position (other properties will be established later, see §26 and §31). The 2-dimensional version of it says that an essential curve can be admissibly deformed into the essential intersection of $F_1$ and $F_2$ if it can be deformed into $F_1$ and $F_2$. This is our basis for the proof of the Enclosing Theorem .

So, for the following lemma let $(X, \underline{\underline{x}})$ denote an I-bundle with projection $p: X \to B$. Define $F = (\partial X - p^{-1}\partial B)^-$ and let $\underline{\underline{f}}$ be the boundary-pattern of $F$ induced from $\underline{\underline{x}}$. Moreover, let $d: (F,\underline{\underline{f}}) \to (F,\underline{\underline{f}})$ be the admissible involution given by the reflections in the I-fibres of $X$. Suppose $G$ is an essential surface in $(F,\underline{\underline{f}})$ which is in a very good position with respect to $dG$, and define $G'$ to be the essential intersection of $G$ and $dG$. Then it is easily checked that $G' = dG' = F \cap p^{-1}pG'$.

<u>11.4 Lemma</u>. <u>Let</u> $f: A \to X$ <u>be an essential singular square or annulus in</u> $(X,\underline{\underline{x}})$ <u>and let</u> $k_1, k_2$ <u>be the two sides of</u> $A$ <u>mapped under</u> $f$ <u>into</u> $F$. <u>Suppose</u> $f(k_1)$ <u>and</u> $f(k_2)$ <u>are contained in</u> $G$. <u>Then</u> $f$ <u>can be admissibly deformed in</u> $(X,\underline{\underline{x}})$ <u>into</u> $p^{-1}pG'$.

<u>Proof</u>. Observe that the restriction of $p$ to any component of $F$ is a covering map. This means that the map $pf: A \to B$ can be lifted to a map $g: A \to F$ with $g|k_1 = f|k_1$. Moreover, $g|k_2$ is either equal to $f|k_2$ or to $df|k_2$. To decide which of these equalities holds, consider an arc, $s$, in $A$ joining $k_1$ with $k_2$. If $g|k_2 = f|k_2$, then $g|s$, together with $f|s$, defines a singular closed curve in $X$ whose composition with $p$ is contractible in the base $B$. Lifting such a contraction to $X$, we find that $f|s$ has to be inessential. But $f$ is essential. Hence $g|k_2 \neq f|k_2$, and so $g|k_2 = df|k_2$.

So we are given an admissible map $g: (A,\underline{\underline{a}}) \to (F,\underline{\underline{f}})$ (with respect to an appropriate boundary-pattern $\underline{\underline{a}}$ of $A$) with $g(k_1) \subset G$ and $g(k_2) \subset dG$. Denote $G_1 = G$ and $G_2 = dG$, and let $S_j = g^{-1}(\partial G_j - \partial F)^-$. Then, without loss of generality, $g(k_j) \subset G_j - (\partial G_j - \partial F)^-$, $j = 1,2$. As in the proof of the transversality lemma (1.1 of [Wa 3]) we see that $g$ can be admissibly deformed (rel $k_1 \cup k_2$) in $(F,\underline{\underline{f}})$ so that afterwards $S_1$ and $S_2$ are two systems of admissible, non-contractible, simple curves in $(A,\bar{\underline{\underline{a}}})$ which are in good position (recall that $G_1$ and $G_2$ are in good position). Moreover, there is no arc in $S_2$ which joins $k_1$ with $k_2$ since $g(k_2) \subset G_2 - (\partial G_2 - \partial F)^-$. Therefore no arc of $S_2$ meets both $k_1$ and $S_1$. Hence we may deform $g$ admissibly so that the above holds and that, in addition, $S_2 \cap k_1 = \emptyset$, and similarly $S_1 \cap k_2 = \emptyset$.

Then $g^{-1}G_1$ and $g^{-1}G_2$ are (non-empty) essential surfaces in $(A,\bar{\underline{\underline{a}}})$. It follows that $g^{-1}G_2$ is in a very good position with respect to $g^{-1}G_1$ since $G_2$ is in such a position to $G_1$ (apply Nielsen's theorem). This in turn implies that $g^{-1}G_1 \cap g^{-1}G_2$ is a non-empty essential surface. So $k_1$ can be admissibly deformed in $(A,\underline{\underline{a}})$ into $g^{-1}G_1 \cap g^{-1}G_2$, and so of course $g|k_1$ can be admissibly deformed in $(F,\underline{\underline{f}})$ into $G_1 \cap G_2$. q.e.d.

## §12. Proof of the enclosing theorem

The object of this paragraph is to prove the enclosing theorem (see 12.5), which asserts that every essential singular square, annulus, or torus $f: T \to M$ can be admissibly deformed into the characteristic submanifold of $(M,\underline{\underline{m}})$.

Here $(M,\underline{\underline{m}})$ denotes a Haken 3-manifold with useful boundary-pattern, and the proof itself will be by an induction on a hierarchy of $(M,\underline{\underline{m}})$. To make the proof more transparent it is quite convenient to first establish a few facts concerning essential surfaces in connection with characteristic submanifolds and essential singular squares, annuli, and tori.

For this fix a connected, essential surface $F$ in $(M,\underline{\underline{m}})$, $F \cap \partial M = \partial F$, which is neither boundary-parallel, nor a 2-sphere. Let $U(F)$ be a regular neighborhood of $F$, and denote $H = (\partial U(F) - \partial M)^-$. Define $(\tilde{M},\tilde{\underline{\underline{m}}})$ to be the 3-manifold obtained from $(M,\underline{\underline{m}})$ by splitting at $F$, i.e. $\tilde{M} = (M - U(F))^-$.

By 4.7, we may suppose that $f$ is admissibly deformed so that $\tilde{f} = f|f^{-1}\tilde{M}$ consists again of essential singular squares, annuli, or tori in $(\tilde{M},\tilde{\underline{\underline{m}}})$. Let $T_1, T_2, \ldots, T_n$, $n \geq 1$, be all the components of $f^{-1}\tilde{M}$, and let the indices be chosen so that $T_i$ and $T_{i+1}$ are neighboring, for all $1 \leq i \leq n$.

<u>12.1 Lemma</u>. <u>Let</u> $\tilde{W}$ <u>be any essential</u> F-<u>manifold in</u> $(\tilde{M},\tilde{\underline{\underline{m}}})$ <u>which contains</u> $\tilde{f}$. <u>Suppose that</u> $f(T_i)$, <u>for some</u> $1 \leq i \leq n$, <u>lies in a component of</u> $\tilde{W}$ <u>which meets</u> $H$ <u>in inner squares or annuli</u>.

<u>Then</u> $\tilde{f}$ <u>can be admissibly deformed into an essential</u> F-<u>manifold in</u> $(\tilde{M},\tilde{\underline{\underline{m}}})$ <u>which meets</u> $H$ <u>in inner squares or annuli</u>.

<u>Proof</u>. Without loss of generality, $f(T_1)$ is contained in a component of $\tilde{W}$ which meets $H$ in inner squares or annuli, and $f(T_2)$ not. Let $X$ be the component of $\tilde{W}$ which contains $f(T_2)$. Then $X$ is an essential I-bundle which meets $H$ in lids. Let $T_1', \ldots, T_m'$, be all the components of $\tilde{f}^{-1}X$, and let $T_2 = T_1'$. Denote by $k_1$, $k_2$ the two components of $(\partial T_1' - \partial T)^- = (\partial T_2 - \partial T)^-$, and suppose that $k_1$ lies near $T_1$. Then $f(k_1)$ is contained in $H \cap X$, i.e. in a lid of $X$.

By our suppositions on $f|T_1$, it follows that $f|k_1$ can be admissibly deformed in $H$ into a non-singular curve $t$, and $f|k_2$ can be admissibly deformed out of $t$. Moreover, for every $f|T_i'$, $1 \leq i \leq m$, at least one side contained in $X \cap H$ can be admissibly deformed out of $t$. All these deformations may be chosen within $X \cap H$ since $X \cap H$ is an essential surface in $H$ (apply the transversality lemma). Now $f|T_i'$, $1 \leq i \leq m$, is essential in $\widetilde{M}$, and so in $X$ (see 4.7). Hence, using 5.12, we see that $f|T_1'$ can be admissibly deformed in $X$ into a regular neighborhood, $U(B_1)$, of some (non-singular) vertical, essential square, annulus, or Möbius band. Moreover, using the additional remark of 5.10, we see that all $f|T_2',\dots,f|T_m'$ can be admissibly deformed into $(X - U'(B_1))^-$. Here $U'(B_1)$ denotes a regular neighborhood of $U(B_1)$ in $X$. Replacing $X$ by $U(B_1) \cup (X - U'(B_1))^-$ we obtain a new essential F-manifold, and so 12.1 follows inductively. q.e.d.

An essential F-manifold, $\widetilde{W}$, in $(\widetilde{M},\underline{\underline{\widetilde{m}}})$ is called a <u>nice submanifold</u>, if

(i) $\tilde{f}$ can be admissibly deformed in $(\widetilde{M},\underline{\underline{\widetilde{m}}})$ into $\widetilde{W}$,

(ii) $\widetilde{W}$ meets $H$ in an essential surface $G$ with the following property: no component $C$ of $(H - G)^-$ is an inner square or annulus in $H$ such that $(\partial C - \partial H)^-$ is contained in components of $G$ which are also inner squares or annuli,

(iii) $\widetilde{W}$ can be admissibly contracted in $(\widetilde{M},\underline{\underline{\widetilde{m}}})$ to every essential F-manifold, satisfying (i) and (ii), which is contained in $\widetilde{W}$.

<u>12.2 Lemma. Let $\widetilde{W}$ be any nice submanifold in $(\widetilde{M},\underline{\underline{\widetilde{m}}})$ which meets $H$ in inner squares or annuli.</u>

<u>Then $\widetilde{W}$ can be admissibly isotoped so that afterwards</u>

$$\widetilde{W} \cap H = d(\widetilde{W} \cap H),$$

<u>where $d\colon H \to H$ is the admissible involution given by the reflections in the I-fibres of the product I-bundle $U(F)$.</u>

Proof. Let $G = \tilde{W} \cap H$. Then we may suppose that $\tilde{W}$ is admissibly isotoped so that $G$ is in a very good position with respect to dG. Define G' to be the essential intersection of $G$ and dG. Let $G_1$ be any component of $G = \tilde{W} \cap H$. Then we still have to show that $G_1 \cap G' \neq \emptyset$ and that $G_1$ can be admissibly contracted to $G_1 \cap G'$.

Observe that we may suppose that $f$ is admissibly deformed so that $\tilde{f}$ is contained in $\tilde{W}$ ($\tilde{W}$ has property (i)). There is at least one component, $k_1$, of $f^{-1}H$ mapped under $f$ into $G_1$: for otherwise, $(\tilde{W} - U(G_1))^-$ contains an essential F-manifold with (i) and (ii) and which is not admissibly isotopic to $\tilde{W}$, but this contradicts our choice of $\tilde{W}$ (here $U(G_1)$ is a regular neighborhood of $G_1$ in $\tilde{M}$).

Let $A$ be the component of $f^{-1}U(F)$ containing $k_1$, and $k_2$ the side of $A$, different from $k_1$, which is mapped into $H$. Then $f|A$ is an essential singular square or annulus in $U(F)$ and $f(k_1)$, $f(k_2)$ are contained in $G$. By 11.4 and our definition of G', $f|k_1$ can be admissibly deformed in $H$ into G'. Since $G' \subset G$ and since $\tilde{W}$ has property (ii), it follows that this homotopy of $f|k_1$ may be chosen within $G_1$. Hence $G_1 \cap G' \neq \emptyset$ and our claim follows from property (ii) of $\tilde{W}$. q.e.d.

<u>12.3 Lemma</u>. <u>Suppose that there is no nice submanifold in</u> $(\tilde{M},\underline{\tilde{m}})$ <u>which meets</u> $H$ <u>in inner squares or annuli</u>.

<u>Then any nice submanifold</u> $\tilde{W}$ <u>in</u> $(\tilde{M},\underline{\tilde{m}})$ <u>can be admissibly isotoped so that afterwards</u>

$$\tilde{W} \cap H = d(\tilde{W} \cap H),$$

<u>where</u> $d$ <u>is given as in</u> 12.2.

Proof. In the notation of the beginning of the proof of 12.2, we still have to show that $G_1 \cap G' \neq \emptyset$ and that $G_1$ can be admissibly contracted to $G_1 \cap G'$.

For this let $X_1$ be the component of $\tilde{W}$ containing $G_1$. It follows from 12.1 and the suppositions of 12.3 that $X_1$ is an I-bundle which meets $H$ in lids. Denote by $p_1 : X_1 \to B_1$ the fibre projection.

Define $F_1^+ = X_1 \cap H$ (possibly $G_1 = F_1^+$), and let $G_1^+ = G' \cap F_1^+$. Let $e: F_1^+ \to F_1^+$ be the admissible involution defined by the reflections in the I-fibres of $X_1$. Without loss of generality, $G_1^+$ is in a very good position with respect to $e(G_1^+)$. Denote by $C$ the essential intersection of $G_1^+$ and $e(G_1^+)$. Then $p_1^{-1}p_1C$ is an essential I-bundle in $X_1$, and so in $\tilde{M}$.

Replacing $X_1$ by $p_1^{-1}p_1C$, we obtain an essential F-manifold, $\tilde{W}'$, with $\tilde{W}' \subset \tilde{W}$. Without loss of generality, $\tilde{W}'$ has property (ii), for in the other case we simply have to add components of $(X_1 - \tilde{W}')^-$ to $\tilde{W}'$ which are I-bundles over the square or annulus (recall that $\tilde{W}$ has property (ii)). So, if $\tilde{f}$ can be admissibly deformed into $\tilde{W}'$, then $\tilde{W}$ can be admissibly contracted to $\tilde{W}'$ ($\tilde{W}$ is nice), and so $G_1$ to $G_1 \cap G'$. Thus it remains to prove that, for every component $T_i'$, $1 \leq i \leq m$, of $\tilde{f}^{-1}X_1$, $\tilde{f}|T_i'$ can be admissibly deformed into $p^{-1}pC$.

Consider $T_1'$. Let $k_1$, $k_2$ be the two sides of $T_1'$ mapped under $\tilde{f}$ into $H$, i.e. into $F_1^+$. By an argument of 12.2, $f|k_1$ and $f|k_2$ can be admissibly deformed in $H$ into $G_1^+$. These homotopies of $f|k_1$, $f|k_2$ may be chosen within $F_1^+$. For otherwise it follows, using the additional remark of 5.10, that $f|T_1'$ can be admissibly deformed into a regular neighborhood of $(\partial X_1 - \partial\tilde{M})^-$, and, by 12.1, this contradicts the suppositions of 12.3. So, by 11.4, $f|T_1'$ can be admissibly deformed in $X_1$ into $p^{-1}pC$ (see our definition of $C$).

q.e.d.

12.4 Lemma. *In the notation above, let the surface $F$ be chosen so that, in addition, the complexity (see 8.1) of $F$ is minimal. Then the two following statements are equivalent:*

1. *$f$ can be admissibly deformed into an essential F-manifold in $(M,\underline{\underline{m}})$.*
2. *$\tilde{f}$ can be admissibly deformed into an essential F-manifold in $(\tilde{M},\underline{\underline{\tilde{m}}})$.*

Proof.

1 implies 2. By 1, we may suppose that $f$ is contained in an essential F-manifold, $W$, in $(M,\underline{\underline{m}})$. Let $W$ be admissibly isotoped

so that the number of curves of $F \cap (\partial W - \partial M)^-$ is as small as possible. Then, by 4.6, $F \cap W$ is an essential surface in $W$. Hence, by 5.6, each component of $F \cap W$ is either vertical or horizontal in $W$, and so, $\tilde{W} = W \cap \tilde{M}$ is an essential F-manifold in $(\tilde{M}, \tilde{\underline{\underline{m}}})$. Let $h$ be an admissible homotopy which pulls $f\colon T \to M$ into $W$. By 4.4, $h$ may be chosen so that $h^{-1}F$ is an essential surface in $T \times I$. Hence it follows that $h$ induces an admissible homotopy in $(\tilde{M}, \tilde{\underline{\underline{m}}})$ which pulls $\tilde{f}$ into $\tilde{W}$ (apply 5.6).

<u>2 implies 1</u>. By 2, we may suppose that $\tilde{f}$ is contained in an essential F-manifold, $\tilde{W}$, in $(\tilde{M}, \tilde{\underline{\underline{m}}})$. $\tilde{W}$ is contained in an admissible F-manifold, $\tilde{W}'$, with property (ii) of a nice submanifold. In fact, it is contained in an essential F-manifold, $\tilde{W}''$, with property (ii). To see this observe that, for every inessential component, $A$, of $(\partial\tilde{W}' - \partial\tilde{M})^-$, the union of $\tilde{W}'$ with the component of $(\tilde{M} - \tilde{W}')^-$ containing $A$ is again an admissible F-manifold with property (ii). By 10.1, this is trivial, if $A$ is not a torus. In the other case, use the irreducibility of $M$. Applying 9.1, it is easily seen that $\tilde{W}''$ contains a nice submanifold. Hence, without loss of generality, $\tilde{W}$ is a nice submanifold. Then, by 12.2 and 12.3, we may suppose that

$$\tilde{W} \cap H = d(\tilde{W} \cap H).$$

This means that there is a system, $Z$, of essential product I-bundles in $U(F)$ with $Z \cap H = \tilde{W} \cap H$. Define

$$W = \tilde{W} \cup Z.$$

Suppose that $F$ is a square, annulus, or torus. Then $Z = U(F)$ since $\tilde{W}$ is nice, and $W$ is an essential F-manifold (see 4.6). $f$ is contained in $W$, and so we are done. Therefore we may suppose in the following that $F$ is neither a square, annulus, or torus.

We claim that $f$ can be admissibly deformed into $W$. For this let $A$ be a component of $f^{-1}U(F)$, and let $k_1$, $k_2$ the two sides of $A$ mapped under $f$ into $H$. Then $f(k_1)$ and $f(k_2)$ are contained in $Z \cap H$. It suffices to show that $f|A$ can be admissibly deformed (rel $k_1 \cup k_2$) in $U(F)$ into $Z$. This follows easily if $f|k_1$, $f|k_2$ cannot be admissibly deformed in $H$ into $(\partial(Z \cap H) - \partial H)^-$ (apply transversality lemma). So, by an argument of 12.3, we may suppose that $\tilde{W}$ meets $H$ in inner squares or annuli. Then no component of $(H - \tilde{W})^- = (H - Z)^-$ is an inner square or annulus since $\tilde{W}$ is nice, and so $f|A$ can be admissibly deformed into $Z$ (apply transversality lemma and Nielsen's theorem). Thus, altogether, we may suppose that $f$ is contained in $W$.

Let $X$ be the component of $W$ which contains $f$. Then, by definition of $W$, $X$ is either an I-bundle, Seifert fibre space, or Stallings manifold, and $(\partial X - \partial M)^-$ consists of essential squares, annuli or tori (see 4.6). If $X$ is a Stallings manifold, $f$ can be admissibly deformed in $X$ into a regular neighborhood $U(\partial X)$ of $\partial X$ in $X$, for otherwise it follows from 6.1 and 6.2 the existence of an essential annulus or torus in $(M,\underline{\underline{m}})$ which is not boundary-parallel. But the latter would contradict our choice of $F$ (recall the minimality condition on $F$). Thus in any case it follows that $f$ can be admissibly deformed into an essential F-manifold in $(M,\underline{\underline{m}})$.

q.e.d.

We are now in the position to prove the enclosing theorem.

<u>12.5 Enclosing theorem</u>. <u>Let</u> $(M,\underline{\underline{m}})$ <u>be a Haken</u> 3-<u>manifold with useful boundary-pattern</u>. <u>Let</u> $V$ <u>be the characteristic submanifold in</u> $(M,\underline{\underline{m}})$.

<u>Then every essential singular square, annulus, or torus in</u> $(M,\underline{\underline{m}})$ <u>can be admissibly deformed in</u> $(M,\underline{\underline{m}})$ <u>into</u> $V$.

<u>Proof</u>. By supposition and by 4.3, $(M_1,\underline{\underline{m}}_1) = (M,\underline{\underline{m}})$ contains a connected, not boundary-parallel, essential surface, $F_1$, $F_1 \cap \partial M_1 = \partial F_1$, which is not a 2-sphere. Suppose $F_1$ is chosen so that its complexity is minimal. Let $(M_2,\underline{\underline{m}}_2)$ be the manifold obtained from $(M_1,\underline{\underline{m}}_1)$ by splitting at $F_1$. Then, by 4.8, $(M_2,\underline{\underline{m}}_2)$ is again an

irreducible 3-manifold with useful boundary-pattern (note that every i-faced disc, $1 \leq i \leq 3$, in an irreducible 3-manifold with useful boundary-pattern is boundary parallel). If there is a component of $M_2$ which is not a ball then there exists a connected, not boundary-parallel, essential surface, $F_2$ in $M_2$, different from a 2-sphere, with $F_2 \cap \partial M_2 = \partial F_2$ and minimal complexity. We repeat the above construction to get $(M_3, \underline{\underline{m}}_3)$ and $F_3$, and so on.

By a result of Haken, the procedure will stop after a finite number of steps [Ha 2, p. 101]. Let

$$F_i \subset M_i, \ (M_{i+1}, \underline{\underline{m}}_{i+1}), \qquad 1 \leq i \leq n$$

be the sequence of data finally obtained.

$M_{n+1}$ is a system of balls. Hence the boundary curves of the surfaces of $\underline{\underline{m}}_{n+1}$ are contractible in $M_{n+1}$. Since $\underline{\underline{m}}_{n+1}$ is a useful boundary-pattern of $M_{n+1}$, they are in fact contractible in the surfaces themselves, and so all surfaces of $\underline{\underline{m}}_{n+1}$ are discs.

Now let $f: (D, \underline{\underline{d}}) \to (M_{n+1}, \underline{\underline{m}}_{n+1})$ be any essential singular square. Then different elements of $\underline{\underline{d}}$ are mapped by $f$ into different elements of $\underline{\underline{m}}_{n+1}$ since $f$ is essential and $M_{n+1}$ is a system of balls. Thus we are able to deform $f$ admissibly in $(M_{n+1}, \underline{\underline{m}}_{n+1})$ so that after the deformation the restriction $f|\partial D$ is an embedding ("straightening the sides"). Since $M_{n+1}$ is a system of balls, $f$ can be deformed (rel $\partial D$) so that $f$ is an embedding. Finally we apply 10.7 to show that $f$ can be admissibly deformed in $(M_{n+1}, \underline{\underline{m}}_{n+1})$ into the characteristic submanifold in $(M_{n+1}, \underline{\underline{m}}_{n+1})$.

Since $(M_{n+1}, \underline{\underline{m}}_{n+1})$ cannot contain any essential singular annulus or torus, we have shown that $(M_{n+1}, \underline{\underline{m}}_{n+1})$ satisfies the conclusion of 12.5. Thus we may start an induction on the hierarchy, and 12.5 follows from 12.4. q.e.d.

Having established the enclosing theorem, we can now give first applications of the theory of characteristic submanifolds.

First of all observe that the two following corollaries imply the annulus- and torus-theorem as announced in [Wa 6].

12.6 Corollary. *Let* $(M,\underline{\underline{m}})$ *be an irreducible* 3-*manifold with useful boundary-pattern. Then the existence of an essential singular square or annulus,* $f\colon T \to M$, *in* $(M,\underline{\underline{m}})$ *implies the existence of an essential non-singular square or annulus,* $A$, *respectively, in* $(M,\underline{\underline{m}})$.

*If* $G$, $g \in \underline{\underline{m}}$, *contains a side of* $f$, *then* $A$ *may be chosen so that a side of* $A$ *lies in* $G$.

Proof. The existence of $f$ shows that $\partial M \neq \emptyset$, i.e., by 4.3, that $(M,\underline{\underline{m}})$ is a Haken 3-manifold. Hence we may apply 12.5. Therefore $f$ can be admissibly deformed in $(M,\underline{\underline{m}})$ into an essential I-bundle or Seifert fibre space, $X$, in $(M,\underline{\underline{m}})$. Applying 5.4, we find $A$ in $X$.

q.e.d.

12.7 Corollary. *Let* $(M,\underline{\underline{m}})$ *be a Haken* 3-*manifold with useful boundary-pattern, and which is not one of the exceptions* 5.1.5 *or* 5.1.6. *Then the existence of an essential singular torus,* $f\colon T \to M$, *implies the existence of an essential non-singular torus,* $A$, *in* $(M,\underline{\underline{m}})$.

*In addition: If* $f$ *cannot be deformed into* $\partial M$, $A$ *may be chosen so that it is not boundary-parallel in* $M$.

Remarks. 1. To see that 12.7 implies the "torus-theorem" in [Wa 6], apply the generalized loop-theorem [Wa 2].

2. If $M$ is the exception 5.1.6, and $\partial M \neq \emptyset$, the conclusion of 12.6 holds, provided $\underline{\underline{m}}$ is well chosen (for example as the set of all boundary components of $M$) (see 5.4).

3. On the other hand, if $M$ is the exception 5.1.6 with $\partial M = \emptyset$, the existence of an essential singular torus in $M$ does not imply, in general, the existence of an essential non-singular torus in $M$ (cf. the example in [Wa 3, paragraph 2].

Proof. Let $V$ be the characteristic submanifold in $(M,\emptyset)$. This exists, by 9.4. The existence of $f$ shows, by 12.5, that $V \neq \emptyset$. By our choice of the new boundary-pattern of $M$, $V \cap \partial M = \emptyset$, i.e. $V$ consists of essential Seifert fibre spaces in $(M,\emptyset)$. Hence either $M$ is a Seifert fibre space and 12.7 follows from 5.4, or

$\partial V \neq \emptyset$ and one component of $\partial V$ is the required torus.

For the additional remark: if there is a component of $\partial V$ which is not boundary-parallel in M, we are done. If not, it follows that V = M since f cannot be deformed into $\partial M$, and 12.7 follows from 5.4. q.e.d.

Let $(M,\underline{\underline{m}})$ be an irreducible 3-manifold with useful boundary-pattern. Suppose M is sufficiently large. Then, for the essential singular squares, annuli, or tori, $f: T \to M$, in $(M,\underline{\underline{m}})$, we define an equivalence relation by

(i) $f_0 \sim f_1$, if $f_0$ is admissibly homotopic in $(M,\underline{\underline{m}})$ to $f_1$,

(ii) $f_1 \sim f_2$, if there exists a covering map $q: T \to T$ with $f_1 = f_2 \cdot q$.

Applying 12.5, 5.10, and 5.13, we obtain the following

<u>12.8 Corollary</u>. <u>Let</u> $(M,\underline{\underline{m}})$ <u>be given as above</u>. <u>Let</u> $f: T \to M$ <u>be an essential singular square, annulus, or torus in</u> $(M,\underline{\underline{m}})$. <u>Then there exists</u> $f' \sim f$ <u>such that</u> $f'$ <u>is an immersion without triple points</u>.

The following conjecture was formulated as a corollary in the announcement [Jo 1]:

<u>Conjecture</u>. <u>Let</u> $(M,\underline{\underline{m}})$ <u>be given as above</u>. <u>Suppose that</u> $\underline{\underline{m}}$ <u>is the set of boundary components of</u> M. <u>Let</u> $f: T \to M$ <u>be an essential singular annulus or torus in</u> $(M,\underline{\underline{m}})$. <u>Then there exists a finite covering map</u> $p: (\tilde{M},\tilde{\underline{\underline{m}}}) \to (M,\underline{\underline{m}})$ <u>and an essential non-singular annulus or torus</u>, $\tilde{f}: T \to \tilde{M}$, <u>in</u> $(\tilde{M},\tilde{\underline{\underline{m}}})$ <u>such that</u> $f \sim p \cdot \tilde{f}$.

But, unfortunately, the assertion is <u>not</u> an immediate consequence of 12.8, and results on surface groups and Fuchsian groups (the so called Fenchel conjecture). New techniques are probably required.

<u>12.9 Corollary</u>. <u>Let</u> $(N,\underline{\underline{n}})$ <u>be a connected, irreducible</u> 3-<u>manifold with useful boundary-pattern</u>. <u>Suppose</u> N <u>is sufficiently large</u>. <u>Let</u> $(M,\underline{\underline{m}})$ <u>be a Seifert fibre space such that</u> $\underline{\underline{m}}$ <u>is a complete boundary-pattern of</u> M. <u>Suppose</u> $(M,\underline{\underline{m}})$ <u>is not one of the exceptional</u>

*cases* 5.1.1-5.1.5. *Let* $p: (M,\underline{\underline{m}}) \to (N,\underline{\underline{n}})$ *be an essential map*. *Then* $(N,\underline{\underline{n}})$ *is also a Seifert fibre space*.

*Remark*. Notice that any admissible covering map is an essential map, and recall that there is a similiar statement for I-bundles, see 5.8.

*Proof*. By 5.2, $\underline{\underline{m}}$ is a useful boundary-pattern of $M$. Hence, by 3.4, and by 6.1 of [Wa 4], we may suppose $p$ is admissibly deformed into an admissible covering map (see also 3.4). In particular, this implies that $\underline{\underline{n}}$ consists of annuli or tori, and so $\partial N$ consists of tori. Let $V$ be the characteristic submanifold in $(N,\underline{\underline{n}})$. This exists, by 9.4. Then $V$ is complete, by 10.4. Hence, by 10.6, $V$ must contain each surface of $\underline{\underline{n}}$ and so $\partial N$. Thus it follows that $V$ consists of Seifert fibre spaces since $\partial N$ consists of tori.

Now consider $G = (\partial p^{-1}V - \partial M)^{-} = p^{-1}(\partial V - \partial M)^{-}$. If $G$ is empty, then $V = N$ (recall $p$ is a covering map). So assume the converse. Let $G_1$ be a component of $G$ and $G_1'$ be that component of $(\partial V - \partial N)^{-}$ which contains $p(G_1)$. Then $G_1$ is a closed surface since $\partial N \subset V$ and so $\partial M \subset p^{-1}V$. Furthermore, $p|G_1: G_1 \to G_1'$ is a covering map since $p$ is, and $p|G_1$ is an essential map. This implies that $G_1$ is essential in $(M,\underline{\underline{m}})$ since $G_1'$ is essential in $(N.\underline{\underline{n}})$. Thus $G$ is essential in $(M,\underline{\underline{m}})$, and so, by 5.6, an admissible fibration of $(M,\underline{\underline{m}})$ can be chosen so that $G$ is either vertical or horizontal. But then we find an essential vertical annulus or torus, $T$, in $(M,\underline{\underline{m}})$ such that $T$ cannot be admissibly deformed in $(M,\underline{\underline{m}})$ into $p^{-1}V$ (this is clear, if $G$ is horizontal; if $G$ is vertical, apply 5.4). $p|T$ is an essential singular annulus or torus in $(N.\underline{\underline{n}})$ since $T$ and $p$ are essential. Hence, by 12.5, $p|T$ can be admissibly deformed in $(N,\underline{\underline{n}})$ into $V$. Lifting this homotopy ($p$ is an admissibly covering map) we see that $T$ can be admissibly deformed in $(M,\underline{\underline{m}})$ into $p^{-1}V$, which is a contradiction. q.e.d.

As a consequence of 12.5 we obtain an algebraic criterion for the existence of a companion of a knot. More precisely, a non-trivial knot k in $S^3$, which is not a torus knot, has a companion (in the sense of [Sch 1]) if and only if the knot group of k contains a knot group as subgroup of infinite index. We leave the proof as an exercise for the interested reader (hint: observe that a map f: N → M between compact manifolds cannot be deformed into a covering map if $f_* \pi_1 N$ has infinite index in $\pi_1 M$, and apply 3.3 and 12.5).

## Chapter V: Singular submanifolds and characteristic submanifolds.

### §13. An extension of the enclosing theorem

The purpose of this paragraph is to prove the following proposition:

**13.1 Proposition.** *Let $(M,\underline{\underline{m}})$ be a Haken 3-manifold with useful boundary-pattern. Let V be the characteristic submanifold in $(M,\underline{\underline{m}})$. Let $(X,\underline{\underline{x}})$ be an I-bundle or Seifert fibre space whose complete boundary-pattern is useful. Suppose $(X,\underline{\underline{x}})$ is not one of the exceptional cases 5.1.1-5.1.5.*

*Then every essential map $f: (X,\underline{\underline{x}}) \to (M,\underline{\underline{m}})$ can be admissibly deformed in $(M,\underline{\underline{m}})$ into V.*

**13.2 Corollary.** *Every essential singular Klein bottle or Möbius band (whose boundary curve is a side) in $(M,\underline{\underline{m}})$ can be admissibly deformed in $(M,\underline{\underline{m}})$ into the characteristic submanifold of $(M,\underline{\underline{m}})$.*

*Proof of 13.2.* There exists an I-bundle or an $S^1$-bundle over the Möbius band which contains a Möbius band or a Klein bottle resp. as an admissible deformation retract. Hence 13.2 follows from 13.1.

q.e.d.

*Proof of 13.1.* Let G be the union of all free sides of $(X,\underline{\underline{x}})$. Then G is non-empty, for otherwise 13.1 is immediate from 5.8 and 12.9. By 12.5 and 4.5, we may suppose that f is admissibly deformed so that $T = f^{-1}(\partial V - \partial M)^-$ consists of essential, vertical squares, annuli, or tori in $(X,\underline{\underline{x}})$ with $T \cap G = \emptyset$. Denote by $(\tilde{M},\tilde{\underline{\underline{m}}})$ and $(\tilde{X},\tilde{\underline{\underline{x}}})$ the manifold obtained from $(M,\underline{\underline{m}})$, resp. $(X,\underline{\underline{x}})$, by splitting along $(\partial V - \partial M)^-$, resp. T.

**13.3 Assertion.** *Let $(X_1,\underline{\underline{x}}_1)$ be a component of $(\tilde{X},\tilde{\underline{\underline{x}}})$, and let $T_1$ be a component of $(\partial X_1 - \partial X)^-$. Then either one of the following holds:*

1. *There is at least one essential, vertical square, resp. annulus, A, in $(X_1,\underline{\underline{x}}_1)$ with $T_1 \cap A \neq \emptyset$ such that $f|A$ is*

<u>essential in</u> $(\tilde{M},\tilde{\underline{m}})$.

2. $f|X_1$ <u>can be admissibly deformed in</u> $(M,\underline{m})$ <u>into</u> $(\partial V - \partial M)^-$, <u>using a homotopy which is constant on</u> $(\partial X_1 - \partial X)^-$.

To see this observe that there is a system, S, of essential, vertical squares, resp. annuli, in $(X_1,\underline{\underline{x}}_1)$ which all meet $T_1$ such that $(X_1 - U(T \cup S))^-$ consists of I- resp. $S^1$-bundles over an i-faced disc, $1 \leq i \leq 3$, where $U(T \cup S)$ denotes a regular neighborhood of $T \cup S$ in $X_1$. If 1 does not hold, we find an admissible homotopy of $f|X_1$ in $(M,\underline{m})$, constant on $(\partial X_1 - \partial X)^-$, which first pulls $f|U$ and then $f|\overline{X - U}$ into $(\partial V - \partial M)^-$, for $(\partial V - \partial M)^-$ is essential and $\tilde{M}$ is aspherical. This proves the assertion.

To continue the proof suppose that f is admissibly deformed so that the above holds, and that, in addition, the number of components of T is minimal.

<u>Case 1</u>. $T = \emptyset$.

Using 12.5, f can be admissibly deformed into a map g so that $g^{-1}(\partial V - \partial M)^-$ is admissibly parallel to G. Let $X_1 = g^{-1}(M - V)^-$, and recall from 10.4 that V is complete. Therefore, by 13.3 and 12.6, we see that either $g|X_1$ can be admissibly deformed (rel $(\partial X_1 - \partial X)^-$) into $(\partial V - \partial M)^-$ and we are done, or that $X_1$ is mapped into a component of $\overline{M - V}$ which itself is the product I- or $S^1$-bundle over the square or annulus, and then obviously f can be admissibly deformed into V.

<u>Case 2</u>. $T \neq \emptyset$.

In this case, $f^{-1}(M - V)^-$ is non-empty, and let $X_1$ be a component of $f^{-1}(M - V)^-$. f maps $X_1$ into a component, $X_1'$, of $(M - V)^-$ and it follows from 13.3 and 12.6 that $X_1'$ is the product I- or $S^1$-bundle over the square or annulus (V is complete). In particular, $(\partial X_1' - \partial M)^-$ consists of two components, $T_1$, $T_2$, since V is a full F-manifold. Let $Y_1$, $Y_2$ (possibly equal) be the components of V which contain $T_1$, $T_2$, respectively. If not both $Y_1$ and $Y_2$ are the

$S^1$-bundles over the Möbius bands, one easily proves, using 1 of 13.3 and 5.10 that the fibrations of $Y_1$ and $Y_2$ coincide via $X_1'$ which contradicts the fact that $\dot{V}$ is a full F-manifold. So we may suppose the converse. Let $X_2$ be a component of $f^{-1}V$ which contains a free side of $(X,\underline{\underline{x}})$. $V = Y_1 \cup Y_2$, and so, without loss of generality, $f(X_2) \subset Y_2$. By 5.5, $f|X_2\colon X_2 \to Y_2$ can be deformed (rel. $(\partial X_2 - \partial X)^-$) into a map $g\colon X_2 \to Y_2$ with $g(\partial X_2) \subset \partial Y_2$. Then, by 6.1 of [Wa 4], either $X_2$ is the $S^1$-bundle over the annulus, or $g$ is a covering map. But a finite covering of $Y_2$ is the $S^1$-bundle over the annulus or Möbius band. To see this recall that $Y_2$ admits also a fibration as I-bundle over the Klein bottle, and lift this fibration to an I-fibration of $X_2$. Hence Case 2 leads to contradictions, for, by our minimality condition on $T$, $X_2$ cannot be the $S^1$-bundle over the annulus and it cannot be that one over the Möbius band since $T \neq \emptyset$. q.e.d.

## §14. Homotopy equivalences between 3-manifolds with torus boundaries

Our aim is to apply the results as developed so far to the study of homotopy equivalences, $f : M \to N$, between 3-manifolds (irreducible, etc.). Before coming to the general case (see §§15-24), we here restrict ourselves to 3-manifolds whose sides (bound or free) are tori, e.g. knot spaces. Indeed, this assumption on the sides, i.e. on the boundaries, simplifies the whole discussion extremely. The underlying reason behind this is threefold, namely:

1. $f|\partial M$ can be studied explicitly, using the enclosing theorem.
2. The characteristic submanifold of $M$ contains all the bound sides of $M$ (see 10.6), and so
3. the characteristic submanifold of $M$ is a system of Seifert fibre space, rather than a mixture of I-bundles and Seifert fibre spaces.

As a first consequence we show that the property: "$\partial M$ consists of tori" is a homotopy invariant. This is based on the following observation:

<u>14.1 Proposition</u>. <u>Let $(M_1,\emptyset),(M_2,\emptyset)$ be connected, irreducible 3-manifolds with non-empty boundaries. Suppose that $M_1$ is boundary-irreducible, and that $\partial M_1$ consists of tori. Let $f : (M_1,\emptyset) \to (M_2,\emptyset)$ be an essential map.</u>

<u>Then there exists a submanifold, $W_2$, in $\mathring{M}_2$ such that $\partial W_2$ consists of essential tori and that $f$ can be deformed into $W_2$.</u>

<u>Proof</u>. Let $V_2$ be the characteristic submanifold of $(M_2,\emptyset)$. This exists, by 9.4. Define $W_2$ to be the union of $V_2$ with all components of $(M_2 - V_2)^-$ which do not meet $\partial M_2$.

Now, by the enclosing theorem and by 4.5, f can be deformed so that (1) $f(\partial M_1) \subset V_2$ and that (2) $f^{-1}\partial V_2$ consists of essential tori. Suppose that f is deformed so that the above holds and that, in addition, the number of components of $f^{-1}\partial V_2$ is minimal.

It remains to show that any component, $N_1$, of $(M_1 - f^{-1}V_2)^-$ is mapped under $f$ into a component, $N_2$, of $(M_2 - V_2)^-$ which does not meet $\partial M_2$. To see this observe that $\partial N_1 \subset f^{-1}V_2$. Then $f(\partial N_1) \subset V_2 \cap \partial N_2 \subset \partial N_2$ and we know, by [Wa 4,6.1] and our minimality condition on $f^{-1}\partial V_2$, that $f|N_1: (N_1, \partial N_1) \to (N_2, \partial N_2)$ can be deformed into a covering map. q.e.d.

14.2 Corollary. Let $(M_1,\emptyset$, $(M_2,\emptyset)$ be connected 3-manifolds with non-empty boundaries which are irreducible and boundary-irreducible. Suppose $(M_1,\emptyset)$ and $(M_2,\emptyset)$ are homotopy equivalent. Then $\partial M_1$ consists of tori if and only if $\partial M_2$ consists of tori.

Proof. Suppose $\partial M_1$ consists of tori. Let $f: M_1 \to M_2$ be a homotopy equivalence and $g: M_2 \to M_1$ be a homotopy inverse of $f$. By 14.1, we may suppose that $f(M_1)$ lies in the interior of some submanifold, $W_2 \subset \overset{0}{M}_2$, whose boundary components are essential tori in $M_2$. Now let $F$ be any component of $\partial M_2$. Then $F \cap W_2 = \emptyset$ since $W_2 \subset \overset{0}{M}_2$. On the other hand, by our supposition on $f$, $fg(F)$ is contained in the interior of $W_2$, and $id|F$ is homotopic to $fg|F$. Applying the transversality lemma ( see 4.5) to such a homotopy and then Nielsen's theorem, it follows that $F$ must be a torus. q.e.d.

We now establish two lemmas which are needed for the main result of this paragraph proved in 14.6.

14.3 Lemma. Let $(M,\emptyset)$ be a connected, irreducible 3-manifold whose boundary is not empty and consists of tori. Let $(N,\emptyset)$ be a Seifert fibre space with non-empty boundary. Let $f: (M,\emptyset) \to (N,\emptyset)$ be an essential map.

Then $(M,\emptyset$ is a Seifert fibre space.

Remark. This result will be generalized in §15.

Proof.

14.4 Assertion. M is a solid torus if N is a solid torus.

Suppose N is a solid torus. $\partial M$ is not empty and consists of tori. $\mathbb{Z} \oplus \mathbb{Z}$ is not a subgroup of $\mathbb{Z}$. Hence, for every boundary component, T, of M, $(f|T)_*: \pi_1 T \to \pi_1 N$ cannot be an injection. Since $f_*$ is an injection, it follows that M is boundary reducible. Therefore, by the loop-theorem, M must be a solid torus since M is irreducible and $\partial M$ consists of tori. This proves 14.4.

To continue the proof of 14.3, note that N is irreducible since it is a Seifert fibre space with boundary. Moreover, $\partial N$ consists of tori. Hence it follows that the complete boundary-pattern, $\bar{\emptyset}$, of $(N,\emptyset)$ is useful if N is not a solid torus. Choose a system of essential vertical annuli, $A_1,\ldots,A_n$, in the Seifert fibre space $(N,\emptyset)$ which splits N into solid tori. Such a system exists since $\partial N \neq \emptyset$ (see 5.4). By 4.5, f can be deformed so that $f^{-1}(\cup A_i)$ is a system of essential annuli or tori in $(M,\bar{\emptyset})$. But $f^{-1}(\cup A_i)$ cannot contain any essential torus since f is essential and since the $A_i$'s are annuli. $f^{-1}(\cup A_i)$ splits M into a system of connected manifolds, $M_1,\ldots,M_m$, whose boundary consists of tori (recall $\partial M$ consists of tori). Thus, by 14.4, and by our choice of $\cup A_i$, $M_i$ is a solid torus, for all $1 \leq i \leq m$. $M_i$ can be fibered as a Seifert fibre space so that $M_i \cap f^{-1}(\cup A_i)$ is vertical. Moreover, since $f^{-1}(\cup A_i)$ consists of annuli, the Seifert fibrations of the $M_i$'s may be chosen so that they coincide on $f^{-1}(\cup A_i)$ [Wa 1, (5.1)], and so they define a fibration of M as a Seifert fibre space. q.e.d.

For the next lemma let $(M,\emptyset)$ be a connected 3-manifold which is aspherical. Let N be a connected, irreducible submanifold in M such that $\partial N$ consists of incompressible surfaces in M and that $(\partial N - \partial M)^-$ consists of closed surfaces.

14.5 Lemma. Let $H: N \times I \to M$ be a homotopy of the embedding $N \to M$. Then the following holds:

1. *If* $H(N \times 1) \subset N$, *either* $H$ *can be deformed* (*rel* $N \times \partial I$) *into* $N$, *or* $N$ *is the product* I-*bundle over a closed orientable surface*.
2. *If* $H(N \times 1) \subset (M - N)^-$, $N$ *is the product* I-*bundle over a closed orientable surface*.

*Proof*. Let $p: \tilde{M} \to M$ be the covering map induced by the subgroup $\pi_1 N$ in $\pi_1 M$. Then there is a component, $\tilde{N}$ of $p^{-1}N$ such that $p|\tilde{N}: \tilde{N} \to N$ is a homeomorphism. We conclude that $\tilde{N}$ is a deformation retract of $\tilde{M}$ since $\pi_1\tilde{N} \cong \pi_1\tilde{M}$ and since $\tilde{M}$ is an aspherical simplicial complex ($M$ is aspherical). Since $H$ is a homotopy of the embedding $N \to M$, there is a lifting $\tilde{H}: N \times I \to \tilde{M}$ of $H$ with $\tilde{H}(N \times 0) \subset \tilde{N}$.

Suppose $\tilde{H}(N \times \partial I) \subset \tilde{N}$. Then $\tilde{H}$ can be deformed (rel. $N \times \partial I$) into $\tilde{N}$ since $\tilde{N}$ is a deformation retract of $\tilde{M}$. Hence also $H = p\tilde{H}: N \times I \to M$ can be deformed (rel $N \times \partial I$) into $N$.

Suppose $\tilde{H}(N \times 1) \subset (\tilde{M} - \tilde{N})^-$. Let $\tilde{M}_1$ be that component of $(\tilde{M} - \tilde{N})^-$ which contains $\tilde{H}(N \times 1)$. Since $\pi_1\tilde{N} \cong \pi_1\tilde{M}$, there is a component, $\tilde{F}$, of $\partial\tilde{N}$ such that the embedding $\tilde{F} \to \tilde{M}_1$ induces an isomorphism of the fundamental groups (we may suppose the basepoint lies in $\tilde{F}$). Thus there is also an isomorphism $\varphi: \pi_1\tilde{M}_1 \to \pi_1\tilde{F}$. Now define a map $g: \tilde{F} \to \tilde{F} \times I$ by $g(x) = (x,0)$. Then clearly $g_*\varphi(\tilde{H}|N \times 1)_*: \pi_1(N \times 1) \to \pi_1(\tilde{F} \times I)$ is an injection. Since $\tilde{F} \times I$ is aspherical, this injection is induced by a map $f: N \to F \times I$, and we suppose $f$ is deformed so that $f(\partial N) \subset F \times \partial I$. Both $N$ and $F \times I$ are irreducible and boundary-irreducible. Thus, by 6.1 of [Wa 4], $N$ must be the product I-bundle over a closed orientable surface. q.e.d.

The following theorem is a special case of the classification theorem, see 24.2.

*14.6 Theorem*. *Let* $(M_1,\emptyset)$, $(M_2,\emptyset)$ *be connected* 3-*manifolds which are irreducible and boundary-irreducible*. *Suppose* $\partial M_i$, $i = 1,2$, *is not empty and consists of tori*. *Let* $V_i$ *be the characteristic submanifold in* $(M_i,\underline{\underline{m}}_i)$, *where* $\underline{\underline{m}}_i$ *is the set of all the boundary components*

<u>of</u> $M_i$.

<u>Then every homotopy equivalence</u> $f\colon (M_1,\emptyset) \to (M_2,\emptyset)$ <u>can be deformed into</u> $\hat{f}$ <u>so that</u>

1. $\hat{f}|\,(M_1 - V_1)^-\colon (M_1 - V_1)^- \to (M_2 - V_2)^-$ <u>is a homeomorphism</u>.
2. $\hat{f}|V_1\colon V_1 \to V_2$ <u>is a homotopy equivalence</u>.

<u>In addition: If</u> $W_i$ <u>denotes the union of all the components of</u> $V_i$ <u>which meet</u> $\partial M_i$, $i = 1,2$, $f$ <u>can be deformed into</u> $\hat{f}$ <u>so that</u> $\hat{f}|\,(M_1 - W_1)^-\colon (M_1 - W_1)^- \to (M_2 - W_2)^-$ <u>is a homeomorphism and that</u> $\hat{f}|W_1\colon W_1 \to W_2$ <u>is a homotopy equivalence</u>.

<u>Proof</u>. By 10.4 and 10.6, $V_i$, $i = 1,2$, contains $\partial M_i$. 13.1 implies that $f|V_1$ can be deformed into $V_2$, and so we may suppose that $f$ is deformed so that $f(V_1) \subset V_2^0$. Moreover, by 4.5, we may suppose that $f$ is deformed (rel $V_1$) so that, in addition, $(\partial f^{-1}V_2 - \partial M_1)^- = f^{-1}(\partial V_2 - \partial M_2)^-$ consists of essential squares, annuli, or tori in $(M_1,\underline{\underline{m}}_1)$. More precisely, we conclude, from the fact that $\partial M_1 \subset V_1 \subset f^{-1}V_2$, that $f^{-1}V_2$ is an essential submanifold of $M_1$ whose boundary consists of tori. Therefore, applying 14.3, it follows that $f^{-1}V_2$ is an essential F-manifold in $(M_1,\underline{\underline{m}}_1)$. Let $\alpha(f)$, $\beta(f)$ be the number of all the components of $f^{-1}V_2$, $(M_1 - f^{-1}V_2)^-$, resp. We suppose $f$ is deformed so that it has the above properties and that, in addition, $(\alpha(f),\beta(f))$ is minimal with respect to the lexicographical order. Notice that this implies, in particular, that $f|Q\colon Q \to Q'$ cannot be deformed (rel. $\partial Q$) into $\partial Q'$, if $Q$ is a component of $f^{-1}V_2$ or $(M_1 - f^{-1}V_2)^-$ with $Q \cap \partial M_1 = \emptyset$, and $Q'$ is the component of $V_2$ or $(M_2 - V_2)^-$ resp. with $f(Q) \subset Q'$. Since $f^{-1}V_2$ is an essential F-manifold which contains the characteristic submanifold, $V_1$, there exists, by 10.4 and 10.6, an admissible ambient isotopy, $\alpha_t$, $t \in I$, of $(M_1,\underline{\underline{m}}_1)$ such that each component of $V_1$ is a component of $\alpha_1 f^{-1}V_2$. $f\alpha_t^{-1}$, $t \in I$, is a homotopy of $f$, and so we may suppose $f$ is deformed so that each component of $V_1$ is a component of $f^{-1}V_2$. Let $g$ be a homotopy inverse of $f$. Then, dually, we may suppose $g$ has the analogous properties as $f$.

For the following we now introduce the following convention: if $Q$ is any component of $f^{-1}V_2$ or $(M_1 - f^{-1}V_2)^-$, we denote by $Q'$ the component of $V_2$ or $(M_2 - V_2)^-$, respectively, which contains $f(Q)$.

<u>14.7 Assertion.</u> $V_1 = f^{-1}V_2$.

Assume the contrary. Then, by our suppositions on $f$, there is at least one component, $Y$, of $f^{-1}V_2$ which lies in $(M_1 - V_1)^-$. Since each component of $V_1$ is a component of $f^{-1}V_2$, it follows from 10.4 and 10.6 that $Y$ is a regular neighborhood of a torus which is parallel to a component of $(\partial V_1 - \partial M_1)^-$. In particular, $Y$ is torus $\times$ I. Let $W_1$, $W_2$ (possibly equal) be the two components of $(M_1 - f^{-1}V_2)^-$ which meet $Y$. Suppose $Y$ is chosen so that $W_1$ meets a component, $N_1$, of $V_1$ and that $W_1$ is torus $\times$ I. Applying 6.1 of [Wa 4] to $f|W_1: W_1 \to W_1'$ and recalling our minimality condition on $(M_1 - f^{-1}V_2)^-$, we see that $f|W_1: (W_1, \partial W_1) \to (W_1', \partial W_1')$ can be deformed into a covering map. This implies (apply 5.8) that $W_1'$ is the I-bundle over the torus or Klein bottle. In any case, $W_1'$ contains an essential annulus, and so $W_1'$ must be torus $\times$ I, since, by 10.4, $V_2$ is complete. Analogously, we conclude that $Y'$ is either torus $\times$ I or the I-bundle over the Klein bottle. Since $V_2$ is a full F-manifold, it follows that $Y'$ must be the I-bundle over the Klein bottle. In particular, $\partial Y'$ is connected. Hence $f(W_2) \subset W_1'$. Moreover, $W_2$ is the preimage of $W_1'$ under $f$, and so, applying 6.1 of [Wa 4], we see that $W_2$ is torus $\times$ I since $W_2'$ is torus $\times$ I. Let $N_2$, $N_2 \neq Y$, be the component of $f^{-1}V_2$ which meets $W_2$. Then, recalling our minimality condition on $(M_1 - f^{-1}V_2)^-$, we see that $f(N_2) \subset N_1'$. Since $V_2$ is a full F-manifold and since $\partial M_2 \neq \emptyset$, we conclude that $N_1'$ cannot be the I-bundle over the torus or Klein bottle. Therefore it follows that $N_i$, $i = 1$ and $2$, cannot be the I-bundle over the torus or Klein bottle with $\partial N_i \cap \partial M_1 = \emptyset$.

Now let $N_3$ be that component of $V_1$ which contains $g(Y') = gf(Y)$. This exists since $f(V_1) \subset V_2$ and $g(V_2) \subset V_1$, by our suppositions on $f$ and $g$. $Y' \cap \partial M_2 = \emptyset$ and so we may suppose $g|Y': Y' \to N_3$ is a covering map (see above). Therefore (apply 5.8)

$N_3$ is also the I-bundle over the Klein bottle (since Y' is). This implies that $N_3$ is neither $N_1$ nor $N_2$. On the other hand, $gf|Y \simeq 1|Y$ since g is a homotopy inverse of f. Hence we conclude that either $N_1$ or $N_2$ must be the I-bundle over the torus or Klein bottle, with $N_i \cap \partial M_1 = \emptyset$, i = 1 or 2 (consider the homotopy of one boundary torus of Y into $N_3$. Apply the transversality lemma, with respect to $((\partial N_1 \cup \partial N_2) - \partial M_1)^-$, and 6.1 of [Wa 4]). We obtain a contradiction and hence 14.7 is proved.

By 14.7, $f(V_1) \subset V_2$ and $f(M_1 - V_1)^- \subset (M_2 - V_2)^-$. Dually, $g(V_2) \subset V_1$ and $g(M_2 - V_2)^- \subset (M_1 - V_1)^-$. Consequently, if $Q_1$ is any component of $V_1$ or $(M_1 - V_1)^-$, there is a component, $Q_2$, of $V_1$ or $(M_1 - V_1)^-$ resp. which contains $gf(Q_1)$. Since $gf \simeq 1$, there is a homotopy $H: Q_1 \times I \to M_1$ such that $H|Q_1 \times 0$ is the embedding $Q_1 \to M_1$ and that $H|Q_1 \times 1 = gf|Q_1$.

Assume $Q_1 \neq Q_2$. Then, by 14.5.2, $Q_1$ is torus $\times$ I. Furthermore, we conclude that there is at least one component, W, of $(M_1 - V_1)^-$ or $V_1$, respectively, which is torus $\times$ I and which meets $Q_1$ in exactly one boundary component (consider the restriction of H to one boundary torus of $Q_1$. Apply the transversality lemma and 6.1 of [Wa 4]). But then, in any case, it follows that $V_1$ is not a full F-manifold in $(M_1, \underline{\underline{m}}_1)$. This is a contradiction.

Thus $gf(Q) \subset Q$, for all components Q of $V_1$ or $(M_1 - V_1)^-$. In particular, it follows that f maps different components of $V_1$ or $(M_1 - V_1)^-$ into different components of $V_2$ or $(M_2 - V_2)^-$.

Suppose Q is a component of $V_1$ or of $(M_1 - V_1)^-$ with $Q \cap \partial M_1 = \emptyset$. Then $f(\partial Q) \subset \partial Q'$ since $f(V_1) \subset V_2$ and $f(V_1 - M_1)^- \subset (M_2 - V_2)^-$. Hence, applying 6.1 of [Wa 4], we see that $f|Q: (Q, \partial Q) \to (Q', \partial Q')$ can be deformed so that either $f|Q$ is a covering map or that $f(Q) \subset \partial Q'$. By our minimality condition on $(\alpha(f), \beta(f))$ the latter case is impossible.

Hence there is a homotopy, $f_t$, $t \in I$, of f with $f_t^{-1}V_2 = f^{-1}V_2$ which deforms f into $\hat{f}$ so that $\hat{f}|Y: Y \to Y'$ is a covering map, for all components, Y, of $(M_1 - V_1)^-$. A corresponding result holds for g. Therefore we see that $\hat{f}|Y$ is in fact a homeomoprhism (recall $gf(Y) \subset Y$). Since f maps different components of $(M_1 - V_1)^-$ into different components of $(M_2 - V_2)^-$, we have shown in this way that $\hat{f}|(M_1 - V_1)^-: (M_1 - V_1)^- \to (M_2 - V_2)^-$ is a homeomorphism. For the additional remark note that, for every component, X, of $V_1$ with $X \cap \partial M_1 = \emptyset$, the restriction of $\hat{f}$ to any component of $\partial X$ is now a covering map. Applying 6.1 of [Wa 4], it follows that $\hat{f}$ can be deformed (rel $(M_1 - V_1)^-$) so that $\hat{f}|(M_1 - W_1)^-$ is a homeomorphism (notation as in the formulation of the theorem).

We still have to show that $\hat{f}|V_1: V_1 \to V_2$ is a homotopy equivalence. For this, let X be a component of $V_1$. Recall $gf(X) \subset X$, i.e. $g(X') \subset X$. If $X \cap \partial M_1 = \emptyset$, then by the preceding argument, $\hat{f}|X: X \to X'$ can be deformed into a homeomorphism, and so $\hat{f}|X$ is, in particular, a homotopy equivalence. So let $X \cap \partial M_1 \neq \emptyset$. Then we distinguish two cases: X is torus $\times$ I, or not. If X is torus $\times$ I, $g|X'$ can be deformed into $\partial X$. Applying 6.1 of [Wa 4], it follows that X' must be torus $\times$ I, too. In this case both $\hat{f}|X: X \to X'$ and $g|X': X' \to X$ can be deformed (rel $X \cap (M_1 - V_1)^-$, resp. rel $X' \cap (M_2 - V_2)^-$) so that they map different boundary components into different boundary components (recall $X \cap \partial M_1 \neq \emptyset$ and so $X' \cap \partial M_2 \neq \emptyset$). Thus, by 6.1 of [Wa 4] both $\hat{f}|X: X \to X'$ and $g|X': X' \to X$ can be deformed into a covering map, and so $\hat{f}|X$ can be deformed into a homeomorphism, i.e. $\hat{f}|X$ is, in particular, a homotopy equivalence. Finally we consider the case that X is not torus $\times$ I. $g\hat{f}(X) \subset X$ and $g|X' \cdot \hat{f}|X \simeq 1|X$ in $M_1$ since $g\hat{f} \simeq 1$. Hence, by 14.5.1, it follows that $g|X' \cdot \hat{f}|X$ is in fact homotopic in X to $1|X$. A corresponding result holds for $\hat{f}|X \cdot g|X'$. Thus $g|X'$ is a homotopy inverse of the map $\hat{f}|X: X \to X'$, and so $\hat{f}|X$ is a homotopy equivalence. Since $\hat{f}$ maps different components of $V_1$ into different components of $V_2$, we have shown that $\hat{f}|V_1: V_1 \to V_2$ is a homotopy equivalence. q.e.d.

We finally apply 14.6 to knot spaces. In view of 14.6 it is interesting to know how the Seifert fibre spaces in knot spaces look like. In general they might be complicated, i.e. they can have a lot of boundary components (of course the orbit surface is always a 2-sphere with holes and the number of exceptional fibres is restricted; apply 5.4 and recall that every torus in $S^3$ bounds a solid torus [Al 1]). However, if the Seifert fibre space contains the boundary component of the knot space, we can say a little more.

14.8 Lemma. *Let k be a prime knot in $S^3$, and U(k) a regular neighborhood in $S^3$. Let X be a Seifert fibre space in the knot space $M = S^3 - \overset{\circ}{U}(k)$ with $\partial M \subset X$ and fixed Seifert fibration. Suppose every component of $(\partial X - \partial M)^-$ is incompressible in M.*

*Then X = M, or X is the Seifert fibre space with the annulus as orbit surface and at most one exceptional fibre.*

*In addition: k is a torus knot or trivial, provided X = M, and a cable knot, provided $X \neq M$ and X is the Seifert fibre space with the annulus as orbit surface and precisely one exceptional fibre (the exceptional fibre is the cable).*

*Proof*. Suppose X = M. Then X has precisely one boundary component. If X is the Seifert fibre space over the disc with at most one exceptional fibre, then X is a solid torus and k is the trivial knot. If not, there exists at least one incompressible vertical annulus in X which is not boundary-parallel (apply 5.4). The boundary curves of this annulus cannot be contracted in U(k) since k is not a product knot, and so they induce a fibration of the solid torus U(k) as a Seifert fibre space. In this way the fibration of X induces a fibration of $S^3$. The Seifert fibrations of $S^3$ have the 2-sphere as orbit surface. Thus the orbit surface of X is orientable, and so $\pi_1 X$ has a non-trivial center. Then, by [BZ 1] k is a torus knot.

Suppose $X \neq M$ and is not torus $\times$ I. As above, the fibration of X induces a fibration of $\bar{X} = X \cup U(k)$ as a Seifert fibre space. $\partial\bar{X} \neq \emptyset$ since $X \neq M$. Let $T_1$ be any boundary component of $\bar{X}$. Then $T_1$ is a torus in $S^3$ and so, by [Al 1], $T_1$ bounds a solid torus in $S^3$. Since the components of $(\partial X - \partial M)^- = \partial\bar{X}$ are incompressible in M and since $\partial M \subset X$, it follows that $\bar{X}$ must be a solid torus. By [Wa 1, [2.3)], every incompressible annulus in a solid torus is boundary parallel. Hence $\bar{X}$ must be the Seifert fibre space over the disc with at most one exceptional fibre (apply 5.4). Since X is not torus $\times$ I, $\bar{X}$ has precisely one exceptional fibre and this does not lie in U(k). Therefore X is the Seifert fibre space over the annulus with precisely one exceptional fibre. Furthermore, we may suppose $k$ is a regular fibre of the Seifert fibration of $\bar{X}$. Hence $k$ is a cable knot with cable the exceptional fibre of $\bar{X}$ (recall $k$ is not a torus knot since $X \neq M$). q.e.d.

In the proof of 14.8, we used the paper of Burde and Zieschang for the fact that only the torus knots have fundamental groups with non-trivial center. On the other hand, this fact also follows from the annulus theorem (see [Si 2]).

Now, finally, with the help of 14.2, 14.6 and 14.8, we may deduce from 6.1 of [Wa 4] the following result on knot spaces.

<u>14.9 Proposition</u>. <u>Let</u> $(M_1,\emptyset)$ <u>be a connected</u> 3-<u>manifold which is irreducible and boundary-irreducible</u>. <u>Let</u> $k$ <u>be a non-trivial prime knot, but neither a torus knot nor a cable knot</u>. <u>Let</u> $U(k)$ <u>be a regular neighborhood in</u> $S^3$ <u>and define</u> $(M_2,\emptyset)$ <u>as the knot space</u> $S^3 - \mathring{U}(k)$.

<u>Then every homotopy equivalence</u> $f\colon (M_1,\emptyset) \to (M_2,\emptyset)$ <u>can be deformed into a homeomorphism</u>.

Recall that every isomorphism $\varphi\colon \pi_1 M_1 \to \pi_1 M_2$ is induced by a homotopy equivalence, provided $M_1$, $M_2$ are aspherical. Hence 14.9 implies (see also [Wa 6]):

<u>14.10 Corollary</u>. <u>If</u> $\pi_1 M_1 \cong \pi_1 M_2$, <u>then</u> $M_1$ <u>is homeomorphic to</u> $M_2$.

This is a special case of a much more general theorem (see 24.2). For more information about the mapping class group of knot spaces see §27.

<u>Remark 1</u>. Let $k$ be a knot in $S^3$ and $U(k)$ a regular neighborhood in $S^3$. Then we say $k$ has the <u>unique embedding property</u> if each embedding $f\colon S^3 - \mathring{U}(k) \to S^3$ extends to a homeomorphism $\bar{f}\colon S^3 \to S^3$. If we suppose the unique embedding property holds for all knots, then it is not difficult to show that every two non-trivial prime knots have homeomorphic knot spaces (and hence are equivalent; up to

orientation). On the other hand, J. Hempel [He 2] constructed cable knots for which this conclusion is false, provided the unique embedding property is false in general.

Remark 2. As pointed out by Gramain [Gr 1], 14.10 implies the result of Simon [Si 1].

Remark 3. In view of 14.10 the questions arises whether or not the isomorphism problem for knot groups is solvable. In §29 we answer this question, in a more general setting, affirmatively. But recall that in contrast to this, it cannot be decided whether or not a finitely generated group is a knot group (see [St 4]). To see this, observe, as pointed out by F. Waldhausen, that a solution of this problem for groups of the form $G*\mathbb{Z}$ would lead to a solution of the triviality problem for finitely presented groups.

## Part III. THE SPLITTING THEOREMS

In contrast to the last paragraph we now drop the condition that the boundaries of 3-manifolds $M_1$, $M_2$ consist of tori. In this general setting the restriction of a homotopy equivalence $f\colon M_1 \to M_2$ to the boundary $\partial M_1$ can be a very complicated singular surface, and it seems hard to obtain any helpful information from this map. Hence, instead of studying the restriction $f|\partial M_1$, we shall study more intensely the behavior of homotopy equivalences $f\colon M_1 \to M_2$ with respect to the characteristic submanifolds itself (chapter VI) and to essential surfaces (chapter VII)--i.e. their splitting properties. We will see later on (see part IV) how to utilize these properties in a proof of the classification theorem.

### Chapter VI. Invariance of the characteristic submanifolds under homotopy equivalences.

Throughout this chapter let $(M_1, \underline{\underline{m}}_1)$ and $(M_2, \underline{\underline{m}}_2)$ be irreducible 3-manifolds whose completed boundary-patterns, $\bar{\underline{\underline{m}}}_1$ and $\bar{\underline{\underline{m}}}_2$, are useful and non-empty (e.g. irreducible and boundary-irreducible 3-manifolds with non-empty boundaries whose boundary-patterns are empty). Furthermore, denote by $\bar{V}_1$ and $\bar{V}_2$ the characteristic submanifolds of $(M_1, \bar{\underline{\underline{m}}}_1)$ and $(M_2, \bar{\underline{\underline{m}}}_2)$, respectively.

The purpose of this chapter is to prove the <u>first splitting theorem</u> which asserts that any given admissible homotopy equivalence $f\colon M_1 \to M_2$ is admissibly homotopic to a map $g$ such that $g|\bar{V}_1\colon \bar{V}_1 \to \bar{V}_2$ and $g|(M_1 - \bar{V}_1)^-\colon (M_1 - \bar{V}_1)^- \to (M_2 - \bar{V}_2)^-$ are admissible homotopy equivalences (see 18.3).

We have tried to make the proof as less restrictive as possible. In particular, we state and prove some results on essential maps and admissible homotopies (see 15.2) which in this generality are not really needed for the above splitting theorem.

## §15. The preimage of an essential F-manifold

Throughout this paragraph suppose that neither $(M_1,\bar{\underline{\underline{m}}}_1)$ nor $(M_2,\bar{\underline{\underline{m}}}_2)$ is a ball with at most four sides. Let X be an essential I-bundle or Seifert fibre space in $(M_2,\bar{\underline{\underline{m}}}_2)$, denote by $\underline{\underline{x}}$ the boundary-pattern of X induced by $\underline{\underline{m}}_2$, and let $\underline{\underline{x}}^+$ be the union of $\underline{\underline{x}}$ with all the components of $(\partial X - \partial M_2)^-$. Now let $f\colon (M_1,\underline{\underline{m}}_1) \to (M_2,\underline{\underline{m}}_2)$ be an essential map, let Y be a component of $f^{-1}X$, and suppose that $(\partial Y - \partial M_1)^-$ consists of essential squares, annuli, or tori in $(M_1,\bar{\underline{\underline{m}}}_1)$. Finally, define the boundary patterns $\underline{\underline{y}}$ and $\underline{\underline{y}}^+$ of Y as above for X.

*15.1 Proposition. Suppose that* $f|G\colon G \to X$ *is essential in* $(X, \bar{\underline{\underline{x}}}^+)$, *for every free side* G *of* $(Y, y^+)$, *and suppose that* X *has no exceptional fibres if* $X \cap \partial M_2 = \emptyset$. *Then one of the following holds:*

1. *Suppose* $(X,\bar{\underline{\underline{x}}}^+)$ *is not the* I- *or* $S^1$*-bundle over the square, annulus, or Möbius band with* $(\partial X - \partial M_2)^- = \emptyset$. *Then* $(Y,\bar{\underline{\underline{y}}}^+)$ *is an* I*-bundle (Seifert fibre space), if* $(X,\bar{\underline{\underline{x}}}^+)$ *is an* I*-bundle (Seifert fibre space).*
2. *Suppose* $(X,\bar{\underline{\underline{x}}}^+)$ *is the* I- *or* $S^1$*-bundle over the square, annulus or Möbius band with* $(\partial X - \partial M_2)^- \neq \emptyset$. *Then there is an essential square, annulus, or torus in* $(Y,\underline{\underline{y}})$ *which splits* $(Y,\bar{\underline{\underline{y}}}^+)$ *into an essential* F*-manifold.*

*In addition: If* X *has exceptional fibres, then* 1 *or* 2 *holds, provided that, in addition,* f *is an admissible homotopy equivalence.*

Applying 4.5 and the proof of 4.10, we obtain the following corollary as an easy consequence of 15.1:

*15.2 Corollary. Let* W *be an essential* F*-manifold in* $(M_2,\bar{\underline{\underline{m}}}_2)$. *Then every admissible homotopy equivalences* $f\colon (M_1,\underline{\underline{m}}_1) \to (M_2,\underline{\underline{m}}_2)$ *can be admissibly deformed so that afterwards* $f^{-1}W$ *is an essential* F*-manifold in* $(M_1,\bar{\underline{\underline{m}}}_1)$.

*Proof of 15.1.* By 4.8.2, the boundary-patterns $\bar{\underline{\underline{x}}}^+$ and $\bar{\underline{\underline{y}}}^+$ are useful, and, by our suppositions on $(M_i,\bar{\underline{\underline{m}}}_i)$, $i = 1,2$, no side of $\bar{\underline{\underline{x}}}^+$ or $\bar{\underline{\underline{y}}}^+$

is an i-faced disc, $1 \leq i \leq 3$.

In the course of the proof we shall have often to distinguish the bound sides of $(Y,\underline{\underline{y}}^+)$ which are components of $(\partial Y - \partial M_1)^-$ from those which lie in bound sides of $(M_1,\underline{\underline{m}}_1)$. So we call the latter ones $\underline{\underline{m}}_1$-bound sides. Similarly, the bound sides of $(X,\underline{\underline{x}}^+)$ which lie in bound sides of $(M_2,\underline{\underline{m}}_2)$ are called $\underline{\underline{m}}_2$-bound sides.

Since the restriction of $f$ to any free side $G$ of $(Y,\underline{\underline{y}}^+)$ is an essential map into $(X,\underline{\underline{x}}^+)$, we may suppose that $f|G$ is either vertical or horizontal in $(X,\underline{\underline{x}}^+)$ (see remark of 5.6).

<u>15.3 Lemma</u>. 15.1 <u>holds</u> <u>if</u> $X$ <u>is</u> <u>an</u> I-<u>bundle</u>.

Fix an admissible fibration of $X$ as I-bundle and define

$$\underline{\underline{g}} = \{G \in \bar{\underline{\underline{y}}} \mid G \text{ is either an } \underline{\underline{m}}_1\text{-bound side mapped under } f \text{ into some lid of } X, \text{ or a free side such that } f|G \text{ is horizontal}\}.$$

Of course, $\underline{\underline{g}}$ is non-empty since $X$ is an I-bundle.

<u>15.4 Assertion</u>. <u>If all sides from</u> $\underline{\underline{g}}$ <u>are</u> $\underline{\underline{m}}_1$<u>-bound, then all sides from</u> $\bar{\underline{\underline{y}}}^+ - \underline{\underline{g}}$ <u>are squares or annuli (with respect to the boundary-pattern induced by</u> $\bar{\underline{\underline{y}}}^+$).

To show the assertion, let $B \in \bar{\underline{\underline{y}}}^+ - \underline{\underline{g}}$. Without loss of generality, $B$ is $\underline{\underline{m}}_1$-bound. Let $B'$ be the side of $(X,\underline{\underline{x}})$ which contains $f(B)$ (this exists since $f$ is admissible). Denote by $\underline{\underline{b}}$, $\underline{\underline{b}}'$ the boundary-patterns of $B$, $B'$ induced by $\underline{\underline{y}}$, $\underline{\underline{x}}$, respectively. Observe that $\bar{\underline{\underline{b}}}$ is equal to the boundary-pattern of $B$ induced by $\bar{\underline{\underline{y}}}^+$, for every side of $\underline{\underline{g}}$ is $\underline{\underline{m}}_1$-bound. Hence it remains to prove that $(B,\bar{\underline{\underline{b}}})$ is a square or annulus. For this let $k$ be any admissible singular arc in $(B,\underline{\underline{b}})$ which is mapped under $f$ into an inessential arc in $(B',\underline{\underline{b}}')$. Then we find near $k$ an inessential arc in $(Y,\underline{\underline{y}})$ since $f$ is essential. This means that $k$ is a side of some admissible singular 2-faced disc in $(Y.\underline{\underline{y}})$. The existence of this 2-faced disc implies that $k$ is inessential in $(B,\underline{\underline{b}})$, for $\bar{\underline{\underline{y}}}^+$

is a useful boundary-pattern. Thus we have proved that the restriction $f|B: (B,\underline{\underline{b}}) \to (B',\underline{\underline{b}}')$ is essential. Now $(B',\bar{\underline{\underline{b}}}'$ is either a square or an annulus, and so also $(B.\bar{\underline{\underline{b}}})$. This proves 15.4.

As an easy consequence of 15.4 we obtain that Y must be an I-bundle, if (1) there is a disc G in $\underline{\underline{g}}$ and if (2) all elements of $\underline{\underline{g}}$ are $\underline{\underline{m}}_1$-bound. To see this observe that, by 15.4, the union $G^+$ of G with all sides of $(Y,\bar{\underline{\underline{y}}}^+)$ meeting G must be a disc again. Moreover, a curve k in $\partial Y - G^+$ near $\partial G^+$ is entirely contained in a side of $(Y,\bar{\underline{\underline{y}}}^+)$. Now k is contractible via $G^+$. Hence $(\partial Y - G^+)^-$ is a disc and a side of $(Y,\bar{\underline{\underline{y}}}^+)$ since $\bar{\underline{\underline{y}}}^+$ is a useful boundary-pattern. In particular, $\partial Y$ is a 2-sphere and so Y a ball since Y is irreducible. Hence, altogether, Y is an I-bundle.

The remainder of the proof of 15.3 will be split into three cases:

Case 1. There is a side $G \in \underline{\underline{g}}$ which is a disc.

If G is $\underline{\underline{m}}_1$-bound, then, for every $G_i \in \underline{\underline{g}}$ and every loop k in $G_i$, some non-trivial multiple of k is contractible in $G_i$. To see this join the base point of k with G by an arc t in Y. Then $f(t^{-1}*k*t)^2$ is inessential in X since X is an I-bundle. Hence our claim follows from the facts that f is essential, G is a disc, and $\bar{\underline{\underline{y}}}^+$ is useful. Therefore every side of $\underline{\underline{g}}$ is a disc if G is $\underline{\underline{m}}_1$-bound. So, without loss of generality, G may be chosen to be a free side (see the remark after 15.4).

If G is a free side of $(Y,\underline{\underline{y}}^+)$, $f|G$ is horizontal and so it follows that X, and so Y, is a ball (f is essential).

Suppose that, in addition, G is not a square (with respect to the boundary-pattern induced by $\underline{\underline{y}}^+$). Using the remark after 15.4, it remains to show that every $\underline{\underline{m}}_1$-bound side B of Y which meets G is a square. For this let $b_1$, $b_2$ be the two sides of B which meet G and let $x_1,x_2$ be the end-points of $b_1$, $b_2$, respectively, which do not lie in G. Of course, $x_1$ and $x_2$ are contained in sides $G_1,G_2 \in \underline{\underline{g}}$, respectively. If $G_1 = G_2$, we are done, for then we find an admissible 2-faced disc whose sides lie in B and $G_1$ and which intersects $G_1 \cap B$ near $x_1$ and $x_2$, and $B$ must be a square since $\bar{\underline{y}}^+$ is useful. So assume that $G_1 \neq G_2$. Since $G$ is not a square, it follows the existence of at least three different $\underline{\underline{m}}_1$-bound sides meeting $G$, say $B_1$, $B_2$, $B_3$. We claim that each $B_i$ has to meet $G_1$. This is clear if $G_1$ is $\underline{\underline{m}}_1$-bound, for then joining $B_i$ with $G_1$ we get an admissible arc in $(Y\ \underline{y})$ mapped to an inessential arc in $(X,\underline{\underline{x}})$ ($X$ is a ball) and our claim follows from the fact that $f$ is essential. If $G_1$ is a free side observe that $f|G_1$ is horizontal, i.e. there must be a side of $G_1$ mapped into the same side of $X$ as $B_i$. By the preceding argument this side of $G_1$ must be a side of $B_i$. Hence every $B_1$, $B_2$, $B_3$ must meet $G_1$, and also $G_2$. But this is impossible in the 2-sphere $\partial Y$.

Suppose that $G$ is a square. Then $X$ is the I-bundle over the square. Two opposite sides of $G$ lie in $\underline{\underline{m}}_1$-bound sides, $B_1$, $B_2$ ($G$ is free). These two sides are different, for $\bar{\underline{y}}^+$ is useful. Consider $B_1$. Every side of $Y$ meeting $B_1$ is $\underline{\underline{m}}_1$-bound or a square. So, either 15.1 follows from the remark after 15.4, or w.l.o.g. at least two disjoint $\underline{\underline{m}}_1$-bound sides, say $B_3$, $B_4$, meet $B_1$. By an argument above, it follows that they also meet $B_2$. Define $A$ to be an essential square whose sides are contained in $B_1$, $B_2$, $B_3$, $B_4$. Let $\tilde{Y}$ be the manifold obtained from $Y$ by splitting at $A$, and denote by $Y_1$ and $Y_2$ the two components of $\tilde{Y}$. It remains to show that $Y_i$ is an I-bundle. This is clear if $Y_i$ contains no $\underline{\underline{m}}_1$-bound side different from $B_1$, $B_2$, $B_3$, $B_4$, for any side different from an $\underline{\underline{m}}_1$-bound side is a square. The other case is impossible since each two $\underline{\underline{m}}_1$-bound sides have a non-trivial intersection (see above).

<u>Case 2. All sides $G \in \underline{\underline{g}}$ are homeomorphic to an annulus (not necessarily admissibly homeomorphic)</u>.

Fix a side $G \in \underline{\underline{g}}$. If all sides of $\underline{\underline{g}}$ are $\underline{\underline{m}}_1$-bound, then by 15.4, all the sides of $Y$ which meet $G$ are squares or annuli. So the union, $G^+$, of $G$ with all these sides is homeomorphic to an annulus. Moreover, the two closed curves in $\partial Y - G^+$ lying near the two components of $\partial G^+$ are contained entirely in sides $G_1, G_2 \in \underline{\underline{g}}$. $G_1 = G_2$. To see this, fix an admissible arc in $Y$ near $G^+$ which joins $G_1$ with $G_2$. We easily find such an arc which is mapped under $f$ to an inessential arc in $X$, and our claim follows from the fact that $f$ is essential. Thus, in particular, $Y$ is a solid torus. Fix a base point $x$ in $G$ and let $k_1$, $k_2$ be loops which generate $\pi_1(G,x)$, resp. $\pi_1(Y,x)$. Then $k_2^m \simeq k_1$ (rel $x$), for some integer $m$, so $f|k_2^m \simeq f|k_1$ (rel $f(x)$), and so $m \leq 2$ since $X$ is an I-bundle. This means that the circulation number of $G$ with respect to $Y$ is at most 2, and so $Y$ must be an I-bundle. Thus, without loss of generality, $G$ may be chosen to be a free side.

If $G$ is a free side of $(Y,\underline{y}^+)$, $f|G$ is horizontal and so it follows that $X$, and so $Y$, is a solid torus. All sides meeting $G$ are either components of $f^{-1}(\partial X - \partial M_2)^-$ or $\underline{\underline{m}}_1$-bound. We are going to show that all these sides are squares or annuli. This is clear for the components of $f^{-1}(\partial X - \partial M_2)^-$. So let $B$ be any $\underline{\underline{m}}_1$-bound side which meets $G$, and observe that $A = (\partial Y - G)^-$ is homeomorphic to an annulus since $Y$ is a solid torus. Now suppose that $B$ is a disc. Let $b_1$, $b_2$ be the two sides of $B$ which meet $G$ and let $x_1$, $x_2$ be the end-points of $b_1$, $b_2$, respectively, which do not lie in $G$. Of course, $x_1$ and $x_2$ are contained in sides $G_1$, $G_2$, respectively, with $G_1, G_2 \in \underline{\underline{g}}$. $G_1$, resp. $G_2$, is an incompressible surface in the annulus $A$ whose homeomorphy type is an annulus and which, moreover, can be joined to <u>one</u> specified boundary curve of $A$ without meeting $G_2$, resp. $G_1$. Hence $G_1$ must be equal to $G_2$. Moreover, we find an admissible 2-faced disc in $Y$ whose sides lie in $B$ and $G_1$ and which intersects $G_1 \cap B$ near $x_1$ and $x_2$. This means that $B$ is a square since $\bar{\underline{\underline{y}}}^+$ is useful. If, on the other hand, $B$ is homeomorphic to an annulus, then observe that both the boundary curves of $B$ have to meet sides $G, G_1 \in \underline{\underline{g}}$. Choosing appropriate admissible 2-faced discs whose sides lie in $B$ and $G$ (resp. $G_1$) we find that each boundary curve of $B$ lies in $B \cap G$ or $B \cap G_1$, i.e. $B$ is an annulus.

Now suppose that $G$ is a free side but not an annulus. By what we have seen so far, every side of $Y$ meeting $G$ must be a square or annulus. By our suppositions on $G$, at least one of these sides, say $B$, must be $\underline{\underline{m}}_1$-bound. Moreover, the union, $G^+$, of $G$ with all sides of $Y$ meeting $G$ must be homeomorphic to an annulus. The two closed curves in $\partial Y - G^+$ lying near the components of $\partial G^+$ are contained entirely in sides $G_1$, $G_2$ of $\underline{\underline{g}}$. It remains to show that $G_1 = G_2$ (see beginning of Case 2). Let the indices be chosen so that $G_1$ meets B, and let A be the component of $(G^+ - G)^-$ which does not meet B. If $G_2$ is an $\underline{\underline{m}}_1$-bound side, then fix an arc near $G \cup A$ which joins B with $G_2$. It is easily seen that such an arc is mapped to an inessential arc in X. Hence B must meet $G_2$, i.e. $G_1 = G_2$, since f is essential. If $G_2$ is a free side, observe that $f|G_2$ is horizontal. This means that $G_2$ must meet a number of $\underline{\underline{m}}_1$-bound sides, $B_1,\ldots,B_n$, mapped into the same side of X as B. Fix an arc near $G \cup A \cup G_2$ which joins B with an appropriate side of $B_1,\ldots,B_n$, then we find as above that B must be equal to this side, i.e. $G_1 = G_2$.

Finally suppose that G is a free side and an annulus. Then X is the I-bundle over the annulus or Möbius band. If Y has no two disjoint $\underline{\underline{m}}_1$-bound sides it easily follows that Y is a Seifert fibre space. So let $B_1$, $B_2$ be two disjoint $\underline{\underline{m}}_1$-bound sides of Y. Let A be an essential annulus in Y whose sides lie in $B_1$ and $B_2$. Define $\tilde{Y}$ to be the manifold obtained from Y by splitting at A, and let $Y_1$, $Y_2$ be the two components of $\tilde{Y}$. It remains to prove that $Y_i$, $i = 1,2$ is an I-bundle or Seifert fibre space. This is clear if $Y_i$ contains an $\underline{\underline{m}}_1$-bound side different from $B_1$ and $B_2$, for each such side has to meet both $B_1$ and $B_2$. It also follows in the other case, for every free side and every component of $f^{-1}(\partial X - \partial M_2)^-$ is a square or annulus.

<u>Case 3</u>. <u>We are neither in Case</u> 1 <u>nor in Case</u> 2.

Since $\underline{\underline{g}} \neq \emptyset$, there is a side $G_1 \in \underline{\underline{g}}$ which is neither a disc nor homeomorphic to an annulus. Let C be a side of Y which meets $G_1$. Then of course two disjoint sides of C lie in sides $G_1, G_2 \in \underline{\underline{g}}$ (possibly $G_1 = G_2$).

We claim that there is an I-bundle Z in Y such that $Z \cap G_i = (G_i - U(\partial G_i))^-$, $i = 1,2$, where $U(\partial G_i)$ denotes a regular neighborhood of $\partial G_i$ in $G_i$. By 12.5, every essential singular annulus in $(Y,\underline{\underline{\bar{y}}}^+)$ can be admissibly deformed in $(Y,\underline{\underline{\bar{y}}}^+)$ into the characteristic submanifold of $(Y, \bar{\underline{y}}^+)$. So it remains to show that (a multiple of) every essential, singular closed curve in $G_1$ (and $G_2$) is a side of some essential singular annulus in $Y$ whose other side is contained in $G_2$ (resp. $G_1$). Fix a base point $x_1$ in $G_1$ near $C \cap G_1$, and let $k$ be any loop in $G_1$ with base point $x_1$. Join $x_1$ with a point $x_2$ in $G_2$ by an arc $t$ in $Y$ near $C$. If $G_2$ is an $\underline{\underline{m}}_1$-bound side, it follows that the loop $(f|t^{-1}*k*t)^2$ is inessential in $X$ since $X$ is an I-bundle (observe that a loop can be considered as a singular arc whose end-points lie in the base point). So the loop $(t^{-1}*k*t)^2$ is inessential in $Y$ since $f$ is essential, and so $k^2$ is a side of some essential singular annulus as required. If $G_2$ is a free side, observe that $f|G_2$ is horizontal, i.e. $pf|G_2$ is a covering map, where $p$ is the fibre projection of the I-bundle $X$ (we consider the base of $X$ as embedded in $X$ as section). Hence there is a loop $k'$ in $G_2$ so that $pf|k' \simeq (pf|t^{-1}*k*t)^m$ (rel base point), for some integer $m$. This implies $t^{-1}*k*t \simeq k'$ (rel $x_2$) since $f$ is essential, and again the existence of the required essential singular annulus follows. This establishes our claim.

In the same way it follows that every component $F$ of $(\partial Y - G_1 \cup G_2)^-$ is homeomorphic to an annulus: (a multiple of) every singular closed curve of $F$ can be deformed in $Y$ into $G_1$, and so into the annulus $(\partial Z - \partial Y)^-$. Near $F$ we find an annulus which is essential in $(Y, \bar{\underline{y}})$ and whose sides lie in $G_1 \cup G_2$. Hence, by our choice of $Z$, this annulus can be admissibly isotoped into $Z$, i.e. it is admissibly parallel to some component of $(\partial Z - \partial Y)^-$.

It remains to prove that every side of $Y$ contained in $F$ is a square or annulus which meets both components of $\partial F$. Observe that $F$ cannot contain any side $G$, $G \in \underline{\underline{g}}$. For such a $G$ must be homeomorphic to a disc or annulus. Since we are not in Case 1, it must be homeomorphic to an annulus. By an argument above we find an I-bundle, $Z_1$, in $Y$ with $Z_1 \cap G = (G - U(\partial G))^-$ and $Z_1 \cap G_1 = (G_1 - U(\partial G_1))^-$. In particular, $G_1$ must be homeomorphic to an annulus since $G$ is, but this contradicts our choice of $G_1$. Now let $B$ be any side of $Y$ contained in $F$. Then, of course, at least two disjoint sides of $B$ lie in sides of $\underline{\underline{g}}$. By what we have seen so far, these sides of $B$ must lie in $\partial F$. Choosing appropriate admissible 2-faced discs and recalling that $\bar{\underline{y}}^{+}$ is useful, we find that every component of $\partial F$ contains at most one side of $B$. By the same argument it follows that $B$ is an annulus if it is homeomorphic to an annulus, and this annulus meets both the components of $\partial F$. Moreover, if $B$ is a disc, it follows in the same way that $(\partial B - \partial F)^{-}$ consist of precisely two components, and so $B$ must be a square which meets both the components of $\partial F$.

Hence 15.3 is established.

<u>15.5 Lemma</u>. 15.1 <u>holds if</u> $X$ <u>is a Seifert fibre space</u>.

Fix an admissible fibration of $X$ as Seifert fibre space and define

$$\underline{g} = \{G \in \bar{\underline{y}}^{+} | G \text{ is a free side of } (Y,\underline{y}^{+}) \text{ such that } f|G \text{ is horizontal}\}.$$

If $\underline{g}$ is empty, the restriction of $f$ to any free side of $Y$ is vertical in $X$. Fix a system $A$ of essential vertical annuli in $(X,\bar{\underline{x}}^{+})$ which split $X$ into a system of solid tori. Observe that $g = f|Y\colon (Y,\underline{y}^{+}) \to (X,\underline{x}^{+})$ is an admissible map. Hence, by 4.4, $g$ can be admissibly deformed so that afterwards $g^{-1}A$ is an essential surface in $(Y,\bar{\underline{y}}^{+})$. Since $A$ and the restriction of $g$ to any free side of $(Y,\underline{y}^{+})$ is vertical in $X$, this homotopy may be chosen so that, in addition, $\partial g^{-1}A$ is contained entirely in sides of $(Y,\bar{\underline{y}}^{+})$. Since $A$ consists of annuli and since $g$ induces an injection on the fundamental groups, $g^{-1}A$ consists of annuli, too. Now $A$ splits $X$ into solid tori, and so it follows that also $g^{-1}A$ splits $Y$ into solid tori whose completed boundary-patterns consist of annuli. Hence $(Y,\bar{\underline{y}}^{+})$ must be a Seifert fibre space.

Thus we may suppose that $\underline{g} \neq \emptyset$ The remainder of the proof of 15.5 will be split into three cases:

<u>Case 1. There is a side of</u> $\underline{g}$ <u>which is a square or annulus.</u>

If there is a side of $\underline{\underline{g}}$ which is a square, X is the I-bundle over the annulus or Möbius band (see 5.10), and so 15.1 follows by 15.3.

If, on the other hand, there is a side of $\underline{\underline{g}}$ which is an annulus, X is the $S^1$-bundle over the annulus or Möbius band (see 5.10). In particular, it follows that all sides of Y are annuli or tori. Hence $\partial Y$ consists of tori. Without loss of generality, Y is not a solid torus, for otherwise we are done. Hence $f|\partial Y$ can be considered as an essential map into the interior of X. So, by 5.5, $f|\partial Y$ can be deformed in X into $\partial X$. Since $f|Y: Y \to X$ induces an injection on the fundamental groups, it follows from [Wa 4, 6.1] that $(Y,\emptyset)$ is the $S^1$-bundle over the annulus or Möbius band. Then of course any essential torus in Y splits Y into Seifert fibre spaces.

Case 2. $X \cap \partial M \neq \emptyset$ and no side of $\underline{\underline{g}}$ is a square or an annulus.

At least one side of X is an $\underline{\underline{m}}_2$-bound side. To see this note that every side $G \in \underline{\underline{g}}$ is a free side. Hence every side of Y which meets G must be either a component of $f^{-1}(\partial X - \partial M_2)^-$ or an $\underline{\underline{m}}_1$-bound side. Since $f|G$ is horizontal in X and since $X \cap \partial M_2 \neq \emptyset$, not every side of Y which meets G can be a component of $f^{-1}(\partial X - \partial M_2)^-$. This means that Y has at least one $\underline{\underline{m}}_1$-bound side, and so our claim follows since f is an admissible map.

Let B be any $\underline{\underline{m}}_2$-bound side of X. Then we may fix a system A of essential vertical annuli, $A_1,\ldots,A_n$, in $(X,\bar{\underline{\underline{x}}}^+)$ such that at least one side of A lies in B and which split $(X,\underline{\underline{x}}^+)$ into a system $(\tilde{\underline{\underline{X}}},\tilde{\underline{\underline{x}}}^+)$ of solid tori. Moreover, A may be chosen so that no component of $\tilde{X}$ is an I-bundle over the annulus or Möbius band (no side of $\underline{\underline{g}}$ is a square or annulus). Recalling 4.5 and the proof of 4.10, one easily checks that f can be admissibly deformed (rel M - Y) so that afterwards it satisfies the same properties with respect to $\tilde{X}$ than with respect to X. Hence, for convenience, we may suppose without loss of generality that X itself is a solid torus. In this case Y is either a ball or a solid torus.

Let $G \in \underline{\underline{g}}$. Then G is a disc since $f|G$ is horizontal.

Since $G$ is not a square, there are at least three $\underline{\underline{m}}_1$-bound sides, $B_1$, $B_2$, $B_3$, of $Y$ meeting $G$.

If $Y$ is a solid torus, we easily find a non-contractible loop $t_i$ in $Y$ whose base point lies in $B_i$ and which is mapped under $f$ to an inessential arc in $(X,\underline{\underline{x}})$ ($(X,\underline{\underline{x}})$ is a solid torus). Since $f$ is essential, this means that $t_i$ is inessential, and this proves that $B_1$, $B_2$, and $B_3$ are annuli. Since $\partial Y$ is a torus and since $B_1$, $B_2$, $B_3$ all meet the disc $G$, we get a contradiction to the fact that $\bar{\underline{\underline{Y}}}^+$ is useful.

If $Y$ is a ball, it again remains to prove that all sides meeting $G$ are squares. If this is not the case, it follows that one $\underline{\underline{m}}_1$-bound side of $Y$, say $B_1$, meets three different sides $G_1, G_2, G_3 \in \underline{\underline{g}}$. Since $f|G_i$ is horizontal, we easily find arcs joining $B_1$, $B_2$, $B_3$ respectively with $\underline{\underline{m}}_1$-bound sides meeting $G_2$ and $G_3$ and which are mapped under $f$ to inessential arcs. Since $f$ is essential, this implies that each disc $B_1$, $B_2$, $B_3$ meets each $G_1$, $G_2$, $G_3$. But this is impossible in the 2-sphere $\partial Y$. So we are done in Case 2.

<u>Case 3</u>. $X \cap \partial M = \emptyset$ <u>and no side of</u> $\underline{\underline{g}}$ <u>is an annulus</u>.

Since $\partial X \neq \emptyset$, we may fix a section $F$ in $X$, i.e. an essential horizontal surface in $X$ (orientable or not) which intersects each fibre precisely once ($X$ has no exceptional fibres).

Suppose first that $f$ maps each $G \in \underline{\underline{g}}$ into $F$. Then let $G_1 \in \underline{\underline{g}}$ and let $C$ be any side $Y$ which meets $G_1$. Since $X \cap \partial M_2 = \emptyset$, it follows that $C$ is an annulus which is a component of $f^{-1}(\partial X - \partial M_2)$. Then of course the boundary components of $C$ are contained in sides $G_1, G_2 \in \underline{\underline{g}}$ (possibly $G_1 = G_2$). Now, for every $G \in \underline{\underline{g}}$, the restriction $f|G\colon (G,\partial G) \to (F,\partial F)$ can be deformed into a covering map since $G$ is not an annulus (Nielsen's theorem). Hence, by an argument used in Case 3 of 15.3, we find an I-bundle $Z$ in $Y$ with $Z \cap G_i = (G_i - U(\partial G_i))$, $i = 1,2$, where $U(\partial G_i)$ denotes a regular neighborhood of $\partial G_i$ in $G_i$. Moreover, it follows that there is no side, $G_3$, of $\underline{\underline{g}}$ which is contained entirely in $(\partial Y - Z)^-$. To see this observe that we may also construct an I-bundle $Z_1$ with

$Z_1 \cap G_1 = (G_1 - U(\partial G_1))^-$ and $Z_1 \cap G_3 = (G_3 - U(\partial G_3))^-$ and recall that $G_3$ cannot be an annulus. Hence all sides of $Y$ different from $G_1$, $G_2$ are components of $f^{-1}(\partial X - \partial M_2)^-$, i.e. annuli, and so we have proved that $Y$ is an I-bundle.

Thus, to show that $Y$ is an I-bundle in general, it suffices to construct an essential map $g\colon (M_1,\underline{\underline{m}}_1) \to (M_2,\underline{\underline{m}}_2)$ with the following properties:

1. $g^{-1}X = f^{-1}X$,

   and $g$ satisfies the suppositions of 15.1

2. $g(G) \subset F$, for all $G \in \underline{\underline{g}}$.

In order to construct such a map $g$, fix a system $A$ of essential vertical annuli in $(X,\underline{\underline{x}}^+)$ which splits $(X,\underline{\underline{x}}^+)$ into a system $(\tilde{X},\tilde{\underline{\underline{x}}}^+)$ of solid tori. Define $\tilde{F} = \tilde{X} \cap F$, and observe that $\tilde{F}$ is a system of meridian discs and that each solid torus of $\tilde{X}$ contains precisely one such disc. Now, by 4.4, $f|Y\colon (Y,\underline{\underline{y}}^+) \to (X,\underline{\underline{x}}^+)$ can be admissibly deformed so that afterwards $f^{-1}A$ consists of essential surfaces in $(Y,\underline{\underline{y}}^+)$. Let $(\tilde{Y},\tilde{\underline{\underline{y}}}^+)$ be the manifold obtained from $(Y,\underline{\underline{y}}^+)$ by splitting at $f^{-1}A$, and define $\tilde{G} = G \cap \tilde{Y}$. Of course, $f|\tilde{Y}\colon (\tilde{Y},\tilde{\underline{\underline{y}}}^+) \to (\tilde{X},\tilde{\underline{\underline{x}}}^+)$ can be admissibly deformed so that afterwards $f(\tilde{G}) \subset \tilde{F}$. Then it is easy to define an admissible map from a regular neighborhood $U(f^{-1}A)$ to $U(A)$ which extends $f|M_1 - U(f^{-1}A)$ to an essential map $g$ as required.

Hence 15.5 is established.

For the additional remark, recall that the hypothesis that X is free of exceptional fibres is used only in Case 3 of 15.5. Hence it suffices to show that Case 3 of 15.5 is impossible, provided f is an admissible homotopy equivalence. Observe that in this case, there is at least one surface $G \in \underline{\underline{g}}$ which is not a disc, annulus, or torus. Hence there is at least one essential singular closed curve, k, in G which is not homotopic G to a simple closed curve. Furthermore, we may suppose that k is not homotopic in G to a non-trivial multiple of another curve. $f \circ k$ can be considered as an essential, based loop in the Seifert fibre space X. So, there is another essential, based loop, t in X (e.g. t a fibre of X) which commutes with $f \circ k$. This implies the existence of a map $h\colon S^1 \times S^1 \to X$ with $h|S^1 \times 0 = f \circ k$, and $h|0 \times S^1$ essential. h is essential, for otherwise it can be extended to a map of a solid torus into X which is impossible since no multiple of $f|k$ is homotopic to a non-trivial multiple of t ($f|G$ is horizontal). Let g be an admissible homotopy inverse of f. Then $g \circ h$ is an essential singular torus in $M_1$. $g \circ h|S^1 \times 0$ can be deformed to k, since $gf \simeq id$. Hence splitting $S^1 \times S^1$ at $S^1 \times 0$, we see that $g \circ h$ defines an essential singular annulus in $(M_1, \bar{\underline{\underline{m}}}_1)$. By 12.5, this singular annulus can be admissibly deformed in $(M_1, \bar{\underline{\underline{m}}}_1)$ into a component Z of the characteristic submanifold of $(M_1, \bar{\underline{\underline{m}}}_1)$. Z also contains the essential singular torus $g \circ h$, up to homotopy. So, by 5.13, Z has to be a Seifert fibre space, i.e. $Z \cap \partial M_1$ consists of annuli and tori. But this is impossible since K cannot be deformed in G to a simple closed curve.

q.e.d.

## §16. Singular characteristic submanifolds

In this paragraph we shall prove that any admissible homotopy equivalence between 3-manifolds (irreducible etc.) can be admissibly deformed so that afterwards the characteristic submanifold is mapped into the characteristic submanifold.

The following lemma is one of the key observations which make the proof of the above result possible:

<u>16.1 Lemma.</u> <u>Let</u> $(M_1,\underline{\underline{m}}_1)$ <u>and</u> $(M_2,\underline{\underline{m}}_2)$ <u>be irreducible</u> 3-<u>manifolds whose completed boundary-patterns are useful and non-empty</u>. <u>Let</u> $F$ <u>be an essential surface in</u> $(M_2,\bar{\underline{\underline{m}}}_2)$ <u>with</u> $F \cap \partial M_2 = \partial F \neq \emptyset$. <u>Let</u> $f\colon (M_1,\underline{\underline{m}}_1) \to (M_2,\underline{\underline{m}}_2)$ <u>be an admissible homotopy equivalence</u>. <u>If</u> $f$ <u>can be admissibly deformed so that afterwards</u> $f^{-1}F = \emptyset$, <u>then</u> $F$ <u>is admissibly parallel in</u> $(M_2,\bar{\underline{\underline{m}}}_2)$ <u>to a free side of</u> $(M_2,\underline{\underline{m}}_2)$.

<u>Remark</u>. This is also true if $F$ is closed.

<u>Proof</u>. Let $(\tilde{M}_2,\tilde{\underline{\underline{m}}}_2)$ be the manifold obtained from $(M_2,\underline{\underline{m}}_2)$ by splitting along $F$, and denote by $F_1$, $F_2$ the two sides of $(\tilde{M}_2,\tilde{\underline{\underline{m}}}_2)$ which are copies of $F$. Assume $f^{-1}F = \emptyset$. Then $f(M_1) \subset \tilde{M}_2$. Let $g\colon (M_2,\underline{\underline{m}}_2) \to (M_1,\underline{\underline{m}}_1)$ be an admissible homotopy inverse of $f$. Then $fg(M_2) \subset \tilde{M}_2$ and $fg$ is admissibly homotopic to the identity. Fixing an appropriate curve we find as a first consequence that $F$ has to be separating. Let $(N_1,\underline{\underline{n}}_1)$ be the component of $(\tilde{M}_2,\tilde{\underline{\underline{m}}}_2)$ which contains $fg(M_2)$, and $(N_2,\underline{\underline{n}}_2)$ be the other one.

In the remainder of this proof we call $N_i$, $i = 1$ or $2$, <u>good</u> if <u>every</u> admissible arc in $(N_i,\underline{\underline{n}}_i)$ with $\partial k \subset F_i$ is inessential. $\bar{\underline{\underline{n}}}_i$ is useful, $i = 1,2$. Hence, if $N_i$ is good, we may conclude that every side of $(N_i,\bar{\underline{\underline{n}}}_i)$ which meets $F_i$ must be a disc or annulus. In the same way we see that, moreover, any two such sides intersect themselves only in sides which meet $F_i$. This implies the existence of a side $G$ of $(N_i,\bar{\underline{\underline{n}}}_i)$ which does not meet $F_i$. If $G$ is the only such side, $(N_i,\bar{\underline{\underline{n}}}_i)$ has to be a product I-bundle whose lids are $F_i$ and $G$. To see the latter observe that every essential singular arc in $G$ is a side of an essential singular square or annulus in $(N_i,\bar{\underline{\underline{n}}}_i)$ whose opposite side lies in $F_i$ ($N_i$ is good), and so our claim follows from 12.5.

Since every closed curve in $M_2$ can be deformed into $N_2$, it follows that $N_i$ is good, for $i = 1$ or $i = 2$. Thus, by what we have seen so far, we are done if there is only one side of $(N_i,\bar{\underline{\underline{n}}}_i)$ which does not meet $F_i$. For this it suffices to show that every bound side of $N_i$ meets $F_i$. This is clear if $N_2$ is good, for $id|N_2$ can be admissibly deformed into $N_1$. If $N_2$ is not good, it follows from the fact that every admissible arc in $(M_2,\underline{\underline{m}}_2)$ can be admissibly deformed into $N_1$. q.e.d.

<u>16.2 Proposition</u>. <u>Let</u> $(M_1,\underline{\underline{m}}_1)$ <u>and</u> $(M_2,\underline{\underline{m}}_2)$ <u>be given as in</u> 16.1 <u>Denote by</u> $\bar{V}_i$, $i = 1,2$, <u>the characteristic submanifold of</u> $(M_i,\bar{\underline{\underline{m}}}_i)$. <u>Then every admissible homotopy equivalence</u> $f\colon (M_1,\underline{\underline{m}}_1) \to (M_2,\underline{\underline{m}}_2)$ <u>can be admissibly deformed so that afterwards</u> $f(\bar{V}_1) \subset \bar{V}_2$.

<u>Remark</u>. The supposition that $f$ is an admissible homotopy equivalence cannot be weakened to the condition that $f$ is an essential map. Here is a counterexample. Let $M_2$ be a Stallings fibration with non-empty boundary, and let $\underline{\underline{m}}_2$ be the set of boundary components of $M_2$. Suppose $(M_2,\underline{\underline{m}}_2)$ is not a Seifert fibre space. Let $M_1$ be a regular neighborhood of some fibre (i.e. a surface) of $M_2$, and $\underline{\underline{m}}_1$ be the boundary-pattern of $M_1$ induced by $\underline{\underline{m}}_2$. Then $(M_1,\bar{\underline{\underline{m}}}_1)$ is an I-bundle, i.e. the characteristic submanifold of $(M_1,\bar{\underline{\underline{m}}}_1)$ is equal to $M_1$. Furthermore, the inclusion $i\colon (M_1,\underline{\underline{m}}_1) \to (M_2,\underline{\underline{m}}_2)$ is an essential map. But of course $i$ cannot be admissibly deformed into the characteristic submanifold of $(M_2,\underline{\underline{m}}_2)$ since $M_2$ is not a Seifert fibre space.

<u>Proof of 16.2</u>. Let $X$ be a component of $\bar{V}_1$ and let $\underline{\underline{x}}$ be the boundary-pattern of $X$ induced by $\underline{\underline{m}}_1$. To prove 16.2 it suffices to show that $f|X\colon (X,\underline{\underline{x}}) \to (M_2,\underline{\underline{m}}_2)$ can be admissibly deformed into $\bar{V}_2$. For this we shall need an admissible homotopy inverse $g\colon (M_2,\underline{\underline{m}}_2) \to (M_1,\underline{\underline{m}}_1)$ of $f$ (see the above remark). By 15.2 we may

suppose that $g^{-1}\bar{V}_1$ is an essential F-manifold in $(M_2,\bar{\underline{\underline{m}}})$.

Case 1. $(X,\underline{\underline{x}})$ is a Seifert fibre space.

If $\bar{\underline{\underline{x}}}$ is a useful boundary-pattern of $X$, our claim follows from 13.1. Thus we suppose the converse. Then, by 5.2, $(X,\bar{\underline{\underline{x}}})$ is the $S^1$-bundle over an i-faced disc, $1 \leq i \leq 3$. Let $G$ be the free side of $(X,\underline{\underline{x}})$. Since $\bar{\underline{\underline{m}}}_1$ is useful, it follows that $G$ contains a component $G_1$ of $(\partial X - \partial M_1)^-$, $G \neq G_1$. In particular, $G_1$ is not a torus and not admissibly parallel to a free side of $(M_1,\underline{\underline{m}}_1)$. By 16.1 and our suppositions on $g$, $g^{-1}G_1$ is a non-empty system of essential squares or annuli in $(M_2,\bar{\underline{\underline{m}}}_2)$. By 10.7, we may suppose that $g^{-1}G_1$ is contained in $\bar{V}_2$. Hence there is an essential curve $k_1, k_1 \subset G_1$, in $(M_1,\underline{\underline{m}}_1)$ and an essential curve $k_2, k_2 \subset \bar{V}_2$, in $(M_2,\underline{\underline{m}}_2)$ such that $g \cdot k_2$ is admissibly homotopic to $k_1^m$, for some $m \geq 1$. This means that $f \cdot k_1^m$ can be admissibly deformed into $\bar{V}_2$.

Suppose for a moment that $f \cdot k_1$ is contained in $\bar{V}_2$. Then $f|X$ can be admissibly deformed (rel $k_1$) so that afterwards $f^{-1}(\partial\bar{V}_2 - \partial M_2)^-$ is a system of incompressible, admissible annuli in $(X,\bar{\underline{\underline{x}}})$ (apply the surgery arguments of the proof of 4.4). Let the homotopy be chosen so that, in addition, the number of components of $f^{-1}(\partial\bar{V}_2 - \partial M_2)^-$ is minimal. Since $\bar{V}_2$ is complete, it follows that $f$ maps each component of $f^{-1}(M_2 - \bar{V}_2)^-$ into a component of $(M_2 - \bar{V}_2)^-$ which is the $S^1$-bundle over the square or annulus. Then it is easy to see that $f|X$ can be admissibly deformed into $\bar{V}_2$.

Now recall that $f \circ k_1^m$ can be admissibly deformed into $\bar{V}_2$. Let $Y$ be a solid torus and let $t$ be a curve on $\partial Y$ which has circulation number $m$ with respect to $Y$. Let $h\colon Y \to M_2$ be a map with $h \circ s = f \cdot k_1$, where $s$ is the core of $Y$. Then $h \cdot t$ can be admissibly deformed into $\bar{V}_2$, and so also $f \circ k_1$, by the above argument.

Case 2. $(X,\bar{\underline{\underline{x}}})$ is an I-bundle.

If $(X,\bar{\underline{\underline{x}}})$ admits an admissible fibration either as Seifert fibre space, or as I-bundle whose lids lie in bound sides of $(M_1,\underline{\underline{m}}_1)$ then 16.2 follows as in Case 1. Thus we suppose the converse. Without loss of generality, $f^{-1}\bar{V}_2$ is an essential F-manifold which is contained in $\bar{V}_1$ (see 15.2 and 10.8).

Suppose $X \cap f^{-1}\bar{V}_2 = \emptyset$. Let $G$ be a horizontal surface in $X$, and let $\underline{\underline{g}}$ be the boundary-pattern of $G$ induced by $\underline{\underline{m}}_1$. $f(G) \subset M_2 - \bar{V}_2$ and $g^{-1}\bar{V}_1 \subset \bar{V}_2$. Hence $gf(G) \subset M_1 - \bar{V}_1$. This means that $G$ can be admissibly deformed out of $X$. Applying the surgery arguments of the proof of 4.4 to this homotopy, we find that $(G,\bar{\underline{\underline{g}}})$ must be a square or annulus, and we obtain a contradiction to our suppositions on $X$.

Suppose $X \cap f^{-1}\bar{V}_2 \neq \emptyset$. Let $X_1$ be a component of $(X - f^{-1}\bar{V}_2)^-$. Without loss of generality, the fibration of $X$ induces an admissible fibration of $X_1$. Fix a horizontal surface $(G,\underline{\underline{g}})$ in $X_1$. As above $(G,\underline{\underline{g}})$ can be admissibly deformed out of $X_1$. This implies that at least one component of $(\partial X_1 - \partial M_1)^-$ separates an I-bundle over the square or annulus from $X$ which contains $X_1$. This is true for every component $X_1$ of $(X - f^{-1}\bar{V}_2)$. Hence it follows that $X$ can be admissibly contracted into a component of $f^{-1}\bar{V}_2$, i.e. $f|X$ can be admissibly deformed into $\bar{V}_2$. q.e.d.

## §17. The preimage of the characteristic submanifold

If $f: M_1 \to M_2$ is an admissible homotopy equivalence, we shall prove that $f$ can be admissibly deformed so that the preimage under $f$ of the characteristic submanifold of $M_2$ is <u>equal</u> to that of $M_1$.

For this we shall use the following:

<u>17.1 Lemma</u>. <u>Denote by $\bar{V}_i'$, $i = 1, 2$, the union of all components of $\bar{V}_i$ which are not regular neighborhoods of free sides of $(M_i, \underline{\underline{m}}_i)$.</u>
<u>If $A$ is a component of $(\partial\bar{V}_1' - \partial M_1)^-$, then $f|A$ can be admissibly deformed in $(M_2, \underline{\underline{m}}_2)$ into $(\partial\bar{V}_2 - \partial M_2)^-$.</u>

<u>Proof</u>. Let $g: (M_2, \underline{\underline{m}}_2) \to (M_1, \underline{\underline{m}}_1)$ be an admissible homotopy inverse of $f$, and suppose that $g$ is admissibly deformed so that $g^{-1}A$ is a system of essential squares, annuli, or tori in $(M_2, \underline{\underline{m}}_2)$ (see 4.5). Observe that no component of $(\partial\bar{V}_1' - \partial M_1)^-$ is admissibly parallel to a free side of $(M_1, \underline{\underline{m}}_1)$ since $\bar{V}_1$ is full. Hence, by 16.1, $g^{-1}A$ is non-empty. Without loss of generality, $g^{-1}A \subset \bar{V}_2$ (see 10.7). By 15.2 and 10.8, $f$ can be admissibly deformed so that afterwards $f^{-1}\bar{V}_2$ is an essential F-manifold contained in $\bar{V}_1$. In particular, $f(A) \subset M_2 - \bar{V}_2$.

If all sides of $A$ lie in bound sides of $(M_1, \underline{\underline{m}}_1)$, then, by the Enclosing Theorem 12.5, $f|A$ can be admissibly deformed into $\bar{V}_2$, and so into $(\partial\bar{V}_2 - \partial M_2)^-$ since $f(A) \subset M_2 - \bar{V}_2$.

We suppose the converse. Let $A$ be an annulus whose sides are all contained in free sides of $(M_1, \underline{\underline{m}}_1)$ (the proof in the other cases is similar). Observe that this choice of $A$ implies that no component of $g^{-1}A$ is a square or torus. Since, by 16.1, $g^{-1}A$ is non-empty it contains at least one annulus, say $B$. Let $a$ and $b$ be essential curves in $A$ and $B$, respectively. Then $g \circ b = a^m$, for some $m \geq 1$. Hence $f \circ a^m = f \circ g \circ b$ which is admissibly homotopic to $b$, by our choice of $g$. Thus $f \circ a^m$, and so, by an argument of Case 1 of 16.2, $f \circ a$ can be admissibly deformed into $\bar{V}_2$. So $f|A$ can be admissibly deformed into $(\partial\bar{V}_2 - \partial M_2)^-$ since

$f(A) \subset M_2 - \bar{V}_2$. q.e.d.

*17.2 Proposition. Let $(M_1,\underline{\underline{m}}_1)$ and $(M_2,\underline{\underline{m}}_2)$ be irreducible and aspherical 3-manifolds whose completed boundary-patterns are useful and non-empty. Denote by $\bar{V}_i$, $i = 1,2$, the characteristic submanifold of $(M_i,\bar{\underline{\underline{m}}}_i)$.*

*Then every admissible homotopy equivalence can be admissibly deformed so that afterwards $f^{-1}\bar{V}_2 = \bar{V}_1$.*

*Proof.* By 16.2, we may suppose that $f(\bar{V}_1) \subset \bar{V}_2$, i.e. $\bar{V}_1 \subset f^{-1}\bar{V}_2$. Denote by $\bar{V}_1'$ the union of all components of $\bar{V}_1$ which are not regular neighborhoods of free sides of $(M_1,\underline{\underline{m}}_1)$, and let $U$ be a regular neighborhood of $(\partial\bar{V}_1' - \partial M_1)^-$ in $(M_1 - \bar{V}_1')^-$. By 17.1, $f|(\partial\bar{V}_1' - \partial M_1)^-$ can be admissibly deformed in $(M_2,\underline{\underline{m}}_2)$ into $(\partial\bar{V}_2 - \partial M_2)^-$. Since $f(\partial\bar{V}_1' - \partial M_1)^- \subset \bar{V}_2$, this homotopy may be chosen within $\bar{V}_2$. This means that $f$ is admissibly homotopic to a map $g$ with $g^{-1}\bar{V}_2 = (f^{-1}\bar{V}_2 - U)^-$. Thus every component of $\bar{V}_1'$ is a component of $g^{-1}\bar{V}_2$. In addition, we may suppose, by 4.5, that $g^{-1}(\partial\bar{V}_2 - \partial M_2)$ consists of essential squares, annuli, or tori.

Let $X$ be a component of $g^{-1}\bar{V}_2$ which is not an essential I-bundle or Seifert fibre space in $(M_1,\bar{\underline{\underline{m}}}_1)$, and let $X'$ be the component of $\bar{V}_2$ which contains $g(X)$. If, in the notation of 15.1, there is a free side $G$ of $X$ such that $g|G$ is inessential in $X'$, then, by 4.9, $G$ has to be a square or annulus. Let $U(G)$ be a regular neighborhood of $G$ in $X$. Then it is easily seen that $g$ can be admissibly deformed into a map $g'$ with $(g')^{-1}\bar{V}_2 = (g^{-1}\bar{V}_2 - U(G))^-$. $G$ cannot meet $\bar{V}_1$, for otherwise, by our choice of $X$, $G$ is a (free) side of $(M_1,\underline{\underline{m}}_1)$ and then $f|G$ is essential in $X'$. Hence $\bar{V}_1 \subset (g')^{-1}\bar{V}_2$. Thus inductively we may suppose that no free side $G$ as above exists. Then it follows easily from 15.1 that $g$ is admissibly homotopic to a map $g'$ such that $(g')^{-1}\bar{V}_2$ is an essential F-manifold which contains $\bar{V}_1$. Hence, by 10.6, we may suppose that $f$ is admissibly deformed so that (1) every component of $\bar{V}_1$ is a component of $f^{-1}\bar{V}_2$, and that (2) $f^{-1}\bar{V}_2 - \bar{V}_1$ is a regular neighborhood of an essential surface in $(M_1,\bar{\underline{\underline{m}}}_1)$ whose components are admissibly parallel to components of $(\partial\bar{V}_1 - \partial M_1)$. Suppose that $f$ is admissibly deformed so that the above holds and that, in addition, the number of components of $f^{-1}(\partial\bar{V}_2 - \partial M_2)^-$ is minimal. Then the following holds:

<u>17.3 Assertion</u>. $\bar{V}_1 = f^{-1}\bar{V}_2$.

Assume the converse. Then, by what we have seen so far, there is a component $W_1$ of $(M_1 - f^{-1}\bar{V}_2)^-$ which meets a component $X_1$ of $\bar{V}_1$ and a component $Y_1$ of $f^{-1}\bar{V}_2 - \bar{V}_1$. Furthermore, we may suppose that both $W_1$ and $Y_1$ are regular neighborhoods of essential squares, annuli, or tori in $(M_1,\bar{\underline{m}}_1)$. If all the sides of these surfaces are contained in bound sides of $(M_1,\underline{m}_1)$ we obtain the required contradiction as in 14.7. So we may suppose that at least one side of $Y_1$ lies in a free side $G$ of $(M_1,\underline{m}_1)$. Let $A$ be a component of $(G - f^{-1}\bar{V}_2)^-$ or of $(G - \bar{V}_1)^-$ which is an inner square or annulus in $G$, and meets $\bar{V}_1$. Then the following holds:

1. $(\partial A - \partial G)^-$ is contained in different components $B_1, B_2$ of $G \cap f^{-1}\bar{V}_2$, resp. $G \cap \bar{V}_1$.
2. Precisely one of $B_1, B_2$ is an inner square or annulus in $G$.

Property 1 follows easily from the fact that $\bar{V}_1$ is full and complete. The same with 2, if $A$ is a component of $(G - \bar{V}_1)^-$. To see 2, if $A$ is a component of $(G - f^{-1}\bar{V}_2)$ but not of $(G - \bar{V}_1)^-$, assume it does not hold. Then, by the above mentioned properties of $f^{-1}\bar{V}_2$, it follows that both $B_1, B_2$ have to be inner squares or annuli in $G$. Let $Z_1, Z_2$ be the components of $f^{-1}\bar{V}_2$ which contain $B_1, B_2$, respectively. Then $Z_1$, say, is a component of $f^{-1}\bar{V}_2$ and $Z_2$ a component of $\bar{V}_1$. Observe that $f|A$ is an essential singular square or annulus in some component $W_1'$ of $(M_2 - \bar{V}_2)^-$ ($f^{-1}(\partial\bar{V}_2 - \partial M_2)^-$ is minimal). Since $\bar{V}_2$ is complete, $W_1'$ has to be an I- or $S^1$-bundle over the square or annulus (see 12.6). This implies that $f|B_1$ is an essential singular square or annulus in some component $Z_1'$ of $\bar{V}_2$, for otherwise $f^{-1}(\partial\bar{V}_2 - \partial M_2)^-$ can be diminished. In the same way it follows that there is at least one essential square or annulus $B$ in $Z_2$ which meets $W_1$ but not $B_1$, so that $f|B$ is essential in some component $Z_2'$ of $\bar{V}_2$ (see the argument in 13.3). By 5.10, we may suppose that $f|B_i$ is vertical in $Z_i'$, $i = 1,2$, and so the fibrations of $Z_1'$ and $Z_2'$

coincide via $W_1'$. But this is impossible since $\bar{V}_2$ is full, and so our claim is established.

1 and 2 above imply in particular that $Y_1$ cannot lie in a component W of $(M_1 - \bar{V}_1)^-$ which is an I- or $S^1$-bundle over the square or annulus with $(\partial W - \partial M_1)^-$ disconnected. Furthermore, recall that $\bar{V}_1$ is complete and that, by 2, $X_1$ cannot be an I- or $S^1$-bundle over the square, annulus, or Möbius band. Hence it follows that there is no essential curve k (closed or not) in $Y_1$ whose end-points lie in bound sides of $(M_1, \underline{\underline{m}}_1)$ and which can be admissibly deformed in $(M_1, \underline{\underline{m}}_1)$ into a component of $\bar{V}_1$ different from $X_1$. But $X_1$ and $Y_1$ are mapped under f into different components $X_1'$ and $Y_1'$ of $\bar{V}_2$ since, in the notation above, $W_1'$ is an I- or $S^1$-bundle over the square or annulus and $f|A: A \to W_1'$ is essential (see above). Now consider an admissible homotopy inverse g of f. Then, by symmetry and an argument of Case 2 of 16.2, we easily see that we may suppose g maps $X_1'$ and $Y_1'$ into different components of $\bar{V}_1$, i.e. at least one into a component Z of $\bar{V}_1$ different from $X_1$. Let k be a curve in $Y_1$ as described above. Then it is easily seen that $gf|k$ can be admissibly deformed in $(M_1, \underline{\underline{m}}_1)$ into Z, and so we obtain the required contradiction since gf is admissibly homotopic to the identity.

q.e.d.

## §18. Splitting a homotopy at the characteristic submanifold

Throughout this paragraph let $(M,\underline{\underline{m}})$ and $\bar{V}$ be given as in the beginning of chapter VI. Furthermore let $H: (M \times I, \underline{\underline{m}} \times I) \to (M,\underline{\underline{m}})$ be an admissible homotopy, and denote $H_t = H|M \times t$, $t \in I$. Suppose that $H_0 = \mathrm{id}$ and that $H_1^{-1}\bar{V} = \bar{V}$. Under this supposition we are going to prove that $H$ can be split along $\bar{V}$, i.e. that $H$ is admissibly homotopic (rel $M \times \partial I$) to $H'$ with $(H_t')^{-1}\bar{V} = \bar{V}$, for all $t \in I$ (see 18.2). Indeed, we shall see that this result easily follows from:

<u>18.1 Proposition</u>. <u>Let</u> $z$ <u>be a point in the interior of</u> $M$ <u>and not contained in</u> $(\partial\bar{V} - \partial M)^-$. <u>Then</u> $H|z \times I$ <u>can be deformed (rel</u> $z \times \partial I$) <u>so that afterwards</u> $H(z \times I) \cap (\partial\bar{V} - \partial M)^- = \emptyset$.

<u>Remark</u>. It will be apparent from the proof that 18.1 remains true if we replace $H_0 = \mathrm{id}$ by the suppositions that $H_0^{-1}\bar{V} = \bar{V}$ and that the two maps $H_0|\bar{V}: \bar{V} \to \bar{V}$ and $H_0|(M - \bar{V})^-: (M - \bar{V})^- \to (M - \bar{V})^-$ are admissible homotopy equivalences.

<u>Proof</u>. Define $h = H|z \times I$, and let $h$ be deformed so that the number of points of $h^{-1}(\partial\bar{V} - \partial M)^-$ is minimal. If $h^{-1}(\partial\bar{V} - \partial M)^-$ is empty, we are done. So we assume the converse and we show that this assumption leads to contradictions. For this let $x$ be the point of $h^{-1}(\partial\bar{V} - \partial M)^-$ which is nearest to $z \times 0$. Denote by $z \times I'$ the arc in $z \times I$ which joins $z \times 0$ with $x$, and let $T_1$ be the component of $(\partial\bar{V} - \partial M)^-$ containing $h(x)$.

$z$ lies in a component $Z$ either of $\bar{V}$ or of $(M - \bar{V})^-$. Let $k$ be any loop in $Z$ with base point $z$. Then, considering $H|k \times I$, we see that $H|z \times I'$ can be extended to a homotopy which moves $k$ into $T_1$. Since $k$ is arbitrarily chosen, this means that there is an injection $\pi_1 Z \to \pi_1 T_1$. Hence $\pi_1 Z$ is isomorphic either to $\mathbb{Z} \oplus \mathbb{Z}$, or $\mathbb{Z}$, or the trivial group. As usual this implies that $Z$ is either torus $\times I$, or a solid torus, or a ball, according to whether $T_1$ is a torus, an annulus, or a square.

Furthermore every bound side of $(Z,\underline{\underline{z}})$ has to meet $T_1$. To see this join $z$ with such a side by an arc $b$ and consider the preimage of $T_1$ under $H|b \times I$.

Now suppose first that $Z$ is a component of $\bar{V}$. Since $T_1$ is a square, annulus, or torus, we find an essential curve $k$ (closed or not) in $Z$ whose end-points lie in bound sides of $(M,\underline{\underline{m}})$, and which contains $z$. $\bar{V}$ is complete. Hence, considering the homotopy $H|k \times I$ and recalling our assumptions on $h = H|z \times I$, it follows from 12.6 that $T_1$ is contained in a component $Y_1$ of $(M - \bar{V})^-$ which is the I- or $S^1$-bundle over the square or annulus. $\bar{V}$ is full. Hence, in particular, $Z$ cannot be the product I- or $S^1$-bundle over the square or annulus. This implies that $(\partial Z - \partial M)^-$ is disconnected (see the above properties of $Z$).

So there must be a component $Y_2$ of $(M - \bar{V})^-$ which contains a component $T_2$ of $(\partial Z - \partial M)^-$ different from $T_1$. Let $z_2$ be a point in $Y_2$, and let $t$ be an arc which joins $z_2$ with $z$, so that $t \cap (\partial\bar{V} - \partial M)^- = t \cap T_2$ is one point. Then, considering the preimage of $(\partial\bar{V} - \partial M)^-$ under $H|t \times I$, we see that $H|z_2 \times I$ cannot be deformed (rel $z_2 \times \partial I$) out of $(\partial\bar{V} - \partial M)^-$ (recall $H_1^{-1}\bar{V} = \bar{V}$).

Thus, in order to get a contradiction we may suppose that $Z$ is a component of $(M - \bar{V})^-$ (for choose $z_2$ instead of $z$ if necessary). Observe that at least one side $C$ of $(Z,\bar{\underline{\underline{z}}}^+)$ (notation is in the beginning of §15) does not meet $T_1$ since $\bar{\underline{\underline{z}}}^+$ is useful (recall the above properties of $Z$). If $Z$ is homeomorphic to torus $\times$ I, this side must be an annulus or a torus. Hence we find an essential annulus in $(Z,\bar{\underline{\underline{z}}}^+)$ which meets $T_1$. Since $\bar{V}$ is complete, the existence of such an annulus implies that $(Z^+,\bar{\underline{\underline{z}}}^+)$ is the $S^1$-bundle over the annulus. If $Z$ is homeomorphic to a ball or a solid torus, we argue similarly. So in any case $Z$ is the I- or $S^1$-bundle over the square or annulus.

Since $\bar{V}$ is full, $(\partial Z - \partial M)^-$ is disconnected and let $T_2$ be a component of $(\partial Z - \partial M)^-$ different from $T_1$. By the same argument, $T_1$ and $T_2$ lie in two different components $Y_1$ and $Y_2$ respectively.

We are now going to show that $Z$ may be chosen to be not an $S^1$-bundle over the annulus. Otherwise $T_2$ is a torus. If $y_2$ is a point in $Y_2$, then, by an argument given above, $H|y_2 \times I$ cannot be deformed (rel $y_2 \times \partial I$) out of $(\partial\bar{V} - \partial M)^-$. So, $Y_2$ has the homeomorphism type of torus $\times$ I, and $(\partial Y_2 - \partial M)^-$ is disconnected (see above). Since $\bar{V}$ is full, $Y_2$ cannot be also an $S^1$-bundle over the annulus, and so at least one component of $(\partial Y_2 - \partial M)^-$ is an annulus. This component lies in a component $Z_2$ of $(M - \bar{V})^-$. Furthermore, if $z_2$ is a point in $Z_2$, then again $H|z_2 \times I$ cannot be deformed (rel $z_2 \times \partial I$) out of $(\partial V - \partial M)^-$. So choosing this $z_2$ instead of $z$, we may suppose that $Z$ is a component of $(M - \bar{V})^-$ which is an I-bundle over the square or annulus.

Assume that at least one side of $T_2$ lies in a free side $G$ of $(M,\underline{\underline{m}})$. Let $A$ be a component of $Z \cap G$, and denote by $G_i$, $i = 1,2$, the component of $G \cap Y_i$ which meets $A$. $G_1$ has to be a square or annulus, for the homotopy $H$ moves a given essential curve in $Z$ along an essential singular square or annulus through $Y_1$, i.e., without loss of generality, either vertically or horizontally (see 5.10 and our suppositions on $h|z \times I$). Now consider the homotopy $H|G_2 \times I$. By our assumption on $h|z \times I$, it follows that $G_2$ can be admissibly deformed into $T_2$. This means that $G_2$ is also a square or annulus. On the other hand, $G_1$ and $G_2$ cannot be both squares or annuli, for $\bar{V}$ is full. This is a contradiction.

So all sides of $T_2$ have to lie in bound sides of $(M,\underline{\underline{m}})$. $\bar{V}$ is full. Hence, considering $H|T_2 \times I$ we find that $T_2$ cannot be a square or annulus (apply 5.8 to see how $Y_1$ otherwise looks like), which is again a contradiction. q.e.d.

<u>18.2 Corollary</u>. <u>Let</u> $H: (M \times I \quad \underline{\underline{m}} \times I) \to (M,\underline{\underline{m}})$ <u>be given as in the beginning of</u> §18. <u>Then</u> $H$ <u>is admissibly homotopic (rel</u> $M \times \partial I$) <u>to</u> $H'$ <u>with</u> $(H'_t)^{-1}\bar{V} = \bar{V}$, <u>for all</u> $t \in I$.

<u>Remark</u>. As for 18.1 also 18.2 remains true under the weaker conditions remarked after 18.1.

<u>Proof</u>. Fix a triangulation $\Delta$ of $(M,\underline{\underline{m}})$ so that $(\partial\bar{V} - \partial M)^-$ and the graph of $(M,\bar{\underline{\underline{m}}})$ are subcomplexes, and denote by $\Delta^i$ the i-skeleton of $\Delta$.

Let $x \in \Delta^0 \cap (\partial\bar{V} - \partial M)^-$. Then $x$ lies in a component $Z_1$ of $\bar{V}$ <u>and</u> in a component $Z_2$ of $(M - \bar{V})^-$. Fix points $z_1, z_2$ in

$Z_1^0, Z_2^0$, respectively, and join $x$ with $z_i$, by an arc $k_i \subset Z_i$. By 18.1, we may suppose that $H(z_i \times I) \cap (\partial\bar{V} - \partial M)^- = \emptyset$. Hence, considering the preimage of $(\partial\bar{V} - \partial M)^-$ under $H|(k_1 \cup k_2) \times I$, we find that $H|x \times I$ can be deformed (rel $x \times \partial I$) into $(\partial\bar{V} - \partial M)^-$. If $x \in \Delta^0 - (\partial\bar{V} - \partial M)^-$, then $x$ lies in a component $Z$ of $\bar{V}$, or of $(M - \bar{V})^-$. If $x$ either lies in the interior of $Z$ or in a free side of $(M,\underline{\underline{m}})$, it follows from 18.1 that $H|x \times I$ can be deformed (rel $x \times \partial I$) into $Z$. If finally $x$ lies in a bound side $B$ of $(M,\underline{\underline{m}})$, join $x$ and a point $z \in Z^0$ with an arc $k \subset Z$. Considering $H|k \times I$ and recalling that $(\partial\bar{V} - \partial M)^-$ is essential in $(M,\bar{\underline{\underline{m}}})$, it follows that $H|x \times I$ can be deformed in $B$ (rel $x \times \partial I$) into $Z$.

Now 18.2 follows by induction on $\Delta^i$ since $(\partial\bar{V} - \partial M)^-$ is essential in $(M,\bar{\underline{\underline{m}}})$ and since $M$ is aspherical. q.e.d.

Combining 17.2 with 18.2, we obtain the following:

18.3 Theorem. Let $f\colon (M_1,\underline{\underline{m}}_1) \to (M_2,\underline{\underline{m}}_2)$ be any admissible homotopy equivalence.

Then $f$ can be admissibly deformed so that afterwards both

$$f|\bar{V}_1\colon (\bar{V}_1,\underline{\underline{v}}_1) \to (V_2,\underline{\underline{v}}_2)$$

and

$$f|(M_1 - \bar{V}_1)^-\colon ((M_1 - \bar{V}_1)^-,\underline{\underline{w}}_1) \to ((M_2 - \bar{V}_2)^-,\underline{\underline{w}}_2)$$

are admissibly homotopy equivalences, with respect to the proper boundary-patterns $\underline{\underline{v}}_i$ and $\underline{\underline{w}}_i$.

Furthermore, if $g$ is any admissible homotopy inverse of $f$, then $g$ can be admissibly deformed so that afterward $g|\bar{V}_2$ and $g|(M_2 - \bar{Y}_2)^-$ are the admissible homotopy inverses of $f|\bar{V}_1$, resp. $f|(M_1 - \bar{V}_1)^-$.

This is our first splitting theorem. Observe that it can be considered as a generalization of 10.9 which in turn tells us that a homeomorphism can be split along the characteristic submanifold.

Chapter VII: Simple 3-manifolds.

A Haken 3-manifold $(M,\underline{\underline{m}})$ whose completed boundary-pattern is useful, is called simple 3-manifold, if every component of the characteristic submanifold of $(M,\bar{\underline{\underline{m}}})$ is a regular neighborhood in $(M,\bar{\underline{\underline{m}}})$ of some side.

Here are some examples:

1. Let $M$ be the knot space of a non-trivial knot, which is not a torus knot and which has no companions (in the sense of [Sch 1]). Then $(M,\{\partial M\})$ is a simple 3-manifold.
2. Let $(M,\underline{\underline{m}})$ be any Haken 3-manifold whose completed boundary-pattern is useful and whose characteristic submanifold is empty (i.e. $(M,\bar{\underline{\underline{m}}})$ contains no essential squares, annuli, or tori--singular or not). Then $(M,\underline{\underline{m}})$ is a simple 3-manifold.
3. Let $(M,\underline{\underline{m}})$ be any Haken 3-manifold whose completed boundary-pattern is useful. Let $(M',\underline{\underline{m}}')$ be the 3-manifold obtained from $(M,\underline{\underline{m}})$ by splitting at the characteristic submanifold of $(M,\bar{\underline{\underline{m}}})$. Then, by 10.4, $(M',\underline{\underline{m}}')$ is the union of simple 3-manifolds with a collection of I-bundles over squares, annuli, or tori.

## §19. Isotopic surfaces in simple 3-manifolds

In this paragraph we are concerned with the problem of isotoping surfaces in irreducible 3-manifolds. To be more precise let $F$ and $G$ be two essential surfaces in an irreducible 3-manifold $(M,\underline{\underline{m}})$, with $F \cap \partial M = \partial F$ and $G \cap \partial M = \partial G$, which are admissibly homotopic. The question is whether or not this implies that they are admissibly isotopic in $(M,\bar{\underline{\underline{m}}})$. This kind of question was attacked in [Wa 4] and solved affirmatively in the case that $\underline{\underline{m}}$ consists of all the boundary components of $M$. However, if $\underline{\underline{m}}$ is not complete, it is easy to construct counterexamples. Hence in order to push the study a bit further we are forced to put some appropriate restrictions on both $F$ and $M$. It will turn out that one can say more if $M$ is simple. The additional information will suffice to attack the splitting problem for surfaces (see §§20, 21).

To begin, we first define a complexity for essential

surfaces $(F,\underline{\underline{f}})$ in a 3-manifold $(M,\underline{\underline{m}})$, $F \cap \partial M = \partial F$. For this denote by $\beta$, resp. $\alpha$, the number of all free sides, resp. the number of all those sides of $(F,\underline{\underline{f}})$ which are contained in squares or annuli of $\underline{\underline{m}}$. Then we define the complexity $d(F,\underline{\underline{f}})$ of $(F,\underline{\underline{f}})$ to be the following triple

$$d(F,\underline{\underline{f}}) = (10 \cdot \beta_1(F) + \operatorname{card}(\bar{\underline{\underline{f}}}), -\alpha, -\beta).$$

Remark. The following observation is crucial. If $G$ is an essential surface in $(F,\underline{\underline{f}})$, and $(\tilde{F},\tilde{\underline{\underline{f}}})$ is the surface obtained from $(F,\underline{\underline{f}})$ by splitting at $G$, then, for each component $(F_1,\underline{\underline{f}}_1)$ of $(\tilde{F},\tilde{\underline{\underline{f}}})$, the integer $10\cdot\beta_1(F_1) + \operatorname{card}(\bar{f}_1)$ is strictly smaller than $10\cdot\beta_1(F) + \operatorname{card}(\bar{\underline{\underline{f}}})$. Except in the case when $G$ is an inner square or annulus admissibly parallel to a side of $(F,\bar{\underline{\underline{f}}})$.

An essential surface $F$ in $(M,\underline{\underline{m}})$, $F \cap \partial M = \partial F$, will be called good, if, for every free side $C$ of $(M,\underline{\underline{m}})$, the following holds:

> If $C_1$ is any inner square or annulus in $C$ such that $(\partial C_1 - \partial C)^-$ is a side of $F$, then the opposite side of $C_1$ does not lie in a bound side of $(M,\underline{\underline{m}})$ which itself is a square or annulus.

Of course this condition is empty if the boundary-pattern of $M$ contains no squares or annuli, e.g. if the boundary-pattern is empty.

For the following proposition let $(M,\underline{\underline{m}})$ be an irreducible 3-manifold whose completed boundary-pattern is useful. Furthermore let $(F,\underline{\underline{f}})$ be a connected, essential surface in $(M,\underline{\underline{m}})$, $F \cap \partial M = \partial F$.

19.1 Proposition. Suppose that the boundary-pattern $\underline{\underline{m}}$ of $M$ is complete. Let $(G,\underline{\underline{g}})$ be an essential surface in $(M,\underline{\underline{m}})$, $G \cap \partial M = \partial G$, which can be admissibly deformed into $F$.

Then $G$ can be admissibly isotoped in $(M,\underline{\underline{m}})$ into $F$.

In addition: If $F \cap G = \emptyset$, then $G$ is admissibly parallel to $F$ in $(M,\underline{\underline{m}})$.

<u>Remark</u>. This is one formulation of 5.5 of [Wa 4] in our language. In particular, 19.1 is equivalent to that result if $\underline{\underline{m}}$ consists of the boundary components of M. The proof is inspired by that of 5.5 of [Wa 4].

<u>Proof</u>. Without loss of generality, G is admissibly isotoped so that G is transversal with respect to F. Now consider an admissible homotopy $\varphi: G \times I \to M$ which pulls G into F, i.e. an admissible homotopy with $\varphi | G \times 0 = \mathrm{id} | G \times 0$ and $\varphi(G \times 1) \subset F$. Observe that the boundary-pattern $\underline{\underline{g}}$ of G induces a canonic boundary-pattern of $G \times I$ which makes it into a product I-bundle. Denote by $r_i \times I$ all the sides of $G \times I$ which are not lids.

Applying the surgery arguments used in the proof of 4.4, we see that $\varphi$ can be admissibly deformed (rel $G \times \partial I$) so that afterwards

(1) the preimage $\varphi^{-1}F$ consists of admissible, connected surfaces in $G \times I$ which are incompressible.

Let $F_1$ be any component of $\varphi^{-1}F$ which is different from $G \times 1$. Since $F_1 \cap G \times 1 = \emptyset$, it follows from 3.2 of [Wa 4] that $F_1$ is parallel to a surface $F_1'$ in $G \times 0 \cup \partial G \times I$. In particular, $F_1 \cup F_1'$ bounds a submanifold $N_1$ and we suppose that $F_1$ is chosen so that $N_1 \cap \varphi^{-1}F = F_1$. Furthermore, observe that every arc k' in $F_1'$ with $k' \cap \partial F_1' = \partial k'$ bounds, together with some arc k in $F_1$, a disc D in $N_1$ with $D \cap \varphi^{-1}F = k$. If, in addition, k' is entirely contained in some $r_i \times I$, we do not find any obstruction to deform $\varphi | D$ admissibly (rel k) into F. This is true, because F is essential, M is aspherical, and $\underline{\underline{m}}$ is useful. Hence as usual $\varphi$ can be admissibly deformed (rel $G \times \partial I$) into $\varphi'$ so that

$$(\varphi')^{-1}F = (\varphi^{-1}F - U(D))^{-} \cup D_1 \cup D_2,$$

where U(D) denotes a regular neighborhood of D in $N_1$ and where $D_1$, $D_2$ are the components of $(\partial U(D) - \partial N_1)^{-}$.

Since we can apply the preceding surgery-procedure if necessary, we may suppose that

(2) for each component $F_1''$ of $F_1' \cap (r_i \times I)$, $(\partial F_1'' - \partial(r_i \times I))^-$ is connected.

Now observe that every 1- or 2-faced disc of $\varphi^{-1}F$ separates a ball $E$ from $G \times I$ which meets $(\partial G \times I) \cup G \times 0$ in a disc. If $E \cap G \times 0 = \emptyset$, there is again no obstruction to deform $\varphi|E$ admissibly into $F$, using a homotopy which is constant on $(\partial E - \partial(G \times I))^-$. Hence all 1- or 2-faced discs of $\varphi^{-1}F$, disjoint to $F \times 0$, can be removed from $\varphi^{-1}F$. Thus, altogether, $\varphi$ can be admissibly deformed so that (1) and (2) above hold and that, in addition,

(3) each component of $F_1' \cap (r_i \times I)$ meets $r_i \times 0$.

It is easily checked that then $F_1$ is admissibly parallel to a surface $F_1''$ in $G \times 0$, i.e. $N_1$ minus the interior of a regular neighbohood of $F_1 \cap (G \times 0)$ has the structure of a product I-bundle.

After a small general position deformation of $\varphi$, $\varphi$ has still the above properties and, in addition, $\varphi^{-1}G$ consists of admissible surfaces. By the above argument, we may suppose that $\varphi$ is admissibly deformed, using a homotopy which is constant outside of $N_1$, so that each component $G_1$ of $N_1 \cap \varphi^{-1}G$ is admissibly parallel in $N_1$ to a surface $G_1'$ in $F_1$. In particular, $G_1$ separates a submanifold $W_1$ from $N_1$ with $W_1 \cap F_1 = G_1'$. Let $G_1$ be chosen so that $W_1 \cap \varphi^{-1}G = G_1$. Now, as before, removing the interior of a regular neighborhood of $G_1 \cap F_1$ from $W_1$ we obtain a submanifold $W_1'$ which has the structure of a product I-bundle. Let $U$ be a regular neighborhood of $F \cap G$ in $M$, and let $W_1^*$ be the closure of the component of $M - (F \cup G \cup U)$ which contains $\varphi(W_1'^{0})$. Then $\varphi|W_1': W_1' \to W_1^*$ can be considered as an essential map. Hence, by 5.8, $W_1^*$ has to be a product I-bundle. This proves that either $F \cap G$ can be diminished, using an admissible isotopic deformation of $G$, or that $G$ is admissibly parallel in $(M,\underline{\underline{m}})$ to $F$.

The additional remark follows immediately from the proof.

q.e.d.

In the next proposition we drop the condition that $\underline{\underline{m}}$ is complete, and replace it by appropriate conditions on $F$ and $M$.

More precisely, let $(M,\underline{\underline{m}})$ be a simple 3-manifold. Furthermore let $(F,\underline{\underline{f}})$ be a connected, essential surface in $(M,\underline{\underline{m}})$,

$F \cap \partial M = \partial F$, which is non-separating and which is chosen so that its complexity is minimal. Then the following holds:

*19.2 Proposition. If* $(G,\underline{\underline{g}})$ *is any good surface in* $(M,\underline{\underline{m}})$ *which can be admissibly deformed in* $(M,\underline{\underline{m}})$ *into* $F$, *then it can be admissibly isotoped in* $(M,\bar{\underline{\underline{m}}})$ *into* $F$, *i.e. with respect to the completed boundary-pattern.*

*In addition: If* $F \cap G = \emptyset$, *then* $G$ *is admissibly parallel to* $F$ *in* $(M,\bar{\underline{\underline{m}}})$.

*Proof.* We suppose that $G$ is admissibly isotoped in $(M,\bar{\underline{\underline{m}}})$ so that $G$ is transversal with respect to $F$, and that, in addition, the number of points of $\partial F \cap \partial G$ contained in free sides of $(M,\underline{\underline{m}})$ is as small as possible. Let $\varphi\colon G \times I \to M$ be any admissible homotopy in $(M,\underline{\underline{m}})$ which pulls $G$ into $F$, i.e. $\varphi|G \times 0 = \mathrm{id}$ and $\varphi(G \times 1) \subset F$.

Now let $r$ be any free side of $(G,\underline{\underline{g}})$ and consider $h = \varphi|r \times I$. After a small admissible general position deformation of $h$ which is constant in $r \times \partial I$, $h^{-1}F$ consists of curves. Suppose that this condition holds and that, in addition, the number of these curves is minimal. $F$ is essential, $M$ is aspherical, and $\bar{\underline{\underline{m}}}$ is useful. Hence one easily checks that, by our minimality conditions on $h^{-1}F$ and $\partial F \cap \partial G$, every curve of $h^{-1}F$ is admissibly parallel in $r \times I$ to $r \times 0$. This means that $h^{-1}F$ splits $r \times I$ into a number of squares or annuli, $A_1,\dots,A_n$, and suppose $A_1 \cap r \times 0 \neq \emptyset$.

Let $k$ be the component of $(\partial A_1 - \partial(r \times I))^-$. Then $h|k$ can be admissibly deformed in $(F,\underline{\underline{f}})$ into a free side of $(F,\underline{\underline{f}})$. To see this denote by $(\tilde{M},\tilde{\underline{\underline{m}}})$ the manifold obtained from $(M,\bar{\underline{\underline{m}}})$ by splitting at $F$, and observe that $h|A_1$ can be considered as an admissible singular square or annulus in $(\tilde{M},\tilde{\underline{\underline{m}}})$. Our claim follows easily if $h|A_1$ is inessential, for $\tilde{\underline{\underline{m}}}$ is useful. If on the other hand $h|A_1$ is essential, then, by 12.5, it can be admissibly deformed into the characteristic submanifold of $(\tilde{M},\tilde{\underline{\underline{m}}})$, i.e. either into an essential I-bundle or Seifert fibre space. Now, $h|r \times 0$ is a side of $h|A_1$ which is a non-singular curve. By 5.10 and since $M$ is simple, this implies that $h|A_1$ can be admissibly deformed in $\tilde{M}$ into a *non-singular* square or annulus, say $B$. Now observe that at least one component F* of $(F - U(B)) \cup (\partial U(B) - \partial\tilde{M})^-$ has to be non-separating in $\tilde{M}$ since F is. Here U(B) denotes a regular neighborhood of B in M. Using a similiar procedure, we obtain from F* an <u>essential</u> surface in $(M,\bar{\underline{\underline{m}}})$ which is non-separating and whose complexity is not bigger than that of F*. Thus, by our choice of F, the complexity of F* cannot be strictly smaller than that of F. Hence the curve $B \cap F$ has to be admissibly parallel in $(F,\underline{\underline{f}})$ to a free side of $(F,\underline{\underline{f}})$. This proves our claim.

After pushing $h|k$ in $(F,\underline{\underline{f}})$ into a free side of $(F,\underline{\underline{f}})$, $h|A_1$ can be considered as an admissible square or annulus in $(M,\bar{\underline{\underline{m}}})$. Then $h|A_1$ can be admissibly deformed (rel $r \times 0 \cup k$) in $(M,\underline{\underline{m}})$ into a free side of $(M,\underline{\underline{m}})$. If $h|A_1$ is inessential in $(M,\bar{\underline{\underline{m}}})$, this follows since both sides $h|r \times 0$ and $h|k$ lie in free sides of $(M,\underline{\underline{m}})$ and since $\bar{\underline{\underline{m}}}$ is useful. If $h|A_1$ is essential, the claim follows since $(M,\bar{\underline{\underline{m}}})$ is a simple 3-manifold and since G is a good surface.

Similarly with the surface $A_2$ meeting $A_1$, etc., and so inductively one proves that $h|r \times 1$ can be admissibly deformed in $(F,\underline{\underline{f}})$ into a free side of $(F,\underline{\underline{f}})$ and that afterwards $h|r \times I$ can be admissibly deformed (rel $r \times \partial I$) into a free side of $(M,\underline{\underline{m}})$.

With other words we have seen that $\varphi$ may be chosen to be an admissible homotopy in $(M,\bar{\underline{\underline{m}}})$, i.e. with respect to the completed boundary-pattern of M. Then 19.2 follows from 19.1. q.e.d.

## §20. Splitting a homotopy equivalence at a surface

Throughout this paragraph let $(M_1,\underline{\underline{m}}_1)$ and $(M_2,\underline{\underline{m}}_2)$ be two simple 3-manifolds. Let $F$ be a connected, essential surface in $(M_1,\bar{\underline{\underline{m}}}_1)$, $F \cap \partial M_1 = \partial F$, which is non-separating and whose complexity is minimal. Let $f\colon (M_1,\underline{\underline{m}}_1) \to (M_2,\underline{\underline{m}}_2)$ be an admissible homotopy equivalence, and let $g$ be an admissible homotopy inverse. Finally suppose that first $g$ and then $f$ are admissibly deformed so that $H = g^{-1}F$ and $G = f^{-1}g^{-1}F$ are essential surfaces whose complexities are as small as possible. All this is possible because of 4.4.

Before we come to the splitting result of this paragraph, we first establish the following property of $G$. This property is crucial in the proof of the next proposition, for it makes the results of §19 available for us.

<u>20.1 Lemma</u>. $G$ <u>is a good surface</u>.

<u>Proof</u>. Assume the converse. We are going to show that this assumption leads to contradictions. For convenience, denote by $(\tilde{M}_2,\tilde{\underline{\underline{m}}}_2)$, resp. $(\tilde{M}_1,\tilde{\underline{\underline{m}}}_1)$, the manifolds obtained from $(M_2,\underline{\underline{m}}_2)$, resp. $(M_1,\underline{\underline{m}}_1)$, by splitting at $H$, resp. $F$.

By our assumption, there is a free side $C$ of $(M_1,\underline{\underline{m}}_1)$ containing an inner square or annulus $A$ with the following properties: $\ell = (\partial A - \partial C)^-$ is a side of $G$, and the side $k$ of $A$ opposite to $\ell$ lies in a bound side $B$ of $(M_1,\underline{\underline{m}}_1)$ which is a square or annulus. We only deal with the case that $A$ is an annulus. The proof in the other case is similiar, only technically a bit more involved, and we leave this to the reader.

$f|A$ is an admissible, singular annulus in $(\tilde{M}_2,\tilde{\underline{\underline{m}}}_2)$. If $f|A$ is inessential, it can be admissibly deformed (rel $\ell$) in $(M_2,\underline{\underline{m}}_2)$ into $H$ (note that $\tilde{\underline{\underline{m}}}_2$ is useful). Choosing this deformation carefully, we see that $f$ is admissibly homotopic to a map $f'$ with $(f')^{-1}H = (f^{-1}H - U(A))^- \cup (\partial U(A) - \partial\tilde{M}_1)^-$, where $U(A)$ is a regular neighborhood in $\tilde{M}_1$. But this contradicts our minimality conditions on $f^{-1}H$. So $f|A$ is essential in $(\tilde{M}_2,\tilde{\underline{\underline{m}}}_2)$.

Observe that the annulus $B$ is mapped under $f$ into a

bound side of $\underline{\underline{m}}_2$, which is an annulus since f is an admissible homotopy equivalence. This implies that $f|A$ can be admissibly deformed in $(\tilde{M}_2,\tilde{\underline{\underline{m}}}_2)$ so that afterwards $f|k$ is an embedding. Moreover, it can be admissibly deformed so that afterwards $f|A$ is an embedding. To see this observe that, by 12.5, $f|A$ can be admissibly deformed into the characteristic submanifold of $\tilde{M}_2$ and that $M_2$ is simple. Denote by A' the essential annulus f(A) in $(\tilde{M}_2,\tilde{\underline{\underline{m}}}_2)$.

If $g|A'$ is essential in $(\tilde{M}_1,\tilde{\underline{\underline{m}}}_1)$, then, as above, we may suppose that g is admissibly deformed so that A" = g(A') is an essential (non-singular) annulus in $\tilde{M}_1$. One side of A" lies in the annulus B (g is admissible homotopy inverse of f), and the opposite side of A" lies in F. By our minimality condition on F, this latter side of A" must be admissibly parallel in F to a side of F. This side of F cannot lie in B, for A" is essential in $\tilde{M}_1$ and $M_1$ is simple. Using again the facts that $M_1$ is simple and that the complexity of F is minimal, we find an annulus A* of $\bar{\underline{\underline{m}}}_1$ neighboring B. $(\partial U(B \cup A^*) - \partial M_1)^-$ is again an admissible annulus in $(M_1,\underline{\underline{m}}_1)$, where $U(B \cup A^*)$ is a regular neighborhood in $M_1$. The existence of this annulus implies that $M_1$ is a solid torus. Then, by our minimality condition on F, F has to be a disc. But G can be admissibly deformed into F since g is an admissible homotopy inverse of f. So each component of G has to be a disc, too. So, by our choice of A, we find that the core of B is the side of an admissible 1-faced disc in $(M_1,\underline{\underline{m}}_1)$. But this is impossible since $\underline{\underline{m}}_1$ is useful.

If finally $g|A'$ is inessential, then g is admissibly homotopic to a map g' with $(g')^{-1}F = (g^{-1}F - U(A'))^- \cup (\partial U(A') - \partial\tilde{M}_2)^-$. By our minimality condition on $g^{-1}F$, we see, by the previous argument, that $M_2$ must be a solid torus. So $M_1$ must be a solid torus, too, since f is a homotopy equivalence. Then we obtain the required contradiction as before. q.e.d.

Given an admissible homotopy equivalence $f: M_1 \to M_2$ between simple 3-manifolds, the following result shows that there is at least one essential, connected, and non-separating surface H in $M_2$ with the property that f is admissibly homotopic to f' such that $(f')^{-1}H$

is an essential and connected surface.

<u>20.2 Proposition.</u> <u>Let</u> F, f, <u>and</u> g <u>be given as in the beginning of this paragraph. Then the following holds:</u>

$G = f^{-1}(g^{-1}F)$ <u>is admissibly isotopic in</u> $(M_1,\bar{\underline{\underline{m}}}_1)$ <u>to</u> F,

<u>and</u> $H = g^{-1}F$ <u>is connected.</u>

<u>Proof.</u> As in 20.1 denote by $(\tilde{M}_2,\tilde{\underline{\underline{m}}}_2)$, resp. $(\tilde{M}_1,\tilde{\underline{\underline{m}}}_1)$, the manifolds obtained by splitting at H, resp. F. Now, for the following observe that any component of G can be admissibly deformed in $(M_1,\underline{\underline{m}}_1)$ into F since g is an admissible homotopy inverse of f.

No component of G is an essential square or annulus in $(M_1,\bar{\underline{\underline{m}}}_1)$. For each such component $G_1$ of G had to be admissibly parallel to a side of $(M_1,\bar{\underline{\underline{m}}}_1)$ since $M_1$ is simple. Therefore all sides of $G_1$ had to be bound since, by 20.1, G is good. Hence it followed from 19.1 that F is admissibly isotopic to $G_1$, and so admissibly parallel to some side of $(M_1,\bar{\underline{\underline{m}}}_1)$. But this is impossible since F is non-separating.

If G is connected, it follows from 20.1 and 19.2 the existence of an admissible ambient isotopy in $(M_1,\bar{\underline{\underline{m}}}_1)$ which moves G to F, i.e., without loss of generality, $f^{-1}g^{-1}F = F$. In this case, H has to be connected. For otherwise there is a component $H_1$ of H with $f(M_1) \cap H_1 = \emptyset$. By 16.1, this means that $H_1$ is admissibly parallel in $(M_2,\bar{\underline{\underline{m}}}_2)$ to a free side of $(M_2,\underline{\underline{m}}_2)$, and so it follows that g is admissibly homotopic to a map g' with $(g')^{-1}F = g^{-1}F - H_1$. But this contradicts our minimality conditions on $g^{-1}F$.

Thus to prove 20.2 it remains to show that G is connected. Applying 20.1 and 19.2 twice, it follows that any two components of G have to be admissibly parallel in $(M_1,\bar{\underline{\underline{m}}}_1)$. This means that there is a connected surface $(G_1,\underline{\underline{g}}_1)$ with complete boundary-pattern and an admissible embedding of $(G_1 \times I, \underline{\underline{g}}_1 \times I)$ into $(M_1,\bar{\underline{\underline{m}}}_1)$ such that $(G_1 \times I) \cap G = G_1 \times \partial I$. Fix an arc k in $G_1 \times I$ joining $G_1 \times 0$ with $G_1 \times 1$, and recall that we may assume $G_1 = F$.

We claim that $gf|k$ is essential in $(\tilde{M}_1,\tilde{\underline{\underline{m}}}_1)$. First of all $f|k$ is essential in $(\tilde{M}_2,\tilde{\underline{\underline{m}}}_2)$. For otherwise there is no obstruction to deform $f|G_1 \times I$ admissibly (rel $G_1 \times \partial I$) into $H$. Choosing this deformation carefully we see that $f$ is admissbly homotopic to a map $f'$ with $(f')^{-1}H = f^{-1}H - (G_1 \cup G_2)$, which contradicts our minimality condition on $f^{-1}H$.

Thus $f|G_1 \times I$ can be considered as an essential, singular I-bundle in $(\tilde{M}_2, \tilde{\underline{\underline{m}}}_2)$. By 13.1, $f|G_1 \times I$ can be admissibly deformed in $(\tilde{M}_2, \tilde{\underline{\underline{m}}}_2)$ into the characteristic submanifold of $(\tilde{M}_2, \tilde{\underline{\underline{m}}}_2)$. More precisely, into an essential I-bundle $Z$ whose lids are contained in copies of $H$ (no component of $G$ is a square or annulus). $k' = f|k$ is an essential arc in $Z$ whose end-points lie in the lids of $Z$. This implies that $gf|k$ is essential in $(\tilde{M}_1, \tilde{\underline{\underline{m}}}_1)$. For if $Z$ is a product I-bundle which meets both the components of $(\partial U(H_1) - \partial M)^-$, of some component $H_1$ of $H$, we find an arc, $\ell'$, in $U(H_1)$ which joins the two end-points of $k'$. Then, without loss of generality, $g|k' \cup \ell'$ meets $F$ in just one point, and our claim follows immediately. If $Z$ is not as above and $gf|k$ is inessential, there is no obstruction to deform $g|Z$ admissibly (rel lids) into $F$. Choosing this deformation carefully, we see that $g$ is admissibly homotopic to a map $g'$ with $(g')^{-1}F = (g^{-1}F - Z)^- \cup (\partial Z - \partial\tilde{M}_2)^-$, and we get a contradiction to our minimality condition on $g^{-1}F$. Thus, in any case, our claim is established.

In particular, $gf|G_1 \times I$ can be considered as an essential singular I-bundle in $(\tilde{M}_1, \tilde{\underline{\underline{m}}}_1)$. $gf|G_1 \times 0$ is admissibly homotopic to $id|G_1 \times 0$. Either this homotopy can be admissibly deformed (rel $G_1 \times \partial I$) into $G_1$, or we consider this homotopy instead of $gf|G_1 \times I$. In any case we may suppose that $gf|G_1 \times 0 = id|G_1 \times 0$. Now, by 13.1, $gf|G_1 \times I$ can be admissibly deformed in $(\tilde{M}_1, \tilde{\underline{\underline{m}}}_1)$ into an essential I-bundle $Z'$ whose one lid is equal to a copy of $F$. Hence $Z'$ cannot be a twisted I-bundle, for $F$ is non-separating. So $Z'$ is a product I-bundle and, without loss of generality, the lids of $Z'$ are equal to the two components of $(\partial U(F) - \partial M_1)^-$. In particular, any free side of $Z'$ is an admissible square in $(M, \bar{\underline{\underline{m}}})$. $(M, \bar{\underline{\underline{m}}})$ is simple. Hence we may suppose that $\tilde{M}_1$ is equal to $Z'$.

By what we have proved so far $G$, splits $(M_1, \bar{\underline{\underline{m}}}_1)$ into a system of product I-bundles, if it is disconnected. Furthermore no such product I-bundle can contain an arc k such that k joins the two lids and that gf|k is inessential in $(\tilde{M}_1, \tilde{\underline{m}}_1)$. But this contradicts the well-known Stallings-trick (see [St 3] [La 1, p.22]) which constructs just such an arc. q.e.d.

## §21. Splitting a homotopy at a surface

In this paragraph we show that a homotopy equivalence between simple 3-manifolds can be split along a surface in such a way that it stays a homotopy equivalence between the split 3-manifolds. For this we still have to show how one splits homotopies along surfaces.

Let $F$ be an admissible, connected surface in a 3-manifold $(M,\underline{\underline{m}})$, $F \cap \partial M = \partial F$. Then $U(F)$ denotes a regular neighborhood of $F$ in $(M,\bar{\underline{\underline{m}}})$. If $h\colon (M,\underline{\underline{m}}) \to (M,\underline{\underline{m}})$ is an admissible map with $h^{-1}F = F$, it is to be understood that $h^{-1}(\partial U(F) - \partial M)^{-} = (\partial U(F) - \partial M)^{-}$.

**21.1 Lemma.** *Suppose $M$ is irreducible and $F$ is essential and not boundary-parallel. Let $h\colon (M,\underline{\underline{m}}) \to (M,\underline{\underline{m}})$ be an admissible map with $h^{-1}F = F$.*

*If $h$ is admissibly homotopic to the identity, then $h$ does not interchange the components of $(\partial U(F) - \partial M)^{-}$.*

*Proof.* Checking the proof of 7.4 of [Wa 4], we see that we still have to show that $\pi_1 M$ is isomorphic to $A \underset{C}{*} B$ in a non-trivial way, if $F$ is separating. To see the latter, let $M_1$, $M_2$ be the components obtained from $M$ by splitting at $F$. Let $F_i$ be the copy of $F$ contained in $\partial M_i$, and let $U_i$ be a regular neighborhood of $\partial F_i$ in $(\partial M_i - F_i)^{-}$. The inclusion $F_i \subset M_i$, $i = 1,2$, induces an injection $\varphi\colon \pi_1 F_i \to \pi_1 M_i$. But $\varphi$ cannot be a surjection. For otherwise there is no obstruction to construct a retraction $r_i\colon M_i \to F_i$ with $f_i^{-1}\partial F_i = U_i$ ($F$ and $M$ are aspherical) and then, using the arguments of 8-12 of [St 2], it is possible to define a homeomorphism $h\colon M \to F_i \times I$ with $h^{-1}(F_i \times \partial I) = (\partial M_i - U_i)^{-}$. This in turn contradicts the fact that $F$ is not boundary-parallel. Thus $\pi_1 M \cong \pi_1 M_1 \underset{\pi_1 F}{*} \pi_1 M_2$ is a free product with a non-trivial amalgam.

q.e.d.

For the following proposition suppose that $(M,\underline{\underline{m}})$ is a simple 3-manifold, and let $(F,\underline{\underline{f}})$ be an essential, connected surface in $(M,\bar{\underline{\underline{m}}})$, $F \cap \partial M = \partial F$, which is non-separating.

21.2 Proposition. *Let* $h\colon (M,\underline{\underline{m}}) \to (M,\underline{\underline{m}})$ *be an admissible map with* $h^{-1}F = F$, *which is homotopic to the identity, using an admissible homotopy* $H\colon M \times I \to M$ *in* $(M,\underline{\underline{m}})$.

*Then there is an admissible ambient isotopy* $\alpha_t$, $t \in I$, *of* $(M,\underline{\underline{m}})$ *and an admissible homotopy* $H'\colon M \times I \to M$ *in* $(M,\underline{\underline{m}})$ *such that the following holds*:

(1) $H'$ *is homotopic to* $H$, *using an admissible homotopy which is constant on* $M \times 1$, *and* $H'|M \times 0 = h \circ \alpha_1$.

(2) $(H')^{-1}F = F \times I$.

Proof. Let $(\tilde{M},\tilde{\underline{\underline{m}}})$ be the manifold obtained from $(M,\underline{\underline{m}})$ by splitting at $F$, and denote by $F_1$, $F_2$ the two sides of $(\tilde{M},\tilde{\underline{\underline{m}}})$ which are copies of $F$. Consider $\varphi = H|F \times I$. Using the surgery arguments of 4.4, we see that $\varphi$ can be admissibly deformed (rel $F \times \partial I$) so that afterwards $\varphi^{-1}F$ is an essential surface in the product I-bundle $F \times I$. By 5.6, $\varphi^{-1}F$ is horizontal, i.e. each component of $\varphi^{-1}F$ is admissibly parallel in $F \times I$ to $F \times 0$. Hence either one of the following holds:

(i) $\varphi(F \times I) \subset \tilde{M}$ and $\varphi\colon F \times I \to \tilde{M}$ is inessential.

(ii) $\varphi(F \times I) \subset \tilde{M}$ and $\varphi\colon F \times I \to \tilde{M}$ is essential.

(iii) there is an admissible homotopy in $(\tilde{M},\tilde{\underline{\underline{m}}})$ which pushes a surface near $F_1$ into a surface near $F_2$.

In case (i) there is no obstruction to deform $\varphi$ admissibly (rel $F \times \partial I$) into $F$. In the other two cases, it follows from 13.1 the existence of an essential I-bundle $Z$ in $\tilde{M}$ with $Z \cap F_1 = F_1$. $Z$ cannot be a twisted I-bundle, for otherwise $(\partial Z - \partial\tilde{M})^-$ consists of admissible squares or annuli in $M$, and this contradicts either the fact that $M$ is simple or that $F$ is non-separating. Thus $Z$ has to be a product I-bundle, and the other lid must lie in $F_2$. More precisely, $Z \cap F_2 = F_2$ since $F_2$ is admissibly homeomorphic to $F_1$. Hence $(\partial Z - \partial M)^-$ consists of admissible squares or annuli in $M$, and since $M$ is simple, it is easily checked that $(\tilde{M},\tilde{\underline{\underline{m}}})$ is a product I-bundle. So $(M,\underline{\underline{m}})$ is a Stallings fibration. This means that we may slide $F$ around $M$.

By what we have seen so far, we find an admissible ambient isotopy $\alpha_t$, $t \in I$, of $M$ (from sliding $F$ around) and an admissible homotopy $H'$ of $h$ which satisfy all the properties of 21.2 except possibly (2), but with $H'(F \times I) \subset F$.

Then we may also suppose that $H'(U(F) \times I) \subset U(F)$. Since, by 21.1, $h = H|M \times 0$ does not interchange the components of $(\partial U(F) - \partial M)^-$, we may even suppose that $H'((\partial U(F) - \partial M)^- \times I) \subset (\partial U(F) - \partial M)^-$. For the final proof of our proposition, it remains to show that $H'$ can be admissibly deformed (rel $U(F) \times I$) so that afterwards $H'((M - U(F))^- \times I) \subset (M - U(F))^-$. To see this choose a triangulation $\Delta$ of $M$ so that the sides of $M$ and $U(F)$ are subcomplexes. Let $z$ be any 0-simplex of $\Delta$ which does not lie in $U(F)$. Then we join $z$ in $(M - U(F))^-$ by an arc $k$ with $(\partial U(F) - \partial M)^-$. Since $H'$ maps $(\partial U(F) - \partial M)^- \times I$ into $(\partial U(F) - \partial M)^-$, we see that $H'|k \times I$ is a map with $H'(\partial(k \times I) - (z \times I))^- \subset (M - U(F))^-$. $F$ is essential, $\underline{\underline{m}}$ is useful, and $M$ is aspherical. Hence it is easily checked that $H'|z \times I$ can be admissibly deformed (rel $z \times \partial I$) into $(M - U(F))^-$. The rest follows in a similar way, using induction on the skeletons of the triangulation $\Delta$. q.e.d.

Combining 20.2 with 21.2 we obtain our <u>second splitting theorem</u>.

To formulate it, let $(M_1, \underline{\underline{m}}_1)$ and $(M_2, \underline{\underline{m}}_2)$ be two simple 3-manifolds. Let $F_1$ be any connected, essential surface in $(M_1, \bar{\underline{\underline{m}}}_1)$, $F_1 \cap \partial M_1 = \partial F_1$, which is non-separating and whose complexity (in the sense of §19) is minimal. Furthermore, let $f: (M_1, \underline{\underline{m}}_1) \to (M_2, \underline{\underline{m}}_2)$ be any admissible homotopy equivalence, and let $g$ be any admissible homotopy inverse of $f$. Then the following holds:

<u>21.3 Theorem</u>. <u>There exists a connected, essential surface</u> $F_2$ <u>in</u> $(M_2, \underline{\underline{m}}_2)$, $F_2 \cap \partial M_2 = \partial F_2$, <u>which is non-separating, and there are admissible deformations which pull</u> $f$ <u>and</u> $g$ <u>so that afterwards</u> $f|\tilde{M}_1: (\tilde{M}_1, \tilde{\underline{\underline{m}}}_1) \to (\tilde{M}_2, \tilde{\underline{\underline{m}}}_2)$ <u>is an admissible homotopy equivalence with admissible homotopy inverse</u> $g|\tilde{M}_2$.
(Here $\tilde{M}_i$, $i = 1,2$, denotes the manifold obtained from $M_i$ by splitting at $F_i$).

<u>Proof.</u> We find the surface $F_2$, using 20.2. This surface $F_2$ has to

be non-separating, too. To see this recall that $F_1$ is supposed to be non-separating, i.e. there is a closed curve $k$ in $M_1$ which meets $F_1$ in just one point, and $f|k$ has the same property with respect to $F_2$. By our choice of $F_2$, $f$ and $g$ can be admissibly deformed so that $f|\tilde{M}_1 : \tilde{M}_1 \to \tilde{M}_2$ and $g|\tilde{M}_2 : \tilde{M}_2 \to \tilde{M}_1$ are admissible maps. Hence and since $f \cdot g$ and $g \circ f$ are both admissibly homotopic to the identity, 21.3 finally follows from 21.2. q.e.d.

## Part IV. THE CONCLUSION OF THE PROOF OF THE CLASSIFICATION THEOREM

### Chapter VIII: Attaching homotopy equivalences.

In this chapter we shall show how an "induction on a great hierarchy" can be utilized to prove our main result (see 24.2). The next paragraph represents the beginning of this induction, which will be completed in §23. Strictly speaking, §22 can be considered as a key--at least from the combinatorial point of view--to our main theorem.

#### §22. The induction beginning

We here study purely the influence of boundary-phenomena to homotopy equivalences--the underlying manifolds being trivial. The result can be formulated as follows:

*22.1 Proposition. Let* $(M_1,\underline{\underline{m}}_1)$ *and* $(M_2,\underline{\underline{m}}_2)$ *be two 3-balls whose completed boundary-patterns are useful and consist of at least four sides. Suppose* $(M_1,\bar{\underline{\underline{m}}}_1)$ *is a simple 3-manifold.*

*Then every admissible homotopy equivalence* $f\colon (M_1,\underline{\underline{m}}_1) \to (M_2,\underline{\underline{m}}_2)$ *is admissibly homotopic to a homeomorphism.*

*Proof.* Since $(M_1,\bar{\underline{\underline{m}}}_1)$ is simple, it follows that no two different free sides of $(M_1,\underline{\underline{m}}_1)$ are admissibly homotopic. In particular, the restrictions of $f$ to different free sides cannot be admissibly deformed into the same free side of $(M_2,\underline{\underline{m}}_2)$. Thus it remains to show that the restriction of $f$ to any free side of $(M_1,\underline{\underline{m}}_1)$ can be admissibly deformed into a free side of $(M_2,\underline{\underline{m}}_2)$ (see 3.4).

We assume the converse, and show that this assumption leads to contradictions. For this it is convenient to use an alternative description of $(M_j,\bar{\underline{\underline{m}}}_j)$, namely one as a simplicial complex. Observe that the boundary-patterns $\underline{\underline{m}}_j$ and $\bar{\underline{\underline{m}}}_j$, $j = 1,2$, induce canonical cell complexes $C_j$ and $\bar{C}_j$: the 2-cells of these are the bound sides of

$(M_j, \underline{\underline{m}}_j)$ resp. $(M_j, \bar{\underline{\underline{m}}}_j)$, the 1-cells are the bound sides of the 2-cells resp., and the 0-cells are the bound sides of the 1-cells resp. Define finally $K_j$ and $\bar{K}_j$ to be the dual complexes of $C_j$ and $\bar{C}_j$ resp. It turns out that $\bar{K}_j$ is a triangulation of the 2-sphere $\partial M_j$ ($\bar{m}_j$ is useful), and so the terms "star" and "link" make sense. By an abuse of language we call the simplices of $K_j$ "bound" and those of $\bar{K}_j - K_j$ "free".

Now observe that the homotopy equivalence $f$ induces a simplicial isomorphism $\varphi: K_1 \to K_2$ in an obvious way. Furthermore, it follows from our assumption on $f$ that there is at least one free 0-simplex $x_1$ such that

(*) $\varphi(\mathrm{link}(x_1, \bar{K}_1)) \neq \mathrm{link}(y, \bar{K}_2)$, for all free 0-simplices $y$ in $\bar{K}_2$.

Since $\varphi$ is a simplicial isomorphism, $\varphi(\mathrm{link}(x_1, \bar{K}_1))$ is a simple closed curve which is simplicial in $K_2$. By the Jourdan courve theorem, this curve splits the 2-sphere $\partial M_2$ into two discs, $D_1'$ and $D_2'$ say. It follows from (*) that each disc $D_j'$, $j = 1,2$, contains at least one bound 1-simplex $t_j' \cap \partial D_j' \neq \emptyset$ and which meets $\partial D_j'$ in points. There is precisely one bound 1-simplex $t_j$ in $\bar{K}_1$ with $\varphi(t_j) = t_j'$, and at least one end-point $z_j$ of $t_j$ lies in $\mathrm{link}(x_1, \bar{K}_1)$. Let $t_1', t_2'$ be chosen so that $z_1$ and $z_2$ lie as near as possible (possibly $z_1 = z_2$) in $\mathrm{link}(x_1, \bar{K}_1)$. Then, by our choice of $t_1'$ and $t_2'$, it follows the existence of at least one free 1-simplex $s$ in $\bar{K}_1$ whose one end-point is either $z_1$ or contained in the shortest arc in $\mathrm{link}(x_1, \bar{K}_1)$ joining $z_1$ with $z_2$. Now observe that, for every free 1-simplex, precisely one end-point is a free 0-simplex. Let $x_2$ be the free end-point of $s$. Then $x_2$ does not lie in $\mathrm{link}(x_1, \bar{K}_1)$, and, without loss of generality, $t_1, t_2 \subset \mathrm{link}(x_2, \bar{K}_1)$. To see the latter observe that the link of a free 0-simplex always consists of bound 1-simplices.

Considering the curve $\varphi(\mathrm{link}(x_2, \bar{K}_1))$ we find that the intersection of $\mathrm{link}(x_1, \bar{K}_1)$ and $\mathrm{link}(x_2, \bar{K}_1)$ is disconnected. Then there is a simple closed and simplicial curve in $\bar{K}_1$ which consists of four 1-simplices (containing $x_1, z_1, x_2, z_2$) and which is neither the boundary of two neighboring 3-simplices, nor the link of a 0-simplex. But it is easily seen that this contradicts the fact that $(M_1, \bar{\underline{\underline{m}}}_1)$ is simple.

q.e.d.

For later use we also consider other small 3-manifolds:

*22.2 Lemma. Let* $(M,\underline{\underline{m}})$ *and* $(M',\underline{\underline{m}})$ *be two irreducible 3-manifolds whose completed boundary-patterns are useful and non-empty. Suppose* $(M,\bar{\underline{\underline{m}}})$ *is the I-bundle over the annulus or torus.*

*Then every admissible homotopy equivalence* $f : (M,\underline{\underline{m}}) \to (M'\underline{\underline{m}}')$ *can be admissibly deformed into a homeomorphism.*

*Proof.* $\pi_1 M' \cong \mathbb{Z}$ or $\mathbb{Z} \oplus \mathbb{Z}$, respectively. Hence it follows by standard arguments (loop-theorem) that M' is either a solid torus or torus × I, respectively. Furthermore, considering the restrictions of f to bound sides, we see that the sides of $(M',\bar{\underline{\underline{m}}}')$ make it to an I-bundle over the annulus or torus, and 22.2 follows immediately (see 3.4).

q.e.d.

## §23. The induction step

In this paragraph we start a study of homotopy equivalences of simple 3-manifolds, which will be continued in §24 and §27. The result of this paragraph is technical and will appear in §24 as the induction step in the proof of the main theorem.

Throughout this paragraph a 3-manifold will always mean an irreducible 3-manifold whose completed boundary-pattern is useful and non-empty. Let $(M,\underline{\underline{m}})$ be such a 3-manifold and specify two sides $F_0$ and $F_1$ of $(M,\bar{\underline{\underline{m}}})$. Although the following definitions are technical, they will play a crucial role in this paragraph and the next.

An essential F-manifold $W$ in $(M,\bar{\underline{\underline{m}}})$ is called a <u>nice submanifold</u> (with respect to $F_0$ and $F_1$) if the following holds:

1. $W$ consists of <u>product</u> I-bundles whose lids are contained in $F_0 \cup F_1$.
2. If $A$ is a component of $(F_i - W)^-$, $i = 1$ and $2$, which is an inner square or annulus in $F_i$, no component of $(\partial A - \partial F_i)^-$ is contained in such a component of $W \cap F_i$.

A 3-manifold $(M,\underline{\underline{m}})$ is called a <u>nice</u> 3-<u>manifold</u> (with respect to $F_0$ and $F_1$) if, for every 3-manifold $(M',\underline{\underline{m}}')$ and every admissible homotopy equivalence $f\colon (M,\underline{\underline{m}}) \to (M',\underline{\underline{m}}')$, there is a nice submanifold $W$ in $(M,\bar{\underline{\underline{m}}})$ (with respect to $F_0$ and $F_1$) and an essential F-manifold $W'$ in $(M',\bar{\underline{\underline{m}}}')$ so that, up to admissible homotopy,

3. $f|W\colon W \to W'$ is an admissible homotopy equivalence, and $f|(M - W)^-\colon (M - W)^- \to (M' - W')^-$ is an admissible homeomorphism,

with respect to the proper boundary-patterns. $W$ is then called a <u>nice</u> <u>submanifold</u> <u>for</u> $f$ (with respect to $F_0$ and $F_1$).

For the following proposition, let $(M,\underline{\underline{m}})$ be a 3-manifold. Furthermore let $F$ be an essential, non-separating surface in $(M,\underline{\underline{m}})$, $F \cap \partial M = \partial F$, whose complexity (in the sense of §19) is minimal. Denote by $(\tilde{M},\tilde{\underline{\underline{m}}})$ the manifold obtained from $(M,\underline{\underline{m}})$ by splitting at $F$, and let $F_0$, $F_1$ be the two sides of $(\tilde{M},\tilde{\underline{\underline{m}}})$ which are copies of $F$.

<u>23.1 Proposition</u>. <u>Suppose that</u> $M$ <u>is</u> <u>a</u> <u>simple</u> 3-<u>manifold</u> <u>and</u> <u>suppose</u> <u>that</u> $\tilde{M}$ <u>is</u> <u>a</u> <u>nice</u> 3-<u>manifold</u>, <u>with</u> <u>respect</u> <u>to</u> $F_0$ <u>and</u> $F_1$. <u>Then</u>, <u>for</u>

<u>every</u> 3-<u>manifold</u> M', <u>every admissible homotopy equivalence</u> $f: (M,\underline{\underline{m}}) \to (M',\underline{\underline{m}}')$ <u>can be admissibly deformed into a homeomorphism</u>.

<u>Remark</u>. In §24 this proposition will be generalized.

<u>Proof</u>. First of all we may draw from 21.3 the existence of an essential surface F' in $(M',\underline{\underline{m}}')$ and an admissible homotopy which pulls f so that afterwards $\tilde{f} = f|\tilde{M}: (\tilde{M},\tilde{\underline{\underline{m}}}) \to (\tilde{M}',\tilde{\underline{\underline{m}}}')$ is an admissible homotopy equivalence, where $(\tilde{M}',\tilde{\underline{\underline{m}}}')$ denotes the manifold obtained from $(M',\underline{\underline{m}}')$ by splitting at F'. Since, by supposition, $\tilde{M}$ is a nice 3-manifold, $\tilde{M}$ contains a nice submanifold W for $\tilde{f}$ with respect to $F_0$ and $F_1$. Let W be chosen to be <u>minimal</u>, in the sense that W can be admissibly contracted in $\tilde{M}$ to any nice submanifold for $\tilde{f}$ contained in W. That such a choice is always possible is a consequence of 9.1.

Now let $F \times I$ be a regular neighborhood of F in M, and let $d: F \times 1 \to F \times 0$ be the admissible homeomorphism defined by reflections in the fibres. Without loss of generality, we may identify $F \times i$ with $F_i$, $i = 0,1$. Denote $G_0 = W \cap F_0$ and $G_1 = d(W \cap F_1)$. Of course we may suppose that W is admissibly isotoped in $\tilde{M}$ so that $G_1$ is in a very good position to $G_0$. Define G to be the essential intersection of $G_0$ and $G_1$. Recall from property 2 of a nice submanifold that the following holds: if A is a component of $(F_i - G_i)^-$, $i = 0$ and 1, which is an inner square or annulus in $F_i$, no component of $(\partial A - \partial F_i)^-$ is contained in such a component of $G_i$. Hence we may suppose that W is admissibly isotoped so that, in addition, the essential intersection G also has this property.

By the very definition, G is contained in $G_0$ as well as in $G_1$. Below we shall prove that furthermore $G_i$ can be admissibly contracted to $G \cap G_i$, for $i = 0$ and 1. This then means that W can be admissibly isotoped in $\tilde{M}$ so that afterwards

$$W \cap F_0 = d(W \cap F_1).$$

Then, by our choice of $W$, $W^+ = W \cup (G_1 \times I)$ is a Stallings fibration and $(\partial W^+ - \partial M)^-$ consists of essential annuli or tori in $(M,\bar{\underline{\underline{m}}})$. Since $M$ is a simple 3-manifold, it follows from 10.7 that either $W^+$, and so $W$, is contained in a regular neighborhood of a side of $M$, resp. $\tilde{M}$, or that $M$ itself is a Stallings fibration. In the first case, 23.1 follows in a straightforward manner from the properties of $W$ (apply 3.4 to the restriction of $f$ to $F \times I$). In the second case observe that $\bar{\underline{\underline{m}}}$ consists of annuli and tori. In fact $\bar{\underline{\underline{m}}}$ must consist of tori since $M$ is a simple 3-manifold. Then 23.1 follows from 14.2 and 14.6.

Thus it remains to prove that $G_i$, $i = 0,1$ can be admissibly contracted to $G \cap G_i$. The proof of this is based on a 2-dimensional result. This result in turn will be established in the appendix. Indeed, the following property of homotopy equivalences between surfaces is an easy consequence of 31.3.

23.2. Let $(F,\bar{\underline{\underline{f}}})$ and $(F',\bar{\underline{\underline{f}}}')$ be two surfaces, different from squares, annuli, and tori, and let $G_0$, $G_2$ resp. $G_0'$, $G_2'$ be essential subsurfaces in $(F,\underline{\underline{f}})$ resp. $(F',\underline{\underline{f}}')$. Suppose that $G_0$, $G_2$ and $G_0'$, $G_2'$ are in a very good position, and define $G$ resp. $G'$ to be the essential intersections of $G_0$, $G_2$ resp. $G_0'$, $G_2'$. Furthermore let $f_0$, $f_2$: $F \to F'$ be two admissible homotopy equivalences such that, for $i = 0$ and $2$,

> $f_i | G_i$: $G_i \to G_i'$ is an admissible homotopy equivalence, and
> $f_i | (F - G_i)^-$: $(F - G_i)^- \to (F' - G_i')^-$ is an admissible homeomorphism,

with respect to the proper boundary-patterns. Then the following holds: If $f_0$ is admissibly homotopic to $f_1$, there is an admissible homotopy $f_t$, $t \in [0,1]$, with $f_t^{-1}(\partial G_0' - \partial F') = (\partial G_0 - \partial F)^-$ so that

> $f_1 | G$: $G \to G'$ is an admissible homotopy equivalence, and
> $f_1 | (F - G)^-$: $(F - G)^- \to (F' - G')^-$ is an admissible homeomorphism,

<u>with respect to the proper boundary-pattern</u>.

To continue the proof let $C$ be any component of $G_0 \cup G_1$. Let $X$ be the component of $W$ containing $C$. Then, by our choice of $W$, $X$ is a <u>product</u> I-bundle with $C$ as one lid. Let $X^*$ be the system of vertical I-bundles in $X$ which meets $C$ in $G \cap C$. By our supposition on $G$, $W^* = (W - X) \cup X^*$ is a nice submanifold, and it furthermore easily follows from 23.2 that $W^*$ is a nice submanifold for $\tilde{f}$. Now by our minimality condition on $W$, $W$ can be admissibly contracted to $W^*$, and this means in particular that $C$ can be admissibly contracted to $C \cap G$. q.e.d.

## §24. The classification theorem

The object of this paragraph is to give the final step in the proof of our main theorem. Later on we shall use this result to study some other aspects of homotopy equivalences (see next chapter).

The results of the preceding paragraph suggest to consider first essential F-manifolds in split 3-manifolds. This will be done in the next proposition. We then finally show how this can be linked with §23 in order to prove the main theorem.

To state the next proposition let $(M,\bar{\underline{\underline{m}}})$ be a simple 3-manifold, let $(F,\underline{\underline{f}})$ be given as in 23.1, and define also $(\tilde{M},\tilde{\underline{\underline{m}}})$ as there. Finally denote by $G_1$, $G_2$ the two sides of $(\tilde{M},\tilde{\underline{\underline{m}}})$ which are copies of $F$.

*24.1 Proposition. Let W be an essential F-manifold in $(\tilde{M},\tilde{\underline{\underline{m}}})$, and let $(X,\underline{\underline{x}})$ be a component of W. Then $(X,\underline{\underline{x}})$ admits an admissible fibration so that one of the following holds:*

1. *X is a solid torus, and $(X,\underline{\underline{x}})$ is a Seifert fibre space.*
2. *$(X,\underline{\underline{x}})$ is an I-bundle over the disc and no lid of $(X,\underline{\underline{x}})$ lies in $G_1 \cup G_2$.*
3. *$(X,\underline{\underline{x}})$ is a product I-bundle and the two lids of $(X,\underline{\underline{x}})$ are contained in $G_1 \cup G_2$.*

*Remark.* Recall from the definition that the characteristic submanifold of a simple 3-manifold $(M,\underline{\underline{m}})$ is trivial. This, however changes if we split $(M,\underline{\underline{m}})$ at a surface, i.e. a simple 3-manifold in general does not stay simple after splitting at a surface. The proposition above describes how the characteristic submanifold must look like for a 3-manifold $(\tilde{M},\tilde{\underline{\underline{m}}})$ split at an essential, non-separating surface whose complexity is minimal. (It is not known whether in a simple 3-manifold the surface can always be chosen so that the split 3-manifold is simple.)

*Proof of 24.1.* Observe that an essential square or annulus A in $\tilde{M}$ separates from $\tilde{M}$ an I- or $S^1$-bundle over a j-faced disc,

$1 \leq j \leq 5$, provided precisely one side of $A$ lies in $G_1 \cup G_2$. To see this denote by $U(A)$ a regular neighborhood of $A$ in $\tilde{M}$, and consider the surface

$$F^* = (F - U(A))^- \cup (\partial U(A) - \partial\tilde{M})^- .$$

At least one component of $F^*$ has to be non-separating since $F$ is. Using a similar procedure, we obtain from $F^*$ an essential surface in $(M,\bar{\underline{\underline{m}}})$ which is non-separating and whose complexity is not bigger than that of $F^*$. Thus, by our choice of $F$, the complexity of $F^*$ cannot be strictly smaller than that of $F$. Hence the curve $A \cap G_1$, say, separates an inner square or annulus $B$ from $G_1$, and $A \cup B$ is an admissible square or annulus in $M$. Since $M$ is simple, our claim follows immediately.

If $X$ is a Seifert fibre space in $(\tilde{M},\tilde{\underline{\underline{m}}})$, every vertical essential torus and annulus in $X$ which does not meet $G_1 \cup G_2$ is admissibly parallel to a side of $(M,\bar{\underline{\underline{m}}})$. This, together with the above observation, implies that $X$ must be a solid torus. In the same way 24.1 follows, if $X$ is an I-bundle which does not meet $G_1 \cup G_2$ in lids.

So we assume that $X$ is an I-bundle which meets $G_1 \cup G_2$ in precisely one lid (without loss of generality it meets $G_1$). We still have to show that $X$ is the I-bundle over the square, annulus, or Möbius band, i.e. that $X$ meets $G_1$ in an inner square or annulus.

<u>Case 1</u>. $X$ <u>is a product</u> I-<u>bundle</u>.

Since $F$ is non-separating, there is at least one component of $(\partial X - \partial\tilde{M})^-$. Let $A_1,\dots,A_n$, $n \geq 1$, be all of them. Then $A_i$, $1 \leq i \leq n$, is an essential square or annulus in $(\tilde{M},\tilde{\underline{\underline{m}}})$. Since we are in Case 1 it follows from our assumptions on $X$, that $A_i$ meets $G_1 \cup G_2$ in precisely one side. By the above observation, each $A_i$ separates a submanifold $Y_i$ from $\tilde{M}$ which is the I- or $S^1$-bundle over a j-faced disc, $1 \leq j \leq 5$. Since $F$ is non-separating, at least one $Y_i$ must contain $X$, and 24.1 follows immediately.

<u>Case 2</u>. X <u>is a twisted</u> I-<u>bundle</u>.

Define

$$F_1 = G_1,\ \hat{F}_1 = (F_1 - X)^-,\ H_1 = F_1 \cap X$$

$$F_2 = (G_1 - X)^- \cup (\partial X - \partial\tilde{M})^-,\ \hat{F}_2 = (F_2 - X)^-,\ H_2 = F_2 \cap X.$$

That $H_1$ is either a square or an annulus as required, will follow by a comparison of the complexities of $F_1$ and $F_2$. Indeed, both $F_1$ and $F_2$ are non-separating, and so, by our choice of $F$, we must have

$$10\ \beta_1(F_1) + \text{card}\ \underline{\underline{f}}_1 \leq 10\ \beta_1(F_2) + \text{card}\ \underline{\underline{f}}_2.$$

(Here $\underline{\underline{f}}_1$, $\underline{\underline{f}}_2$ denote boundary-patterns of $F_1$, resp. $F_2$, induced by $\underline{\underline{m}}$.) In the following we may forget the embedding types of $F_1$ and $F_2$, i.e. we may simply consider $F_1$ and $F_2$ as surfaces. $H_1$ and $H_2$ are essential subsurfaces in $F_1$ resp. $F_2$, and, moreover, $H_2$ is a system of inner squares or annuli in $F_2$. An immediate consequence of this is that card $\underline{\underline{f}}_1 \geq$ card $\underline{\underline{f}}_2$. Thus $\beta_1(F_1) \leq \beta_1(F_2)$, and so $\chi F_1 \geq \chi F_2$ (Euler characteristics). Observe that

$$\chi H_1 \leq \chi H_2$$

and that the equality holds if $H_1$ is either a square or an annulus. To see this recall that $H_1$ is connected, i.e. $\chi H_1 \leq 1$, that the Euler characteristic of each disc is +1, and that each pair of arcs of $H_1 \cap H_2$ corresponds to one component of $H_2$. Now using a well known formulae for the Euler characteristic, we may compute:

$$\chi H_1 = \chi F_1 - \chi\hat{F}_1 + \chi(\hat{F}_1 \cap H_1) \geq \chi F_2 - \chi\hat{F}_2 + \chi(\hat{F}_2 \cap H_2) = \chi H_2.$$

Thus $\chi H_1 = \chi H_2$, and, by what we have seen above, this implies that $H_1$ is either an inner square or an inner annulus. q.e.d.

We now come to our main theorem. In the remainder of this

paragraph we always mean by a 3-manifold, an irreducible 3-manifold whose completed boundary-pattern is useful and non-empty. If $(M,\underline{\underline{m}})$ is any such 3-manifold, we denote by $\bar{V}$ the characteristic submanifold of $(M,\bar{\underline{\underline{m}}})$. With these notations in mind, 18.3 can be refined as follows:

*24.2 Classification theorem. Let* $(M,\underline{\underline{m}})$ *be a 3-manifold. Then, for every 3-manifold* $(M',\underline{\underline{m}}')$, *every admissible homotopy equivalence* $f\colon (M,\underline{\underline{m}}) \to (M',\underline{\underline{m}}')$ *can be admissibly deformed so that afterwards:*

> $f|\bar{V}\colon (\bar{V},\underline{\underline{v}}) \to (\bar{V}',\underline{\underline{v}}')$ *is an admissible homotopy equivalence,*
>
> *and* $f|(M-\bar{V})^-\colon (M-\bar{V},\underline{\underline{w}}) \to ((M'-\bar{V}')^-,\underline{\underline{w}}')$ *is an admissible homeomorphism,*

*with respect to the proper boundary-patterns,* $\underline{\underline{v}}$, $\underline{\underline{v}}'$, $\underline{\underline{w}}$, $\underline{\underline{w}}'$.

*24.3.* The proof of this theorem is by induction on a great hierarchy of $(M,\underline{\underline{m}})$. Here a *great hierarchy* is defined to be a sequence of 3-manifolds,

$$(M,\underline{\underline{m}}) = (M_o,\underline{\underline{m}}_o),(M_1,\underline{\underline{m}}_1),\ldots,(M_n,\underline{\underline{m}}_n), \quad n \geq 1,$$

where $(M_{i+1},\underline{\underline{m}}_{i+1})$ is obtained from $(M_i,\underline{\underline{m}}_i)$ by the following device:

If the index $i$ is even, take the characteristic submanifold $\bar{V}_i$ of $(M_i,\bar{\underline{\underline{m}}}_i)$, and define $M_{i+1} = (M_i - \bar{V}_i)^-$. If we denote by $\underline{\underline{m}}_{i+1}$ the proper boundary-pattern of $M_{i+1}$, it follows that $\bar{\underline{\underline{m}}}_{i+1}$ is useful (see 4.8). Furthermore, $(M_{i+1},\underline{\underline{m}}_{i+1})$ consists of simple 3-manifolds and I-bundles over the square, annulus, or torus (apply 10.4).

If, on the other hand, the index $i$ is odd, fix a non-separating, essential surface $(F_i,\underline{\underline{f}}_i)$ in some component of $(M_i,\bar{\underline{\underline{m}}}_i)$ which is a simple 3-manifold, $F_i \cap \partial M_i = \partial F_i$. Such a surface always exists, by 4.3. Furthermore, suppose that $F_i$ is chosen so that its complexity is minimal. Define $(M_{i+1},\underline{\underline{m}}_{i+1})$ to be the manifold obtained from $(M_i,\underline{\underline{m}}_i)$ by splitting at $F_i$. Then, by 4.8, $\bar{\underline{\underline{m}}}_{i+1}$ is again a useful boundary-pattern.

By a result of Haken [Ha 2, pp. 101] a great hierarchy always exists such that $(M_n, \underline{\underline{m}}_n)$ consists of balls, or I-bundles over the square, annulus, or torus.

Proof of 24.2. First of all fix a great hierarchy of $(M, \underline{\underline{m}})$:

$$(M,\underline{\underline{m}}) = (M_o,\underline{\underline{m}}_o),\ (M_1,\underline{\underline{m}}_1),\dots,(M_n,\underline{\underline{m}}_n),\quad n \geq 1.$$

To start the induction, recall from 22.1 and 22.2 that $(M_n, \underline{\underline{m}}_n)$ satisfies the conclusion of 24.2 Thus we suppose that $(M_{2i+1}, \underline{\underline{m}}_{2i+1})$ satisfies 24.2, and we are done if we show that $(M_{2i-1}, \underline{\underline{m}}_{2i-1})$ satisfies 24.2 (apply 18.3). For this we still have to prove that, for every 3-manifold M', every admissible homotopy equivalence $f: (M_{2i-1}, \underline{\underline{m}}_{2i-1}) \to (M', \underline{\underline{m}}')$ can be admissibly deformed into a homeomorphism. Without loss of generality, $M_{2i-1}$ is connected and a simple 3-manifold. Hence we see finally from 23.1 that it remains to construct a nice submanifold in $M_{2i}$ for a given homotopy equivalence $\tilde{f}: M_{2i} \to M''$.

The characteristic submanifold is a good candidate to start with, for, by our induction assumption, $\bar{V}_{2i}$ satisfies already property 3 of a nice submanifold for $\tilde{f}$ (apply 18.3).

Define $G = (\partial U(F_{2i-1}) - \partial M_{2i-1})^-$. Since $M_{2i-1}$ is a simple 3-manifold, every component of $\bar{V}_{2i}$ which does not meet $G$ is a regular neighborhood of a side of $(M_{2i-1}, \bar{\underline{\underline{m}}}_{2i-1})$ (recall that $\bar{V}_{2i}$ is complete, by 10.4). Hence, by our induction assumption, it is easily seen that, removing all these components, we obtain from $V_{2i}$ an essential F-manifold $W$ in $(M_{2i}, \bar{\underline{\underline{m}}}_{2i})$ which still satisfies property 3 of a nice submanifold for $\tilde{f}$.

Let $W^*$ be the union of all those components of $W$ which are either Seifert fibre spaces, or I-bundles which do not meet $G$ in lids. Denote by $U$ a regular neighborhood of $(\partial W^* - \partial M_{2i})^-$. Now, observe that the components of $W$ look as described in 24.1. In particular, every component, $X$, of $W^*$ is either a solid torus or an I-bundle over the disc, and moreover, at most one side of $X$, different from a lid, is contained in a free side of $(M_{2i-1}, \underline{\underline{m}}_{2i-1})$ ($M_{2i-1}$ is a simple 3-manifold). Observe also that, by our choice of F, every solid torus of $W^*$ which meets a free side of $(M_{2i-1}, \underline{\underline{m}}_{2i-1})$ has to be in fact a product $S^1$-bundle (see the proof of 24.1). Hence it is easily checked that $(W - W^*)^- \cup U$ satisfies property 3 since W does. Define $\hat{W} = (W - W^*)^- \cup U$. Then $\hat{W}$ is an essential F-manifold which already satisfies 3 and 1 of a nice submanifold for $\tilde{f}$ (see 24.1).

Now finally let A be any component of $(G - \hat{W})^-$ which is an inner square or annulus in G. Then at least one component of $(\partial A - \partial G)^-$ is contained in a component B of $G \cap \hat{W}$ which is an inner square or annulus. To see this observe that it holds for $\bar{V}_{2i}$, since, by 10.4, $\bar{V}_{2i}$ is complete and, since, by the very definition, a characteristic submanifold is a full F-manifold. Let Y and X be the components of $(M_{2i} - \hat{W})^-$ resp. $\hat{W}$ which contain A resp. B. Then X is an I-bundle over the square or annulus and $X \cup Y$ is a product I-bundle over the disc (see the properties of $W^*$). Hence it follows that $\hat{W} - X$ satisfies property 3 and 1 of a nice submanifold for $\tilde{f}$ since $\hat{W}$ does (use homotopies which are constant on $(\partial X - \partial M_{2i})^- - Y$). This means that we obtain a nice submanifold for $\tilde{f}$ from $\hat{W}$ by removing appropriate components. q.e.d.

## Part V. LOCAL CONSTRUCTIONS FOR HOMOTOPY EQUIVALENCES

So far we have established the main properties of the characteristic submanifolds, culminating in the Enclosing Theorem and the First Splitting Theorem. With respect to homotopy equivalences we have seen furthermore, by the Classification Theorem, that the study of homotopy equivalences can now be splitted (1) into that of Seifert fibre spaces and I-bundles, and (2) that of homeomorphisms of simple 3-manifolds. This will be our starting point in Part V.

In Chapter IX we take up the second problem. Having decided to consider homeomorphisms, we may equally well study the mapping class group of arbitrary Haken 3-manifolds. It turns out that for this purpose it is convenient to introduce a 3-dimensional version of the Dehn twists, well known for surfaces. With this notion we find that the mapping class group contains a subgroup of finite index generated by Dehn twists. Indeed, in §25, we prove this for Seifert fibre spaces and I-bundles and arrive moreover at a device for actually computing the mapping class group of such 3-manifolds (based on results of Hatcher-Thurston for surfaces). Furthermore, we show that the mapping class group of simple 3-manifolds is finite. For Stallings manifolds this follows immediately from results of Hemion and Zieschang. In the other case we show this by establishing a great hierarchy which is left invariant under the isotopy classes of all, but finitely many, homeomorphisms. Since, by 10.9, the characterisitc submanifold is unique, up to isotopy, the forementioned result on Dehn twists follows easily.

In Chapter X we turn to the first problem, and show in §28 that the homotopy equivalences of "most" Seifert fibre spaces can be deformed into a fibre preserving map. This reduces the study of homotopy equivalences of Seifert spaces to that of Fuchsian complexes (recall that, by its very definition, such a complex is obtained from a surface by attaching to some boundary components $r_i$ a disc $D_i$ via a covering map $\partial D_i \to r_i$). We stop here and do not push further the study of homotopy equivalences in this direction. In §29, we finally show how the results of this paper can be used for the isomorphism problem of 3-manifold groups, in particular knot groups. We do this by showing that the whole homotopy type of a Haken 3-manifold can be obtained by applying finitely many so called Dehn flips (see §29).

## *Chapter IX*: *Dehn twists of 3-manifolds*

By an *admissible Dehn twist* of a Haken 3-manifold $(M,\underline{\underline{m}})$ we mean an admissible homeomorphism of $(M,\underline{\underline{m}})$ which is the identity outside a regular neighborhood of an essential annulus or torus in $(M,\underline{\underline{m}})$. Furthermore, the *mapping class group* $H(N,\underline{\underline{n}})$ of a 2- or 3-manifold $(N,\underline{\underline{n}})$ is the group of all admissible homeomorphisms of $(N,\underline{\underline{n}})$, modulo admissible isotopy. Observe that for an essential torus in $(M,\underline{\underline{m}})$ the Dehn twists along T define a free abelian subgroup of $H(M,\underline{\underline{m}})$ of rank two, if and only if T does not separate a Seifert fibre space from M .

### §25. On the mapping class group of Seifert fibre spaces

We expect that the mapping class group of Seifert fibre spaces can be described by that of the orbit surface plus a certain limited contribution in the vertical, i.e. fibre direction. It is the aim of this paragraph to make this impression precise (see 25.3)

For convenience we begin by introducing some notations. So let $(M,\underline{\underline{m}})$ be a Seifert fibre space with complete and useful boundary-pattern, and with a fixed Seifert fibration. Let $p\colon (M,\underline{\underline{m}}) \to (F,\underline{\underline{f}})$ be the fibre projection onto the orbit surface $(F,\underline{\underline{f}})$. Then -- by definition -- p maps each fibre of M to a point of F, and so any admissible fibre preserving homeomorphism $h\colon (M,\underline{\underline{m}}) \to (M,\underline{\underline{m}})$ induces, in a well-defined way, an admissible homeomorphism $\bar{h}\colon (F,\underline{\underline{f}}) \to (F,\underline{\underline{f}})$.

Denote by $H^+(M,\underline{\underline{m}})$ the subgroup of $H(M,\underline{\underline{m}})$ generated by all orientation-preserving admissible homeomorphisms, and by $H^0(M,\underline{\underline{m}})$ that one generated by all orientation- and fibre-preserving admissible homeomorphism $h\colon (M,\underline{\underline{m}}) \to (M,\underline{\underline{m}})$ with $\bar{h} = \mathrm{id}$. For later application we also need a somewhat refined version of the above notions, namely: if X is any subset of M denote by $H_X(M,\underline{\underline{m}})$ the group generated by all admissible homeomorphisms of $(M,\underline{\underline{m}})$ which are the identity on X, modulo isotopy which is constant on X.

A fibre preserving admissible isotopy $h_t$, $t \in I$, of $(M,\underline{\underline{m}})$ with $\bar{h}_t = \mathrm{id}$, for all $t \in I$, will be called a *fibrewise* isotopy. By a *vertical Dehn twist* we mean a fibre preserving admissible Dehn twist, g, along a vertical annulus or torus with $\bar{g} = \mathrm{id}$.

25.1 *Lemma*. *Let* $(M,\underline{\underline{m}})$ *be the* $S^1$*-bundle over the Möbius band with* $\underline{\underline{m}} = \{\partial M\}$. *Then* $H^0(M,\underline{\underline{m}}) \cong \mathbb{Z}/2\mathbb{Z}$, *and the non-trivial homeomorphism* h *of* $H^0(M,\underline{\underline{m}})$ *may be chosen to be fibre preserving such that* $\bar{h} = \mathrm{id}$

<u>and</u> $h|\partial M = id|\partial M$.

<u>In addition</u>. 1. <u>No ambient isotopy</u> $\alpha_t$, $t \in I$, <u>of</u> $M$ <u>with</u> $\alpha_1 = h^2$ <u>is constant on</u> $\partial M$.

2. <u>Every ambient isotopy</u> $\beta_t$, $t \in I$, <u>of</u> $\partial M$ <u>which slides</u> $\partial M$ <u>once around along the fibres, can be extended to an ambient isotopy</u>, $\sigma_t$, $t \in I$, <u>of</u> $M$ <u>with</u> $\sigma_1 = h^2$.

<u>Proof</u>. Let $K$ be a Klein bottle. Recall from [Li 2], that $K$ contains (up to isotopy) precisely three oriented, 2-sided, simple closed curves, say $\pm k_1$, $+k_2$ and four oriented, 1-sided, simple closed curves, say $\pm t_1$, $\pm t_2$. Moreover, $k_2$ is non-separating and intersects $t_1$, resp. $t_2$, in precisely one point. Let $h^*$ be the (non-trivial) Dehn twist along $k_2$. Then it is easy to see that $h^*(t_1) \simeq t_2$ and $(h^*)^2(+t_1) \simeq +t_1$.

Consider $M$ as the I-bundle over the Klein bottle $K$, and let $q: M \to K$ be the projection. $h^*$ lifts to a homeomorphism $h'$ of $M$, which is I- as well as $S^1$-fibre preserving, and which can also be described as Dehn twist along the annulus $A = q^{-1}k_2$.

Of course, $h'$ is not isotopic to the identity, for otherwise $h^*(t_1) \simeq t_1$. In order to see that $(h')^2$ is isotopic to the identity, observe that $(h^*)^2$ is isotopic to the identity and that every isotopy of $(h^*)^2$ can be lifted to an I-fibre-preserving isotopy of $(h')^2$. By our definition of $H^0(M,\underline{\underline{m}})$, it follows that every homeomorphism of $H^0(M,\underline{\underline{m}})$ can be isotoped into an I-fibre preserving homeomorphism which is the identity on $A$. Hence we see that every homeomorphism of $H^0(M,\underline{\underline{m}})$ is isotopic either to the identity or to $h'$. Without loss of generality, $\bar{h}' = id$ and so $h'$ is a generator of $H^0(M,\underline{\underline{m}})$, i.e. $H^0(M,\underline{\underline{m}}) \cong \mathbb{Z}/2\mathbb{Z}$.

Denote by $a_1$ one boundary curve of the annulus $A$, and let $B_i$, $i = 1,2$, be the Möbius band $q^{-1}t_i$. Then $b_i = \partial B_i$ is a simple closed curve which meets $a_1$ in precisely one point. By the definition of $h'$, $h'(B_1) \simeq B_2$ (modulo boundary) and since $B_1 \cap B_2 = \emptyset$ it follows that $h'(b_1) \simeq b_1$ in the torus $\partial M$. Since also $h'|a_1 \simeq id|a_1$ in $\partial M$, this means that $h'$ induces the identity on the first homology of $\partial M$, and so it follows that $h'|\partial M \simeq id|\partial M$ in $\partial M$ Thus, by Baer's theorem, $h'|\partial M$ is isotopic in $\partial M$ to the identity. More precisely, since $h'$ maps each $S^1$-fibre to itself (i.e. $\bar{h}' = id$), there is an isotopy $h_t'$, $t \in I$, of $M$ with $h_0' = h'$, $h_1'|\partial M = id|\partial M$, and $h_t'(c) = c$, for all $t \in I$ and all $S^1$-fibres $c$ in $M$. We define $h = h_1'$, and $h$ satisfies the conditions of 25.1.

In order to prove the first additional remark, observe that, by $h(A) = A$, it suffices to show that $h|A$ is not isotopic in $A$ to $id|A$, by an isotopy which is constant on $\partial A$. But this is true, for otherwise $h$ could be isotoped (rel $\partial M$) such that afterwards $h|U(A) = id|U(A)$, where $U(A)$ is a regular neighborhood of $A$. $V = (M - U(A))^-$ is a solid torus, and every homeomorphism of a solid torus which is the identity on the boundary, can be isotoped (rel boundary) to the identity. Hence $h$ would be isotopic to the identity, which is a contradiction (see above).

For the second remark, recall the definitions of the homeomorphism $h'$ and the isotopy $h_t'$. If we denote by $T_1$ and $T_2$ the two annuli of $(\partial M - U(A))^-$, it follows that, without loss of generality, $h_t'$ is constant on $T_1$ while it slides $T_2$ once around the $S^1$-fibres. Fix an essential vertical annulus $A'$ in $M$ which does not meet $U(A)$, and let $k'$ be an essential arc in $A'$. Then, by $h(A') = A'$ and the above property of $h_t'$, it follows that the composition of $k'$ with $h(k')$ is a (possibly singular) closed curve which has circulation number one with respect to $A'$.

Now, extend $\beta_t|\partial A'$ to an ambient isotopy $\sigma_t'$ of $A'$. By our choice of $M$, we obtain $M$ from $V = (M - U(A))^-$ by attaching the two copies of $A$ in $V$ via a homeomorphism $A \to A$ which does not reverse the orientation of $A$, but which interchanges the boundary curves of $A$. Hence it follows that $\sigma_t'$, $t \in I$, slides the boundary curves of $A'$ in different directions, and so the composition of $\sigma_1'(k')$ with $k'$ is a singular closed curve which has circulation number two with respect to $A'$.

Hence, altogether, $h^2|A' \simeq \sigma_1'|A'$ (rel $\partial A'$). Now, $(M - U(A))^-$ is a solid torus, and so $\sigma_t'$ can be extended to an ambient isotopy $\sigma_t$ of $M$ with $\sigma_1 = h^2$. q.e.d.

25.1 shows, in particular, that, for a certain Seifert fibre space $M$, $H^0(M,\underline{\underline{m}})$ is isomorphic to the first relative homology group of the orbit surface. The following lemma gives a generalization of this fact (see also [Wa 7]).

<u>25.2 Lemma</u>. <u>Let</u> $(M,\underline{\underline{m}})$ <u>be a Seifert fibre space, with fibre projection</u> $p: M \to F$, <u>but not one of the exceptions</u> 5.1.1-5.1.5. <u>Suppose that</u> $\underline{\underline{m}}$ = {<u>components of</u> $\partial M$} <u>and that</u> $M$ <u>is not an</u> $S^1$-<u>bundle over the torus or Klein bottle</u>.

<u>Then</u> $H^0(M,\underline{\underline{m}}) \cong H_1(F,\partial F)$ (= <u>first relative homology</u>).

<u>In addition</u>. 1. <u>If</u> $F$ <u>is orientable</u>, $H^0(M,\underline{\underline{m}})$ <u>is generated by a finite set of vertical Dehn twists</u>.

2. <u>If</u> $F$ <u>is non-orientable, there is a finite set of vertical Dehn twists which generate a subgroup of finite index in</u> $H^0(M,\underline{\underline{m}})$.

<u>Remark</u>. 1. It will be apparent from the proof that in the additional remarks $H^0(M,\underline{\underline{m}})$ may be replaced by $H^0_A(M,\underline{\underline{m}})$, for any union $A$ of sides of $(M,\underline{\underline{m}})$.

2. For $S^1$-bundles over the torus or Klein bottle we refer to 25.6 and 25.7.

<u>Proof</u>.

<u>Case 1</u>. <u>The orbit surface</u> $F$ <u>is orientable</u>.

In this case we may specify the embedding of a 2-sphere $G$ with holes in $F$ such that $G$ contains $\partial F$, and that the complement of $G$ consists of discs and tori with one hole (classification of surfaces). Suppose that $G$ is chosen so that there is precisely one exceptional point, $x_i$, $i \geq 1$, in each disc $D_i$ of $(F - G)^-$ and that all exceptional points lie in discs of $(F - G)^-$. For every other component $B_i$, $i \geq 1$, of $(F - G)^-$ fix a pair, $s_i$, $t_i$, of simple closed, non-isotopic curves in $B_i$ which intersect themselves in precisely one point. For every boundary curve, $r_j$, $j \geq 2$, of $F$, fix a simple arc $a_j$ in $G$ which joins $r_j$ with $r_1$, and suppose that the $a_j$'s are pairwise disjoint.

Define $S_i = p^{-1}s_i$, $T_i = p^{-1}t_i$, and $A_j = p^{-1}a_j$. Let $\sigma_i$, resp. $\tau_i$, resp. $\alpha_j$, be the vertical Dehn twists which generate the subgroups of all vertical Dehn twists along $S_i$, resp. $T_i$, resp. $A_j$.

We claim that $\sigma_i$, $\tau_i$, and $\alpha_j$ generate a subgroup of $H^0(M,\underline{\underline{m}})$ which is isomorphic to $H_1(F,\partial F)$. To see this, note first that we may fix an orientation of $s_i$, $t_i$, and $a_j$, so that these oriented curves generate $H_1(F,\partial F)$. Moreover, observe that, by the definitions of $\sigma_i$, $\tau_i$, and $\alpha_j$, all these homeomorphisms commute. Thus it remains to show that, for every fibre preserving homeomorphism $h = \Sigma m_i\sigma_i + \Sigma n_j\tau_j + \Sigma p_k\alpha_k$, with $\bar{h} = \mathrm{id}$, where $m_i$, $n_j$, $p_k \in \mathbb{Z}$, the fact that $h$ is isotopic to the identity implies that $m_i = n_j = p_k = 0$, for all $i,j,k$. Fix a closed curve $t$ in $T_i$ which intersects each fibre in precisely one point. Then $h(t)$ is also a closed curve in $T_i$, for $\bar{h} = \mathrm{id}$, and $t \simeq h(t)$ in $M$, for $h \simeq \mathrm{id}$. Let $f\colon t \times I \to M$ be a homotopy with $f|t \times 0 = t$ and $f|t \times 1 = h(t)$. Then, by 5.10, it follows that $f$ can be deformed (rel $t \times \partial I$) into $T_i$, for $M - \mathring{U}(T_i)$ is not the $S^1$-bundle over the annulus or Möbius band (recall that $T_i$ is non-separating and see our suppositions on $M$). Hence $t \simeq h(t)$ in $T_i$, for all $i$, and so it follows that $n_j = 0$, for all $j$. In the same way, we prove that $m_i = 0$. To show that $p_k = 0$, consider a curve in the boundary component of $M$ different from $p^{-1}r_1$ which meets $A_k$. This completes the proof of our claim.

To prove 25.2 in Case 1, it remains to show that $\sigma_i$, $\tau_i$, and $\alpha_j$ generate $H^0(M,\underline{\underline{m}})$. For this let $h\colon M \to M$ be any orientation- and fibre-preserving homeomorphism with $\bar{h} = \mathrm{id}$. Then, in particular, $h$ preserves $S_i$, $T_i$, and $A_j$. The restrictions $h|S_i$ and $h|T_i$ are fibrewise isotopic to Dehn twists along fibres, for $h$ is orientation-preserving and $\bar{h} = \mathrm{id}$. Hence $h|S_i$ and $h|T_i$ can be extended to Dehn twists along $T_i$ and $S_i$, respectively. Let $R_j$ be the boundary component of $M$ which $A_j$ joins with $p^{-1}r_1$. Then $h|R_j$ is also a Dehn twist along a fibre, and so $h|R_j$ can be extended to a Dehn twist along $A_j$. Hence, altogether, there is a product, $\varphi$, of the $\sigma_i$'s, $\tau_i$'s, and $\alpha_j$'s such that $\varphi \cdot h$ is fibrewise isotopic to a homeomorphism $g$ with $g|p^{-1}B_i = \mathrm{id}$, for all $i$, and $g|R_j = \mathrm{id}$, for all

$j \neq 1$. Moreover, without loss of generality, $g|p^{-1}D_i = \mathrm{id}$. Now recall that, by our choice of $G$, $p^{-1}G$ is a product $G \times S^1$. Embed $G$ in a disc $D$ with $r_1 = \partial D$. Then, by the properties of $g$, $g|p^{-1}G$ can be extended to a fibre preserving homeomorphism $g'$ of $D \times S^1$. Of course, $g'$ is fibrewise isotopic to the identity, and so there is a fibrewise isotopy of $g|p^{-1}G$ in $p^{-1}G$ to the identity. This isotopy extends to an isotopy of $M$, and so $g$ is isotopic to the identity. This proves that $h$ is isotopic to $\varphi^{-1}$.

Case 2. The orbit surface $F$ is non-orientable.

In this case we may specify the embedding of a 2-sphere $G$ with holes in $F$ such that $G$ contains $\partial F$ and that the complement consists of discs and Möbius bands. As in Case 1, suppose that $G$ is chosen so that there is precisely one exceptional point, $x_i$, $i \geq 1$, in each disc $D_i$ of $(F - G)^-$ and that all exceptional points lie in discs of $(F - G)^-$. For every Möbius band $B_i$ of $(F - G)^-$, let $b_i$ be the core of $B_i$, and define $W_i = p^{-1}B_i$. For every boundary curve $r_j$, $j \geq 2$, of $F$, fix a simple arc $a_j$ in $G$ which joins $r_j$ with $r_1$, and suppose that the $a_j$'s are pairwise disjoint.

Let $\alpha_j$ be the vertical Dehn twist which generates the subgroup of all vertical Dehn twists along $A_j$. According to 25.1, there is a fibre preserving homeomorphism $h_i^*: W_i \to W_i$ with $\bar{h}_i^* = \mathrm{id}$ and $h_i^*|\partial W_i = \mathrm{id}|\partial W_i$ and which is not isotopic to the identity. $h_i^*$ can be extended to a fibre preserving homeomorphism $h_i$ of $M$ with $\bar{h}_i = \mathrm{id}$, and $h_i|(M_i - W_i)^- = \mathrm{id}|(M_i - W_i)^-$.

We claim that the $\alpha_j$'s and $h_i$'s generate a subgroup of $H^0(M,\underline{\underline{m}})$ which is isomorphic to $H_1(F,\partial F)$. To see this, note that we may fix an orientation of the $a_j$'s and $b_i$'s so that they generate $H_1(F,\partial F)$. Moreover, observe that all the homeomorphisms, $\alpha_j$ and $h_i$, commute with each other, for their supports are disjoint. Then it remains to show the following: a homeomorphism $h = \Sigma m_j\alpha_j + \Sigma n_i h_i$ with $\bar{h} = \mathrm{id}$, where $m_j, n_i \in \mathbb{Z}$, is isotopic to the identity if and only if $m_j = 0$, for all coefficients $m_j$, and all coefficients $n_i$ are equal and even (here the sum $\Sigma n_i h_i$ is taken over all $h_i$).

If $m_j = 0$ and all $n_i$ are equal and even, then it follows from the second additional remark of 2.3 that $h$ is isotopic to the identity.

For the other direction it follows, by an argument of Case 1, that all coefficients $m_j$ have to be zero. Hence we may suppose that $h = \Sigma n_i h_i$. Let $A_i$ be an essential, vertical annulus in $W_i$ and denote by $B_i$ one of the components in which $A_i$ separates $\partial W_i$. Define $T_i = A_i \cup B_i$. Then $T_i$ is a vertical and essential Klein bottle in $M$. Fix a simple closed curve $t_i$ in $T_i$ which meets each fibre in precisely one point. Since $h \simeq id$, there is a homotopy $f\colon t \times I \to M$ with $f|t_i \times 0 = t_i$ and $f|t_i \times 1 = h(t_i)$. $h(t_i)$ is contained in $T_i$ (see our choice of the $h_i$'s), and $M_i - \dot{U}(T_i)$ is not an $S^1$-bundle over the annulus or Möbius band (note that $T_i$ is non-separating and see our suppositions on $M$). Hence, by 5.10, it follows that $f$ can be deformed (rel $t_i \times \partial I$) into $T_i$. This implies the existence of an isotopy of $h|T_i$ which moves $h|T_i$ in $T_i$ into the identity. This isotopy may be chosen to be fibrewise, and so it extends to an isotopy of $h|A_i \cup \partial W_i$ into the identity. Since $(W_i - U(A_i))^-$ is a solid torus it follows that $h|W_i$ is isotopic to the identity. Since $h_i^*$ is not isotopic in $W_i$ to the identity, it follows from 25.1 that $h|W_i = h_i^{n_i}|W_i = h_i^{2n}|W_i$, for some $n \in \mathbb{Z}$. This proves that $n_i$ is even. Assume that not all $n_i$ are equal, and let $n_1$ be the smallest coefficient of $\Sigma n_i h_i$. Define $h' = \Sigma n_1 h_i$. Then $h(h')^{-1}$ is fibrewise isotopic to a homeomorphism $g$ with $g|W_1 = id|W_1$ and $g|W_2 = h_2^{2n}|W_2$, say, with $n \neq 0$. Let $C_1$ and $C_2$ be two disjoint vertical annuli in $p^{-1}G$ with $\partial C_1 \cup \partial C_2 = \partial A_1 \cup \partial A_2$ and such that $C_1 \cup C_2 \cup A_1 \cup A_2$ is connected. Then $T = C_1 \cup C_2 \cup A_1 \cup A_2$ is a torus (and not a Klein bottle). Let $t$ be a closed curve in $T$ which meets each fibre in precisely one point. Then $g(t)$ is a curve in $T$, for $\bar{g} = id$. Now, $g|A_2$ is not isotopic (rel $\partial A_2$) to the identity, for otherwise $g|W_2$, and so $h_2^{2n}|W_2$, is isotopic (rel $\partial W_2$) to the identity (recall that $(W_2 - U(A_2))^-$ is a solid torus), and this contradicts the first additional remark of 25.1. Hence $g(t)$ cannot be homotopic to $t$ in $T$, for $g|(T - A_2)^- = id$. By an argument used above, this leads to a contradiction, for $g \simeq id$ and $(M - U(T))^-$ is not the $S^1$-bundle over the annulus or Möbius band.

Thus all $n_i$ have to be equal.

It remains to show that the $\alpha_j$'s and $h_i$'s generate $H^0(M,\underline{\underline{m}})$. So, let $h: M \to M$ be any orientation- and fibre-preserving homeomorphism with $\bar{h} = id$. Multiplying $h$ with $\alpha_j$'s (if necessary), we may suppose that $h|\partial M = id|\partial M$ (see Case 1). By 25.1, $h|W_i$ is isotopic in $W_i$ to $h_i^{n_i}|W_i$, for all $i \geq 1$ and some $n_i \in \mathbb{Z}$, and since $\bar{h}_i = \bar{h} = id$, the isotopy may be chosen to be fibrewise (note that $G$ is not an annulus, for $M$ is not the $S^1$-bundle over the Klein bottle). Hence, multiplying $h$ with $h_i$'s, we obtain a homeomorphism $g$ with $\bar{g} = id$ such that $g|W_i$ is fibrewise isotopic to the identity. Since $p^{-1}G$ is a product $G \times S^1$, this implies that $g$ itself is fibrewise isotopic to the identity (see the second additional remark of 25.1). Hence $h$ is isotopic to a product of $\alpha_j$'s and $h_i$'s.

For the additional remark note that $h_i^2$ is isotopic to a Dehn twist along $\partial W_i$ and that any conjugate of a Dehn twist is again a Dehn twist. q.e.d.

Let $(M,\underline{\underline{m}})$ be a Seifert fibre space, but not one of the exceptions 5.1.1-5.1.5. Suppose that $\underline{\underline{m}} = \{$components of $\partial M\}$, and let $p: (M,\underline{\underline{m}}) \to (M,\underline{\underline{m}})$ be the fibre projection. Let $x_1,\ldots,x_n$, $n \geq 0$, be all the exceptional points in the orbit surface $F$. Associate to each $x_i$, $1 \leq i \leq n$, the $(\mu_i,\nu_i)$-value of the exceptional fibre above $x_i$. Denote by $H^*(F,\underline{\underline{f}})$ the mapping class group generated by the admissible homeomorphisms of $(F,\underline{\underline{f}})$ which map exceptional points to exceptional points with the same value, modulo admissible isotopies which are constant on the exceptional points.

Let $h$ be any homeomorphism from $H^+(M\ \underline{\underline{m}})$ and suppose that $M$ is not an $S^1$-bundle over the annulus, torus, Möbius band or Klein bottle. Then, by 5.9, $h$ can be admissibly isotoped into a fibre preserving homeomorphism, say $h'$. Associate to $h$ the element $\varphi(h)$ of $H^*(F,\underline{\underline{f}})$ represented by $\bar{h}'$.

$\varphi$ is well-defined, for if $h''$ is another fibre preserving homeomorphism isotopic to $h$, then $h'$ and $h''$ are fibre preserving isotopic (see the remark on p. 85 of [Wa 4]). Such an isotopy induces an isotopy of $\bar{h}'$ to $\bar{h}''$ which is constant on the exceptional points.

Altogether, the above rule defines a homomorphism $\varphi: H^+(M,\underline{\underline{m}}) \to H^*(F,\underline{\underline{f}})$.

<u>25.3 Proposition.</u> <u>Suppose that</u> $(M,\underline{\underline{m}})$ <u>is given as above. Then there is a short exact sequence</u>

$$1 \to H^0(M,\underline{\underline{m}}) \to H^+(M,\underline{\underline{m}}) \overset{\varphi}{\to} H^*(F,\underline{\underline{f}}) \to 1,$$

<u>and that</u> $H^0(M\,\underline{\underline{m}}) \cong H_1(F,\partial F)$, <u>where</u> $H_1(F.\partial F)$ <u>denotes the first relative homology group.</u>

<u>In addition. If</u> $\partial M \neq \emptyset$, <u>the sequence is split exact.</u>

<u>Proof.</u> By 25.2, $\ker \varphi = H^0(M\,\underline{\underline{m}}) \cong H_1(F,\partial F)$. Hence it remains to show that $\varphi$ is surjective. For this let $D$ be a small disc in $F$ disjoint from all the exceptional points $x_1,\dots,x_n$. Define $F' = F - (D^\circ \cup \mathring{U}(x_1) \cup \dots \cup \mathring{U}(x_n))$, where $U(x_i)$ denotes a regular neighborhood of $x_i$. Then $M' = p^{-1}F'$ is an $S^1$-bundle over $F'$ with non-empty boundary. Hence we may fix a section $s': F' \to M'$. Denote by $k_0$ the curve $s'(F') \cap p^{-1}(\partial D)$, and let $k_i = s'(F') \cap p^{-1}(\partial U(x_i))$. If $h_i$ is any Dehn twist along a vertical annulus in $M'$ which joins $p^{-1}\partial D$ with $p^{-1}\partial U(x_i)$, it follows that $h_i \circ s'$ is again a section. Multiplying $s'$ with appropriate Dehn twists $h_i$ if necessary, we may suppose that $s'$ is chosen so that the meridian discs of each $p^{-1}U(x_i)$ meets $\partial M'$ in a curve which is isotopic in $\partial M'$ to $\mu_i k_i + \nu_i b_i$, where $b_i$ is a fibre in $p^{-1}\partial U(x_i)$ and $(\mu_i,\nu_i)$ describes the exceptional fibre above $x_i$.

To prove that $\varphi$ is surjective, let $\bar{g}$ be any homeomorphism of $H^*(F,\underline{\underline{f}})$. Without loss of generality, $\bar{g}(D) = D$. Then observe (use a hierarchy) that $\bar{g}|s'F'$ can be extended to a fibre preserving homeomorphism $g': M' \to M'$. $g'|\partial M'$ is the identity on $p^{-1}\partial D$ and permutes the coordinate systems $(k_i,b_i)$. Moreover, $g'$ maps exceptional points to the exceptional points of the same $(\mu_i,\nu_i)$-value. Hence, by the properties of $s'$, $g'$ extends to a fibre preserving homeomorphism $g: M \to M$ such that $\varphi(g)$ has the same isotopy class as $\bar{g}$, modulo isotopies which are constant on

$x_1,\dots,x_n$.

For the additional remark note that, by supposition on $M$, the manifold $M^* = p^{-1}F^*$, where $F^* = F - \dot{U}(x_1) \cup \dots \cup \dot{U}(x_n)$, has a boundary component different from $p^{-1}\partial U(x_i)$, for all $1 \leq i \leq n$. Hence, as above, we find a section $s^*: F^* \to M^*$ so that every homeomorphism of $s^*(F^*)$, i.e. of $F^*$, can be extended to a fibre preserving homeomorphism of $M$. It is easy to see that this defines a homomorphism $\psi: H^*(F,\underline{\underline{f}}) \to H^+(M,\underline{\underline{m}})$ with $\varphi \cdot \psi = \text{id}$, for every isotopy of the base $F^*$ of the $S^1$-bundle $M^*$ can be lifted to a fibre preserving isotopy of $M^*$ which also preserves $s^*(F^*)$. q.e.d.

25.3 may be considered as a device for the computation of the mapping class group of certain Seifert fibre spaces. For this, recall that Hatcher and Thurston [HT 1] have recently given a presentation of the mapping class group of closed orientable surfaces. As far as Dehn twists are concerned it has been known for a long time that, for a given closed surface $F$ (orientable or not), there is a (finite) homeomorphism $f$ with the property that every homeomorphism $h: F \to F$ is isotopic to a product $\varphi \circ f$, where $\varphi$ is a product of Dehn twists (see [De 1], [Li 1,2,3,4]. This means that the mapping class group of a closed surface contains a subgroup of finite index (more precisely: of index two) generated by Dehn twists. It is not difficult to generalize this fact to the group $H^*(F,\underline{\underline{f}})$ (for a discussion of this aspect of the mapping class group see also [Bi 1]). Then we have, as an easy consequence of 25.3, the following corollary.

<u>25.4 Corollary</u>. <u>Let</u> $(M,\underline{\underline{m}})$ <u>be a Seifert fibre space, but not one of the exceptions</u> 5.1.1-5.1.5. <u>Suppose that</u> $M$ <u>is not an</u> $S^1$-<u>bundle over the annulus, torus, Möbius band, or Klein bottle</u>.

<u>Then the admissible Dehn twists of</u> $(M,\underline{\underline{m}})$ <u>generate a subgroup of finite index in</u> $H^+(M,\underline{\underline{m}})$, <u>and so in</u> $H(M,\underline{\underline{m}})$.

<u>Remark</u>. Using the remark of 25.2, it is apparent that $H(M,\underline{\underline{m}})$ may be replaced by $H_A(M,\underline{\underline{m}})$, for any union $A$ of sides of $(M,\underline{\underline{m}})$.

We still have to consider the $S^1$-bundles over the annulus, torus, Möbius band, and Klein bottle. The $S^1$-bundles over the annulus or Möbius band are also I-bundles over the torus or Klein bottle. Hence their mapping class groups are known since those of the torus and Klein bottle are known [Li 2]. In particular, the assertion of 25.4 holds for them. Hence it remains to study the $S^1$-bundles over the torus or Klein bottle.

<u>25.5 Proposition</u>. <u>Let</u> $M$ <u>be an</u> $S^1$<u>-bundle over the torus</u>. <u>Then every orientation-preserving homeomorphism of</u> $M$ <u>is isotopic to some product of Dehn twists of</u> $M$.

<u>25.6 Corollary</u>. <u>The Dehn twists of</u> $M$ <u>generate a subgroup of index two in the mapping class group of</u> $M$.

<u>Proof of 25.5</u>. Let $p\colon M \to T$ be the projection of the $S^1$-bundle. In $T$ we fix two simple closed, essential curves which are not isotopic in $T$ and which intersect themselves in precisely one point. Let $T_1$ and $T_2$ be defined to be the preimages under $p$ of these curves, respectively. Then $T_1$ and $T_2$ are essential tori in $M$ and $T_1 \cap T_2$ consists of precisely one curve.

Let $h$ be any orientation-preserving homeomorphism. Then we have to show that $h$ is isotopic to a product of Dehn twists.

For this we are first going to show that there is a product $\varphi$ of Dehn twists such that $\varphi h(T_1)$ is isotopic to $T_1$. So let $h$ be isotoped so that $h(T_1)$ intersects $T_1$ in a minimal number of essential curves (innermost-disc-argument). $T_1 \cap h(T_1) \neq \emptyset$, for otherwise our claim follows immediately. Observe that $T_1$ splits $M$ into a product $S^1$-bundle over the annulus, say $\widetilde{M}$, and that, by 4.6, $h(T_1) \cap \widetilde{M}$ consists of essential annuli in $\widetilde{M}$. Hence the $S^1$-fibration of the annuli of $h(T_1) \cap \widetilde{M}$ estends to an $S^1$-fibration of $\widetilde{M}$ which makes $\widetilde{M}$ to an $S^1$-bundle over the annulus. Moreover, this fibration of $\widetilde{M}$ fits to an $S^1$-fibration of $M$ over the torus such that both $T_1$ and $h(T_1)$ are vertical. Let $q\colon M \to F$ be the fibre projection of this new fibration. Then $s_1 = qT_1$ and $s_2 = qhT_1$ are two essential curves in the torus $F$. By the first part of the proof of lemma 2 in [Li 1] (since $F$ is a torus only this part occurs), there is a product $\varphi^*$ of Dehn twists of $F$ such that $\varphi^*(s_2)$ is isotopic in $F$ to $s_1$. $\varphi^*$ can be lifted to a product $\varphi$ of fibre preserving Dehn twists of $M$, and note that every isotopy in $F$ lifts to a fibre preserving isotopy in $M$. Hence it follows that $\varphi h(T_1)$ is isotopic to $T_1$. Thus we may suppose that $h(T_1) = T_1$.

Consider the torus $T_1$. By our choice of $T_2$, the intersection $T_1 \cap T_2$ consists of precisely one curve $t_2$ and, moreover, $t_2$ is a fibre of the $S^1$-fibration $p$ of $M$. By our suppositions on $h$, $h(t_2)$ is again an essential curve in $T_1$. Furthermore, we may suppose that $h$ is isotoped so that, in addition, $T_2 \cap hT_2$ consists of essential curves in $T_2$ and that $t_2 \cap ht_2$ is minimal with respect to isotopies of $t_2$ in $T_1$ (innermost-disc-argument). Since $T_1$ is a torus, we then may fix orientations of $t_2$ and $h(T_2)$ such that $h(t_2)$ intersects $t_2$ in all points of $t_2 \cap ht_2$ in the same direction, with respect to the direction of $t_2$.

Define $T_2' = h(T_2)$ and let $t_2'$ be the curve $T_2' \cap T_1$.

<u>Case 1</u>. <u>$t_2 \cap t_2'$ is empty.</u>

Note that $t_2'$ is equal to $h(t_2)$, for $h(T_1) = T_1$ and $T_2' = h(T_2)$. Hence and since we are in Case 1, $h|T_1 \colon T_1 \to T_1$ can be isotoped into a fibre preserving homeomorphism with respect to $p$. This means that $h|T_1$ is isotopic in $T_1$ to a Dehn twist of $T_1$ along a fibre. Of course, such a Dehn twist can be extended to a fibre preserving Dehn twist $\sigma$ of $M$. So we may suppose that $\sigma^{-1}h|T_1 = \mathrm{id}|T_1$. Since $h$ and $\sigma$ are orientation preserving, we even may suppose that $\sigma^{-1}h|U(T_1) = \mathrm{id}|U(T_1)$, where $U(T_1)$ is a regular neighborhood of $T_1$ in $M$. Now $M' = (M - U(T_1))^-$ is the $S^1$-bundle over the annulus and so it follows that $\sigma^{-1}h|M'$ is isotopic (rel boundary) to a Dehn twist. This complete the proof in Case 1.

<u>Case 2</u>. <u>$t_2 \cap t_2'$ is non-empty.</u>

Observe that we obtain $M$ from $\tilde{M}$ by attaching the two boundary components of $\tilde{M}$ under a homeomorphism, say $\psi$. Furthermore, under $\psi$ the system $T_2 \cap \tilde{M}$ as well as $T_2' \cap \tilde{M}$ fits together to the torus $T_2$, resp. $T_2'$. Since we are in Case 2 this implies that $\psi$ has to be the trivial homeomorphism, i.e. M is the 3-torus $S^1 \times S^1 \times S^1$. By our suppositions on $t_2 \cap t_2'$ and since $T_1$ is a torus, $t_2'$ meets $t_2$ in each point of $t_2 \cap t_2'$ in the same direction. Hence, by lemma 2 of [Li 1], there is a product of Dehn twists $\varphi$ of $T_1$ such that $\varphi(t_2')$ can be isotoped in $T_1$ out of $t_2$. Each Dehn twist of $T_1$ extends to a Dehn twist of M since M is the 3-torus. Hence 25.5 follows by an argument of Case 1. q.e.d.

<u>25.7 Proposition</u>. <u>Let</u> M <u>be an</u> $S^1$-<u>bundle over the Klein bottle</u>. <u>Then the Dehn twists of</u> M <u>generate a subgroup of finite index in the mapping class group of</u> M.

<u>Proof</u>. For the proof, we heavily utilize the fact that the Klein bottle has very few essential curves. Indeed, recall from [Li 2] the following facts for the Klein bottle, K:

1. There is, up to isotopy, just one 2-sided, non-separating, simple closed and essential curve in K, say $t_1$. $t_1$ splits K into an annulus.
2. There is, up to isotopy, just one 2-sided, separating, simple closed and essential curve in K, say $t_2$. $t_2$ separates K into two Möbius bands, say $B_1$, $B_2$.
3. If we denote by $k_1$, $k_2$ the cores of $B_1$, $B_2$, respectively, every 1-sided, simple closed curve in K is isotopic either to $k_1$, or to $k_2$.
4. $t_1$ intersects $k_1$, as well as $k_2$, in precisely one point and $t_2$ in precisely two points.

M is an $S^1$-bundle over K, and let $p: M \to K$ be the projection. Define

$$T_i = p^{-1}t_i \quad \text{and} \quad K_i = p^{-1}k_i, \quad i = 1,2.$$

Then $T_1$ is a non-separating, essential torus, $T_2$ is a separating essential torus, and $K_1$, $K_2$ are essential Klein bottles in M. Denote by $t_i'$, $k_i'$, $i = 1,2$, the essential curves in the Klein bottle $K_1$ as described above. Observe that $T_1 \cap K_1$ is a non-separating, 2-sided, essential curve in $K_1$. Hence we may suppose that $t_1' = T_1 \cap K_1$, i.e. $t_1'$ is a fibre with respect to $p$.

Consider $M_i = p^{-1}B_i$, $i = 1,2$. $M_i$ is the $S^1$-bundle over the Möbius band, i.e. also the I-bundle over the Klein bottle, $q_i: M_i \to I_i$. $T \cap M_i$ is an essential annulus in $M_i$, and let $A_i$ be an essential annulus in $M_i$ which is not isotopic to $T \cap M_i$.

We obtain $M$ from $M_1$ and $M_2$, by attaching the two boundary components $\partial M_1$ and $\partial M_2$ via a homeomorphism $h: \partial M_1 \to \partial M_2$. Now it might happen that $h(\partial A_1) = \partial A_2$ (at least up to isotopy). In this case $A_1$ and $A_2$ fit together and give a surface $T_3$ in $M$ (note that this construction is not unique).

For each two $T_i, T_j$, $i \neq j$, of the three surfaces $T_1$, $T_2$, $T_3$ we find a Seifert fibration of $M$ with the property that both $T_i$ and $T_j$ are vertical with respect to this Seifert fibration. For convenience, we call such a Seifert fibration a $(T_i,T_j)$-fibration. In the following we are going to consider two of these Seifert fibrations, namely:

1) the original $S^1$-fibration $p$ of $M$ which is of course a $(T_1,T_2)$-fibration;
2) the $S^1$-fibration of the annulus $A_i$, $i = 1,2$, extends to a Seifert fibration of $M_i$ and this makes $M_i$ to a Seifert fibre space over the 2-disc with two exceptional fibres. These Seifert fibrations of $M_1$ and $M_2$ can be isotoped so that afterwards they give a $(T_2,T_3)$-fibration (provided $T_3$ exists).

Let $r: M \to S$ be the fibre projection of the $(T_2,T_3)$-fibration. Then $S$ is a 2-sphere with four exceptional points, say $x_1$, $x_2$, $x_3$, $x_4$. Define $S' = S - \cup \mathring{U}(x_i)$, where $U(x_i)$, $1 \leq i \leq 4$, is a regular neighborhood of $x_i$. $r(T_2) = s_2$ and $r(T_3) = s_3$ are two non-isotopic essential curves in $S'$ which intersect themselves in two points. There are two other non-isotopic curves in $S'$ not isotopic to $s_2$ and $s_3$ and which intersect both $s_2$ and $s_3$ in two points. Let $s_4$ be one of them, and define $T_4 = r^{-1}s_4$.

We are now going to classify essential tori in $M$, up to isotopy and up to Dehn twists.

25.8 Assertion. *Every essential torus* $T$ *can be isotoped so that afterwards it is vertical either with respect to the* $(T_1,T_2)$*-fibration, or to the* $(T_2,T_3)$*-fibration.*

If $T$ is contained in $M_i$, for $i = 1$ or $2$, note that it cannot be isotoped in $M_i$ into a horizontal surface with respect to the $(T_1,T_2)$-fibration. Hence, by 5.6, $T$ can be isotoped in $M_i$, and so in $M$, into a vertical surface with respect to the $(T_1,T_2)$-fibration.

If, on the other hand, $T$ cannot be isotoped out of $T_2$, then $T$ can be isotoped so that $C_i = T \cap M_i$, $i = 1,2$, consists of essential annuli in $M_i$ (see 4.6). Hence $C_i$ is isotopic to a vertical surface in $M_i$, with respect to the I-fibration $q_i$. Now $K_i$ has only two 2-sided curves, up to isotopy, and, by our choice of $T_2$ and $T_3$, the preimages of these two curves are equal to $T_1 \cap M_i$ and $T_3 \cap M_i$, respectively. Therefore every component of $C_i$ is isotopic either to $T_1 \cap M_i$, or to $T_3 \cap M_i$, and so it follows that $T$ can be isotoped into a surface which is vertical either with respect to the $(T_1,T_2)$-fibration, or with respect to the $(T_2,T_3)$-fibration.

25.9 Assertion. *Let* $T$ *be an essential torus in* $M$. *Then there is a product* $\varphi$ *of Dehn twists of* $M$ *such that* $\varphi(T)$ *is isotopic either to* $T_1$, $T_2$, $T_3$ *or* $T_4$.

If $T$ is vertical with respect to the $(T_1,T_2)$-fibration, then $T$ is isotopic either to $T_1$ or to $T_2$, for $t_1$, $t_2$ are the only 2-sided curves in the Klein bottle $K$ (up to isotopy). So, by 25.8, we may suppose that $T$ is vertical with respect to the $(T_2,T_3)$-fibration. Then $t = r(T)$ is a simple closed essential curve in $S'$ which is not boundary-parallel. By [Li 1, lemma 2], there is a product $\varphi_1$ of Dehn twists of $S'$ such that $\varphi_1 t$ is isotopic in $S'$ to a curve which either does not meet $s_2$ at all, or in precisely two points. Since $S'$ is the 2-sphere minus four holes, one easily sees that there is a product $\varphi_2$ of Dehn twists such that $\varphi_2\varphi_1 t$ is isotopic in $S'$ either to $s_3$ or to $s_4$. Lifting the Dehn twists to Dehn twists of $M$ the claim follows.

Having classified tori in $M$, we are now ready to complete the proof of 25.7. For this keep in mind that $T_1$ is the only non-separating torus in $M$ (see 25.9).

Note that, by lemma 5 of [Li 2], the mapping class group of $K$ is isomorphic to $\mathbb{Z}/2\mathbb{Z} \oplus \mathbb{Z}/2\mathbb{Z}$. Let f', g' be the two homeomorphisms of $K$ which generate this group. Furthermore, if a $(T_2,T_3)$-fibration exists, let $h_3^*,h_4^*$ be the homeomorphisms of S' which map $s_2$ to $s_3$, resp. $s_4$, and let $h_3',h_4'$ be their extensions to S.

Now fix the following homeomorphisms of $M$ (which are not products of Dehn twists):

1. Let $f$ and $g$ be liftings of f' and g', resp. (see the proof of 25.3 for their existence).
2. Let $h_3$ and $h_4$ be liftings of $h_3'$ and $h_4'$ which map $T_2$ to $T_3$ or $T_4$, resp.
3. Let $h_1$ be the homeomorphism of $M$ which is the identity on $M_2$ and which on $M_1$ is the non-trivial homeomorphism as described in 25.1. Define $h_2$ similarly, but to be the identity on $M_1$.

Let $h$ be any homeomorphism. By 25.9, $h(T_1)$ is isotopic to $T_1$ ($T_1$ is the only non-separating torus) and there is a product, $\varphi$, of Dehn twists such that $\varphi h(T_2)$ is isotopic to $T_2$, $T_3$ or $T_4$. Hence there are integers $\varepsilon_3,\varepsilon_4 = 0$ or -1 such that

$$h_4^{\varepsilon_4}h_3^{\varepsilon_3}\cdot\varphi h$$

preserves both $T_1$ and $T_2$. Such a homeomorphism can be isotoped into a fibre preserving homeomorphism with respect to the $(T_1,T_2)$-fibration (see the proof of 5.9). Then, by our choice of $f$ and $g$, there are integers $\eta_1,\eta_2 = 0$ or 1, such that

$$g^{\eta_2}f^{\eta_1}h_4^{\varepsilon_4}h_3^{\varepsilon_3}\cdot\varphi h$$

induces the identity on the base $K$ (up to isotopy in $M$). Hence, by our choice of $h_1$ and $h_2$, it follows that, up to isotopy,

$$\varphi\cdot h = h_3^{-\varepsilon_3}h_4^{-\varepsilon_4}f^{-\eta_1}g^{-\eta_2}h_2^{\alpha}h_1^{\beta},$$

for some integers $\alpha$ and $\beta$. By the second additional remark of 25.1, it follows that $h_i^2$, $i = 1,2$, is isotopic to a Dehn twist along $p^{-1}\partial B_1 = \partial M_1$. Therefore, modulo Dehn twists, $h$ is isotopic to

$$h_3^{-\varepsilon_3} h_4^{-\varepsilon_4} f^{-\eta_1} g^{-\eta_2} h_2^{\varepsilon_2} h_1^{\varepsilon_1}, \qquad (*)$$

for some integers $\varepsilon_1,\ldots,\varepsilon_4,\eta_1,\eta_2 = 0$ or $\pm 1$ (observe that the Dehn twists generate a normal subgroup). This completes the proof since there are certainly only finitely many homeomorphisms of the form (*).

q.e.d.

As a result of the foregoing discussion we may summarize:

<u>25.10 Corollary</u>. *Let* $(M,\underline{\underline{m}})$ *be a Seifert fibre space, but not one of the exceptions* 5.1.1-5.1.5. *Then the admissible Dehn twists of* $(M,\underline{\underline{m}})$ *generate a subgroup of finite index in* $H^+(M,\underline{\underline{m}})$ *and so in* $H(M,\underline{\underline{m}})$.

<u>Remark</u>. Again we may replace $H(M,\underline{\underline{m}})$ by $H_A(M,\underline{\underline{m}})$, for any union $A$ of sides of $(M,\underline{\underline{m}})$.

## §26. Homeomorphisms of I-bundles

Before we are able to prove in the next paragraph our main theorem on the mapping class group of 3-manifolds, we need a certain technical result on homeomorphisms of I-bundles. This will be established in 26.3. This result might have some interest in its own since it leads to the definition of a geometric obstruction for a homeomorphism to be the identity.

Let $(X,\underline{\underline{x}})$ denote an I-bundle (twisted or not) with complete boundary-pattern, and with projection $p: X \to B$. Let $F$ be the union of the lids of $(X,\underline{\underline{x}})$, i.e. $F = (\partial X - p^{-1}\partial B)^-$, and let $\underline{\underline{f}}$ be the boundary-pattern of $F$ induced by $\underline{\underline{x}}$. Finally, denote by $d: (F,\underline{\underline{f}}) \to (F,\underline{\underline{f}})$ the fixpoint-free, admissible involution given by the reflections in the I-fibres.

<u>26.1 Lemma</u>. *Let $G$ be an essential surface in $(F,\underline{\underline{f}})$. Suppose that $G$ is in a very good position with respect to $dG$ (for the definition see §11). Let $h: (X,\underline{\underline{x}}) \to (X,\underline{\underline{x}})$ be an admissible homeomorphism with $h|(F - G)^- = \mathrm{id}|(F - G)^-$.*

*Then there is an admissible isotopy $h_t$, $t \in I$, of $h = h_0$, with $h_t(G) = G$, for all $t \in I$, such that $h_1|(\partial dG - \partial F)^- = \mathrm{id}|(\partial dG - \partial F)^-$ and $h_1|(F - G)^- = \mathrm{id}|(F - G)^-$.*

<u>Proof</u>. Denote by $k_1,\dots,k_n$, $n \geq 1$, all the components of $(\partial G - \partial F)^-$. Suppose that $h|dk_1 \cup\dots\cup dk_j = \mathrm{id}$, $j \geq 0$, and consider $k = k_{j+1}$. It suffices to show the existence of an admissible isotopy $h_t$, $t \in I$, of $h = h_0$, with $h_t(G) = G$, for all $t \in I$, such that $h_1|dk_1 \cup\dots\cup dk_j \cup dk = \mathrm{id}$ and $h_1|(F - G)^- = \mathrm{id}|(F - G)^-$.

$k$ is an essential curve (closed or not) in $(F,\underline{\underline{f}})$ since $G$ is an essential surface in $(F,\underline{\underline{f}})$. The preimage $p^{-1}(pk)$ is, in general, not a square or annulus, for $k \cap dk$ need not be empty. But it is easy to see that there is always an I-fibre preserving immersion $g_k: k \times I \to X$ with $g_k(k \times I) = p^{-1}pk$, and $g_k(k \times 0) = k$ and $g_k(k \times 1) = dk$.

Define $\ell = h^{-1}(dk)$. Observe that $h^{-1}|k = \mathrm{id}|k$, for

$h|(F - G)^- = id|(F - G)^-$, and that the immersion $g_k$ and the homeomorphism $h$ are both essential maps. Hence $h^{-1}g_k$ is an essential singular square or annulus in $(X,\underline{\underline{x}})$ with $k$ as one side. This implies that $h^{-1}g_k$ can be admissibly deformed (rel $k \times 0$) in $(X,\underline{\underline{x}})$ into a vertical map, i.e. into $g_k$. To see this observe that $p \circ h^{-1}g_k$ can be admissibly contracted (rel $k \times 0$) in the base $B$ into $p(k \times 0)$ and lift such a contradiction to an admissible homotopy of $h^{-1}gk$. The restriction of this homotopy to $(k \times 1) \times I$ defines an admissible deformation $f: k \times I \to F$ with $f|k \times 0 = \ell$ and $f|k \times 1 = dk$.

<u>Case 1.</u> $dk \cap (\partial G - \partial F)^-$ <u>is empty</u>.

In this case dk does not meet $\cup_{1 \leq i \leq j} dk_i$ or $(\partial G - \partial F)^-$. The same holds for $\ell$: that $\ell$ does not meet $\cup_{1 \leq i \leq j} dk_i$ follows from $h|\cup_{1 \leq i \leq j} dk_i = id$, $h(\ell) = dk$, and $dk \cap \cup_{1 \leq i \leq j} dk_i = \emptyset$, and that $\ell$ does not meet $(\partial G - \partial F)^-$ follows from $h|(F - G)^- = id$. Hence, and since $G$ is in a very good position with respect to dG it follows that $f$ can be admissibly deformed (rel $k \times \partial I$) so that afterwards $S = f^{-1}((\partial G - \partial F)^- \cup \cup_{1 \leq i \leq j} dk_i)$ is a system of pairwise disjoint curves which are admissibly parallel to the side $k \times 0$ of $k \times I$.

If $S$ is empty and dk lies in $(F - G)^-$, nothing is to show since $h|(F - G)^- = id|(F - G)^-$. If $S$ is empty and dk lies in $G$, the existence of the required isotopy $h_t$, $t \in I$, follows from Baer's theorem.

Thus we may suppose that $S$ is non-empty. Then $S$ splits $k \times I$ into squares or annuli. Let $A'$ be the one of them which contains $k \times 1$. Recall that dk does not meet $\cup_{1 \leq i \leq j} dk_i$ or $(\partial G - \partial F)^-$. Hence, by Nielsen's theorem, the existence of the map $f|A'$ implies the existence of an inner square or annulus $A$ in $(F,\underline{\underline{f}})$ with $(\partial A - \partial F)^- = t \cup dk$, where $t$ is either $dk_i$, for some $1 \leq i \leq j$, or a component of $(\partial G - \partial F)^-$. Moreover, it follows from our choice of $A'$ that $A^0 \cap ((\partial G - \partial F)^- \cup \cup_{1 \leq i \leq j} dk_i) = \emptyset$. Consider $h^{-1}A$, and note that $h^{-1}t = t$, for $h|\cup_{1 \leq i \leq j} dk_i = id$ and $h|(F - G)^- = id$. Hence $h^{-1}A$ is also an inner square or annulus in $(F,\underline{\underline{f}})$ with

$(h^{-1}A)^0 \cap ((\partial G - \partial F)^- \cup \cup_{1 \leq i \leq j} dk_i) = \emptyset$, and
$(\partial h^{-1}A - \partial F)^- = h^{-1}dk \cup h^{-1}t = \ell \cup t$. Now $\ell$ is admissibly isotopic, via $h^{-1}A$, to $t$ and then, via $A$, to $dk$. Extending these isotopies in the obvious way, we get the required isotopy $h_t$, $t \in I$, provided $h$ does not interchange the components of $(\partial U(t) - \partial F)^-$, where $U(t)$ is some regular neighborhood of $t$ with $h(U(t)) = U(t)$. But the latter must be true, for otherwise $h$ reverses the orientation of $F$ which would imply that $G = F$ since $h|(F - G)^- = id|(F - G)^-$.

<u>Case 2</u>. $dk \cap (\partial G - \partial F)^-$ <u>is non-empty</u>.

$G$ is an essential surface in $(F,\underline{\underline{f}})$ which is in a very good position with respect to $dG$. Hence we may suppose that $f$ is admissibly deformed (rel $k \times \partial I$) so that $f^{-1}(\partial G - \partial F)^-$ is a system of curves which join $k \times 0$ with $k \times 1$.

We first consider the subcase that $f^{-1}(\cup_{1 \leq i \leq j} dk_i)$ is empty. Let $a_1$ be a component of $(dk - G)^-$, and let $F_1$ be the component of $(F - G)^-$ which contains $a_1$. Then $a_1$ is an essential arc in $F_1$, for $G$ is in a very good position with respect to $dG$. $f|a_1 \times I$ is an admissible homotopy of $a_1$ in $F_1$, and since $h|(F - G)^- = id|(F - G)^-$ we have $a_1 = f(a_1 \times 0) = f(a_1 \times 1)$. If $f|a_1 \times I$ cannot be admissibly deformed (rel $a_1 \times \partial I$) into $a_1$, then, by Nielsen's theorem, $F_1$ has to be an inner annulus in $(F,\underline{\underline{f}})$. Moreover, it follows that $F_1 \cap \cup_{1 \leq i \leq j} dk_i = \emptyset$ since $f^{-1}(\cup_{1 \leq i \leq j} dk_i) = \emptyset$. Sliding $a_1$ around $F_1$ (if necessary), we may suppose that $h$ is isotoped so that $f|a_1 \times I$ now can be admissibly deformed (rel $a_1 \times \partial I$) into $a_1$, for all components $a_1$ of $(dk - G)^-$. In this situation, the existence of the homotopy $f$ shows that every component $b_2$ of $\ell \cap G$ can be admissibly deformed in $G$ into a component $a_2$ of $k \cap G$, using a deformation which is constant on $\partial b_2$ and which does not meet $\cup_{1 \leq i \leq j} dk_i$. In fact, by Baer's theorem, these deformations may be chosen as isotopies. Extending all these isotopies in the obvious way, we get the required isotopy $h_t$, $t \in I$.

Now let us suppose that $f$ cannot be admissibly deformed so that afterwards $f^{-1}(\cup_{1 \leq i \leq j} dk_i) = \emptyset$ and that $f^{-1}(\partial G - \partial F)^-$ consists of curves which join $k \times 0$ with $k \times 1$. Then $f$ can be admissibly deformed so that $f^{-1}(\cup_{1\leq i\leq j} dk_i)$ is a non-empty system of curves which are parallel to $k \times 0$. This system splits $k \times I$ into squares or annuli. Let A' be the one of them which contains $k \times 1$. Using this A' the existence of the required isotopy $h_t$ follows by a similar argument as in Case 1. q.e.d.

For the next lemma let $G$ be again an essential surface in $(F,\underline{\underline{f}})$, and suppose that $G$ is in a very good position with respect to dG. Let $U$ be a regular neighborhood of $(\partial G - \partial F)^-$ in $(F,\underline{\underline{f}})$. Denote by $C$ the <u>essential union</u> of $U$ and dU, i.e. the smallest essential surface in $(F,\underline{\underline{f}})$ containing $U \cup dU$.

<u>26.2 Lemma</u>. <u>Suppose that</u> $(X,\underline{\underline{x}})$ <u>is not the</u> I-<u>bundle over the annulus, torus, Möbius band, or Klein bottle</u>. <u>Let</u> $h\colon (X,\underline{\underline{x}}) \to (X,\underline{\underline{x}})$ <u>be an admissible homeomorphism with</u> $h|(F - G)^- = id|(F - G)^-$.

<u>Then there is an admissible isotopy</u> $h_t$, $t \in I$, <u>of</u> $h = h_0$, <u>with</u> $h_t(G) = G$, <u>for all</u> $t \in I$, <u>such that</u> $h_1|p^{-1}pC = id$ <u>and</u> $h_1|(F - G)^- = id$.

<u>Proof</u>. By 26.1, we may suppose that $h|(\partial G - \partial F)^- = id$ and $h|(\partial dG - \partial F)^- = id$. $h|F$ is orientation preserving, for $h|(F - G)^- = id$. Hence we may suppose that $h|U = id$ and $h|dU = id$, and hence also $h|C = id|C$. Observe that, by definition, $dC = C$.

Denote $N = p^{-1}pC$, and let $\underline{\underline{n}}$ be the boundary-pattern of $N$ induced by $\underline{\underline{x}}$. Then the fibration of $(X,\underline{\underline{x}})$ induces an admissible fibration of $(N,\underline{\underline{n}})$ as a system of I-bundles. $C$ is then the union of all the lids of these I-bundles.

By its very definition, $C$ is an essential surface in $(F,\underline{\underline{f}})$. Hence it follows that each component $A$ of $(\partial N - \partial X)^-$ is an essential square or annulus in $(X,\underline{\underline{x}})$. Since $h|F \cap \partial A = id|F \cap \partial A$ and by our suppositions on $(X,\underline{\underline{x}})$, it follows that $h|A$ can be admissibly deformed (rel $F \cap \partial A$) in $(X,\underline{\underline{x}})$ into $A$. By 5.5 of [Wa 4] (see 19.1), the deformation may be chosen as an isotopy, and this isotopy can be extended to an admissible isotopy of $h$, which is constant on $F$. Therefore it follows that $h$ can be admissibly isotoped (rel $F$) so that afterwards, $h(N) = N$.

Let $(N_1,\underline{\underline{n}}_1)$ be any component of $(N,\underline{\underline{n}})$, and let $\bar{\underline{\underline{n}}}_1$ be the completed boundary-pattern of $(N_1,\underline{\underline{n}}_1)$. Then $h|N_1\colon (N_1,\bar{\underline{\underline{n}}}_1) \to (N_1,\bar{\underline{\underline{n}}}_1)$ is an admissible homeomorphism with $h|F \cap N_1 = \mathrm{id}|F \cap N_1$. If $(N_1,\bar{\underline{\underline{n}}}_1)$ is not the I-bundle over the annulus or Möbius band, it follows, by an argument of 5.9, that $h|N_1\colon (N_1,\bar{\underline{\underline{n}}}_1) \to (N_1,\bar{\underline{\underline{n}}}_1)$ can be admissibly isotoped into the identity, using an isotopy which is constant on $N_1 \cap F$. This is also true if $(N_1,\bar{\underline{\underline{n}}}_1)$ is the I-bundle over the Möbius band. To see this,note that in this case $N_1$ is a regular neighborhood of a vertical Möbius band. Moreover, every homeomorphism of the Möbius band which is the identity on the boundary is isotopic (rel boundary) to the identity.

Let $\bar{N}$ be a union of components of $N$ such that $h|\bar{N} = \mathrm{id}|\bar{N}$. By what we have seen so far, we may suppose that $\bar{N}$ is chosen so that $N - \bar{N}$ consists of I-bundles over the annulus.

So let $N_1$ be any component of $N - \bar{N}$. Then $(N_1,\bar{\underline{\underline{n}}}_1)$ is an I-bundle over the annulus and we may suppose that $h|N_1\colon (N_1,\bar{\underline{\underline{n}}}_1) \to (N_1,\bar{\underline{\underline{n}}}_1)$ cannot be admissibly isotoped to the identity, using an isotopy which is constant on $N_1 \cap F$. It remains to show that there is an admissible isotopy $h_t$, $t \in I$, of $h = h_0$, with $h_t(G) = G$ and $h_t(N) = N$, such that $h_1|\bar{N} \cup N_1 = \mathrm{id}|\bar{N} \cup N_1$ and $h_1|(F - G)^- = \mathrm{id}|(F - G)^-$.

For this consider $N_1$ as a regular neighborhood of a vertical annulus $A_1$ in $(X,\underline{\underline{x}})$. Without loss of generality, one boundary component of $A_1$, say $k_1$, is a component of $(\partial G - \partial F)^-$ and the other one, say $k_2$, is contained either in $G$ or in $(F - G)^-$ without meeting $(\partial G - \partial F)^-$ (recall our choice of $N_1$).

If $k_2$ lies in $G$, observe that $h|A_1\colon A_1 \to A_1$ is isotopic to the identity, using an isotopy which is constant on $k_1$. Extending such an isotopy to an admissible isotopy of $h$ which is constant outside a regular neighborhood of $N_1$, we find the required isotopy $h_t$.

If $k_2$ lies in $(F - G)^-$, then $\partial A_1$ lies in $(F - G)^-$. It follows that, for one component $X_1$ of $(X - N)^-$ which meets $N_1$, all lids are contained in $(F - G)^-$. Let $B$ be an essential vertical square in $(X_1,\bar{\underline{\underline{x}}}_1)$ which meets $N_1$, where $\bar{\underline{\underline{x}}}_1$ denotes the completed (boundary-pattern of $(X_1,\underline{\underline{x}}_1)$. Since $h|(F - G)^- = \mathrm{id}|(F - G)^-$, we have that $h|B$, together with $\mathrm{id}|B$, defines an admissible singular annulus in $(X_1,\bar{\underline{\underline{x}}}_1)$. By our suppositions on $h|N_1$, this singular annulus is essential in $(X_1,\underline{\underline{x}}_1)$ and cannot be admissibly deformed into a vertical map. Hence, applying Nielsen's theorem to the composition of this singular annulus and the projection p, we find that $(X_1,\bar{\underline{\underline{x}}}_1)$ has to be the I-bundle over the annulus or Möbius band. But it cannot be the I-bundle over the Möbius band, for $h|X_1\colon (X_1,\bar{\underline{\underline{x}}}_1) \to (X_1,\bar{\underline{\underline{x}}}_1)$ cannot be admissibly isotoped (rel F) into the identity since, by supposition, $h|N_1$ cannot.

By what we have seen so far, $(X_1,\bar{\underline{\underline{x}}}_1)$ has to be the I-bundle over the annulus, Moreover, $h|X_1\colon (X_1,\bar{\underline{\underline{x}}}_1) \to (X_1,\bar{\underline{\underline{x}}}_1)$ cannot be admissibly isotoped into the identity, using an isotopy which is constant on $X_1 \cap F$. Thus, in particular, $X_1$ cannot meet $\bar{N}$. So, either $(\partial X_1 - \partial X)^-$ is connected or $X_1$ meets a component $N_2$ of N which is also an I-bundle over the annulus. Again, consider $N_2$ as a regular neighborhood of vertical annulus $A_2$ in $(X,\underline{\underline{x}})$. Without loss of generality, one boundary component, say $\ell_1$, of $A_2$ is a component of $(\partial G - \partial F)^-$ and so the other one, say $\ell_2$, is a component of $(\partial dG - \partial F)^-$. Since $X_1$ is an I-bundle over the annulus, it follows that $k_1$ and $\ell_1$, resp. $k_1$ and $\ell_2$, bound an inner annulus in $(F,\underline{\underline{f}})$. Since G is in a very good position to dG, it follows that $k_1$ and $\ell_1$ bound an inner annulus, i.e. $k_1$ lies in a component of $(F - G)^-$ which is an inner annulus in $(F,\underline{\underline{f}})$. Let H be the lid of $N_1 \cup X_1 \cup N_2$ which contains $k_2$. Observe that $h|N_1 \cup X_1 \cup N_2$ is admissibly isotopic in $N_1 \cup X_1 \cup N_2$ to the identity, using an isotopy which is constant on H. Extending this isotopy to an admissible isotopy of h which is constant outside of a regular neighborhood of $N_1 \cup X_1 \cup N_2$ we find the required isotopy $h_t$. q.e.d.

Again let G be an essential surface in $(F,\underline{\underline{f}})$, and suppose that G is in a very good position with respect to dG.

<u>26.3 Proposition.</u> <u>Suppose that</u> $(X,\underline{\underline{x}})$ <u>is not the I-bundle over the annulus, Möbius band, torus, or Klein bottle.</u> <u>Let</u> $h\colon (X,\underline{\underline{x}}) \to (X,\underline{\underline{x}})$ <u>be an admissible homeomorphism with</u> $h|(F - G)^- = \mathrm{id}$.

<u>Then there is an admissible isotopy</u> $h_t$, $t \in I$, <u>of</u> $h = h_0$, <u>with</u> $h_t(G) = G$, <u>for all</u> $t \in I$, <u>such that</u> $h_1|p^{-1}pH = id|p^{-1}pH$, <u>where</u> $H$ <u>is the essential union of</u> $(F - G)^-$ <u>and</u> $(F - dG)^-$.

<u>Remark</u>. In the appendix we establish a similar result for homotopy equivalences of I-bundles (see 31.1).

<u>Proof</u>. Let $U$ be the regular neighborhood of $(\partial G - \partial F)^-$ in $(F, \underline{\underline{f}})$, and define $C$ to be the essential union of $U$ and $dU$. Then, by 26.2, we may suppose that $h|p^{-1}pC = id|p^{-1}pC$ and $h|(F - G)^- = id|(F - G)^-$. Let $H'$ be the union of $C$ with all the components of $(F - C)^-$ which lie either in $(F - G)^-$ or in $(F - dG)^-$. Then observe that $H$ is contained in $H'$.

Let $X_1$ be any component of $(X - p^{-1}pC)^-$ with $X_1 \cap F \subset H'$. Then, by the definition of $H'$, at least one lid of $X_1$ lies in $(F - G)^-$ It suffices to show that $h|X_1 \colon X_1 \to X_1$ is admissibly isotopic to the identity, using an isotopy which is constant on $(\partial X_1 - \partial X)^-$ and all the lids of $X_1$ which lie in $(F - G)^-$. By our suppositions on $h$, this follows by an argument of 5.9. q.e.d.

The above result leads us in a natural way to a construction of a certain <u>geometric obstruction for homeomorphisms of surfaces</u>. Let us close this paragraph with a brief description of this obstruction. This is based on the following observation. Let $G_0$ and $G_1$ be two (not necessarily disjoint) essential surfaces in a surface $(F, \underline{\underline{f}})$. Let $h$ be an admissible homeomorphism of $(F, \underline{\underline{f}})$ with $h|(F - G_0)^- = id$ and which can be admissibly isotoped into $h_1$ with $h_1|(F - G_1)^- = id$. The isotopy in question may be considered as an admissible homeomorphism of $F \times I$. So it follows from 26.3 that $h$ can be admissibly isotoped into a homeomorphism which is the identity on the essential union of $(F - G_0)^-$ and $(F - G_1)^-$. In fact, since $F$ is compact, one can prove that for any given admissible homeomorphism $h$ of $(F, \underline{\underline{f}})$ there is an essential surface $F(h)$ in $F$ with

1. $h$ can be admissibly isotoped so that afterwards $h|(F - F(h))^- = id$.

2. F(h) can be admissibly isotoped into every essential surface with 1.

3. F(h) minus a component of F(h) does not satisfy 1.

The surface F(h) is unique (up to admissible isotopy), and it can be considered as a geometric obstruction for the homeomorphism h. It measures whether or not h is admissibly isotopic to the identity, and how far away it is from being the identity.

In §30 we shall describe a similar geometric obstruction for homotopy equivalences of surfaces with boundary. Indeed, see the proof of 30.15 for a rigorous argument concerning the existence of the above mentioned surface F(h).

## §27. On the mapping class group of 3-manifolds

The aim of this paragraph is to study to which extent the Dehn twists along embedded annuli and tori generate the whole mapping class group of a sufficiently large 3-manifold. The key result in this direction is that <u>the mapping class group of a simple 3-manifold is always finite</u> (see 27.1). Combining this with the results of §25, we shall see that the Dehn twists generate a subgroup of finite index in the whole mapping class group (see 27.6).

As an application we shall show that the 2-sphere is the only closed surface whose homeomorphisms are all extendable to a 3-manifold (see 27.9).

Furthermore recall from 24.2 that every homotopy equivalence $f: M_1 \to M_2$ between simple 3-manifold can be deformed into a homeomorphism. Moreover, every isomorphism $\varphi: \pi_1 M_1 \to \pi_1 M_2$ is induced by a homotopy equivalence. Hence we have the following:

<u>Corollary</u>. <u>If</u> $M$ <u>is a simple</u> 3-<u>manifold, then the outer automorphism group of</u> $\pi_1 M$ <u>is a finite group</u>.

To give a concrete example, let $k$ be any non-trivial knot in $S^3$ which is not a torus knot and which has no companions. Then the outer automorphism group of the knot group of $k$ is a finite group, and the knot space of $k$ admits only finitely many homeomorphisms, up to isotopy.

On the other hand observe that not every knot has a finite outer automorphism group. E.g., consider a knot $k$ which is <u>not</u> a Neuwirth knot. Imagine $k$ as a closed braid contained in the standard solid torus $V$ in $S^3$. Now map $V$ homeomorphically onto the regular neighborhood $U$ of $k$. The image of $k$ is a knot $\ell$ contained in $U$. The knot space of $\ell$ has infinite mapping class group. For we find Dehn twists $h$ of the knot space of $\ell$ such that no power of $h$ is isotopic to the identity.

After these general remarks we study the mapping class group more closely.

<u>27.1 Proposition.</u> <u>Let</u> $(M,\underline{\underline{m}})$ <u>be</u> <u>a</u> <u>simple</u> 3-<u>manifold</u> <u>with</u> <u>complete</u> <u>and</u> <u>useful</u> <u>boundary-pattern</u>. <u>Then</u> $H(M,\underline{\underline{m}})$ <u>is</u> <u>a</u> <u>finite</u> <u>group</u>.

<u>Proof</u>. The proof is based on the following two finiteness theorems:

1. In a simple 3-manifold there are, up to admissible isotopy, only finitely many essential surfaces of a given admissible homeomorphism type. See [Ha 1].
2. The proposition is true for Stallings fibrations which are simple 3-manifolds. See [He 1].

As a first consequence of these two facts, we show that the mapping class group of all simple Stallings manifolds is finite. Here a Stallings manifold means a 3-manifold $(M,\underline{\underline{m}})$ which contains an essential surface $F$ such that $(M - U(F))^-$ consists of I-bundles, where $U(F)$ denotes a regular neighborhood of $F$ in $(M,\underline{\underline{m}})$. By 2 above, we may suppose that $(M - U(F))^-$ consists of two twisted I-bundles, say $M_1$, $M_2$. $M_1$ and $M_2$ have product I-bundles $\tilde{M}_1$, $\tilde{M}_2$, respectively, as 2-sheeted coverings. Attaching the lids of $\tilde{M}_1$ and $\tilde{M}_2$ in the obvious way, we obtain a manifold $\tilde{M}$ and a 2-sheeted covering $p: \tilde{M} \to M$. By 1 above, it suffices to show that the subgroup of $H(M,\underline{\underline{m}})$ generated by all admissible homeomorphisms $h: (M,\underline{\underline{m}}) \to (M,\underline{\underline{m}})$ with $h(F) = F$ is finite. Since $\underline{\underline{m}}$ is a finite set, we may restrict ourselves to the case that $\underline{\underline{m}}$ is the set of all boundary components of $M$. $h|M_i$ is an admissible homeomorphism of $M_i$, and so, as a well-known fact, it can be lifted to an admissible homeomorphism $\tilde{h}_i$ of $\tilde{M}_i$. The two liftings $\tilde{h}_1$ and $\tilde{h}_2$ define a lifting $\tilde{h}: \tilde{M} \to \tilde{M}$. By construction, $\tilde{M}$ is a Stallings fibration, and, by 12.6 and 12.7, it is a simple 3-manifold. Hence, by 2 above, there are only finitely many homeomorphisms $\tilde{h}: \tilde{M} \to \tilde{M}$, up to isotopy. Hence it remains to prove that $\tilde{h}$ is isotopic to the identity if and only if $h$ is. This in turn follows from (7) of [Zi 1]. Indeed, all suppositions of (7) of [Zi 1] are satisfied: a homeomorphism of $\tilde{M}$ is isotopic to the identity if and only if it is homotopic to the identity [Wa 4]. Moreover, the centralizer of $p_*\pi_1\tilde{M}$ is trivial in $\pi_1 M$. For otherwise $\pi_1\tilde{M}$ has non-trivial center since $\pi_1 M$ is torsion-free [Ep 1] and since $p_*\pi_1\tilde{M}$ has finite index in $\pi_1 M$. Then, by [Wa 3], $\tilde{M}$ has to be a Seifert fibre space, and so $M$ (see 12.9 and 6.8). But this is a contradiction to the fact that $M$ is a simple 3-manifold.

Now we come to the proof of the general case. It is by induction on a great hierarchy

$$(M,\underline{\underline{m}}) = (M_o,\underline{\underline{m}}_o),(M_1,\underline{\underline{m}}_1),\ldots,(M_n,\underline{\underline{m}}_n), \quad n \geq 1,$$

(see 24.3 for the definition of a great hierarchy).

If $j \geq 1$ is an odd integer, denote by $H(M_j,\underline{\underline{m}}_j,F_j)$ the subgroup of $H(M_j,\underline{\underline{m}}_j)$ generated by all the admissible homeomorphisms of $(M_j,\underline{\underline{m}}_j)$ which preserve $U(F_j)$. Of course, $H(M_n',\underline{\underline{m}}_n')$ is a finite group, for all components $M_n'$ of $M_n$ which are simple 3-manifolds. So by the facts quoted in the beginning of the proof, it suffices to prove the following:

<u>27.2 Lemma</u>. <u>If</u> $H(M_{2i+1},\underline{\underline{m}}_{2i+1})$ <u>is finite</u>, <u>and if</u> $M_{2i-1}$ <u>is not a Stallings manifold</u>, <u>then</u> $H(M_{2i-1},\underline{\underline{m}}_{2i-1},F_{2i-1})$ <u>is finite</u>.

To begin with we simplify the notations somewhat, and we write $(N_0,\underline{\underline{n}}_0) = (M_{2i-1},\underline{\underline{m}}_{2i-1})$, $(N_1,\underline{\underline{n}}_1) = (M_{2i},\underline{\underline{m}}_{2i})$ and $(N_2,\underline{\underline{n}}_2) = (M_{2i+1},\underline{\underline{m}}_{2i+1})$. Moreover, denote $F = F_{2i-1}$ and $H = (\partial U(F) - \partial N_0)^-$. Observe that $\underline{\underline{n}}_0$, together with the components of $H$, induces a boundary-pattern of the regular neighborhood $U(F)$ which makes $U(F)$ into a product I-bundle.

By 10.9, the characteristic submanifold of a 3-manifold is unique, up to admissible ambient isotopy. This means that every admissible homeomorphism of $(N_1,\underline{\underline{n}}_1)$ can be admissibly isotoped so that it preserves the characteristic submanifold $V_1$ of $(N_1,\underline{\underline{n}}_1)$. This, together with the suppositions of lemma 27.2, implies the following: there are finitely many admissible homeomorphisms $g_1,\ldots,g_m$ of $(N_0,\underline{\underline{n}}_0)$ with $g_j(U(F)) = U(F)$, for all $1 \leq j \leq m$, such that for a given admissible homeomorphism $g$, $g \in H(N_0,\underline{\underline{n}}_0,F)$, $g|N_1$ can be admissibly isotoped in $(N_1,\underline{\underline{n}}_1)$ so that afterwards

$$g|(N_1 - V_1)^- = g_j|(N_1 - V_1)^-, \quad \text{for some} \quad 1 \leq j \leq m.$$

We claim that even $g$ is admissibly isotopic to $g_j$. Since $g$ is arbitrarily given, this would prove 27.2.

Define $h = g_j^{-1}g$. Then $h(N_1) = N_1$ and $h|(N_1 - V_1)^- = \text{id}$. It remains to show that $h$ is admissibly isotopic to the identity. By the following assertion, it suffices to prove that the restriction $h|H$ can be admissibly isotoped in $H$ into the identity.

<u>27.3 Assertion.</u> <u>Suppose that</u> $h|H$ <u>is admissibly isotopic in</u> $H$ <u>to the identity.</u> <u>Then</u> $h$ <u>is admissibly isotopic in</u> $(N_0, \underline{\underline{n}}_0)$ <u>to the identity.</u>

Since $(F, \underline{\underline{f}})$ is not an annulus or torus, it is easily seen that there is an admissible isotopy $\varphi_t$, $t \in I$, of $h|H$ with $\varphi_t(H) = H$ and $\varphi_t(V_1 \cap H) = V_1 \cap H$, for all $t \in I$, and $\varphi_1 = \text{id}|H$ (apply the theorems of Nielsen and Baer).

Recall that $(N_0, \underline{\underline{n}}_0)$ is a simple 3-manifold. Hence every component of $(\partial V_1 - \partial N_1)^-$ has to meet $U(F)$, and this in turn implies that every component of $V_1$ and every component of $(N_1 - V_1)^-$ meets $U(F)$. More precisely, we have a partition of $N_0$ consisting of the following parts:

1. the regular neighborhood of $F$, $U(F)$,
2. components of $(N_1 - V_1)^-$ which are not I-bundles over the square or annulus,
3. I-bundles of $V_1$ which meet $U(F)$ in lids, but which are not I-bundles over the square or annulus,
4. I-bundles over discs which do not meet $U(F)$ in lids, and Seifert fibre spaces over discs with at most one exceptional fibre (i.e. solid tori).

By 10.4, the parts described in 2 meet $H$ in an essential surface whose components are different from inner squares or annuli.

$h$ is an admissible homeomorphism which preserves this partition, and, of course, $\varphi_t$, can be extended to an admissible isotopy $h_t$, $t \in I$, of $h$ which preserves the partition and which is constant outside a neighborhood of $H$. In fact, $h_t$ may be chosen so that, in addition, $h_1$ is the identity on $U(F)$ and on all parts of the partition described in 2. To see this note first that $U(F)$ is a product I-bundle and that the regular neighborhood of $H$

intersects every part of the partition in a system of product I-bundles. Then recall that $h|(N_1 - V_1)^-$ is the identity, and observe that every admissible homeomorphism of an I-bundle which is the identity on the lids can be admissibly isotoped into the identity (compare the proof of 3.5 of [Wa 4]), and this isotopy may be chosen to be constant on the lids provided the base of the I-bundle is not an annulus or a torus. Moreover, this isotopy may be chosen to be constant on all the sides of the I-bundle on which the homeomorphism is already the identity. Hence and since every part of the partition meets U(F), this implies that $h_t$ may be chosen so that, in addition, $h_1$ is the identity on all the parts as described in 3. Therefore we may suppose that $h_1$ is the identity on all parts except those described in 4.

So, let X be a submanifold of the partition as described in 4. Let A be the union of all the sides of X which are contained in parts of the partition different from X. Then it follows from the properties of X that A is connected, for otherwise we find an essential square or annulus in X which does not meet U(F), which is impossible since $(N_0, \underline{\underline{n}}_0)$ is simple. Hence and by the properties of X, every admissible homeomorphism of X which is the identity on A can be admissibly isotoped to the identity, using an isotopy which is constant on A. By the suppositions on the isotopy $h_t$, $t \in I$, this implies the assertion.

In order to prove the supposition of 27.3, i.e. that $h|H$ is admissibly isotopic in H to the identity, we introduce the concept of "good submanifolds".

An essential F-manifold W in $(N_1, \underline{\underline{n}}_1)$ is called a <u>good submanifold</u>, if

(i) W meets H in an essential surface G with the property: no component of $(H - G)^-$ is an inner square or annulus in H which meets a component of G which is also an inner square or annulus,

(ii) there is an admissible isotopy of h which preserves U(F) and which moves h so that afterwards $h(W) = W$ and $h|(H - G)^- = id|(H - G)^-$.

In the remainder of the proof the property (i) of an essential surface in $H$ will be called the square- and annulus-property.

27.4 Assertion. There is at least one good submanifold in $(N_1, \underline{\underline{n}}_1)$.

We obtain a good submanifold by modifying the characteristic submanifold $V_1$ of $(N_1, \underline{\underline{n}}_1)$. Indeed, by what we have seen so far, $V_1$ satisfies (ii), and $V_1 \cap H$ is an essential surface in $H$. Suppose that there is a component $A$ of $(H - V_1)^-$ which is an inner square (resp. annulus) in $H$ and which meets a component $B$ of $V_1 \cap H$ which itself is also an inner square (resp. annulus) in $H$. Let $U(B)$ be a regular neighborhood of $B$ in $(N_1, \underline{\underline{n}}_1)$, and define $V_1' = (V_1 - U(B))^-$. Then $V_1'$ satisfies (ii), for $V_1$ satisfies (ii) and since $h|B \cup A$ is isotopic to the identity, by an isotopy in $B \cup A$ which is constant on $\partial B - A$. Thus, after finitely many steps, we obtain an admissible F-manifold with (i) and (ii). Removing trivial components from this F-manifold, we finally get a good submanifold. This completes the proof of 27.4.

To continue the proof, let $W$ be any good submanifold in $(N_1, \underline{\underline{n}}_1)$. A moments reflection shows that, by 9.1, we may suppose that $W$ is in fact chosen so that for every good submanifold $W'$ with $W' \subset W$, the essential surface $W \cap H$ can be admissibly isotoped in $H$ into $W' \cap H$. With this choice of $W$ we can show:

27.5 Assertion. $W$ can be admissibly isotoped in $(N_1, \underline{\underline{n}}_1)$ so that afterwards

$$W \cap H = d(W \cap H),$$

where $d: H \to H$ is the involution given by the reflections in the fibres of the product I-bundle $U(F)$.

To prove this, define $G = W \cap H$, and suppose that $G$ is in a very good position to $dG$. Of course, this position can always be obtained, using an admissible isotopic deformation of $W$ in $(N_1, \underline{\underline{n}}_1)$.

Denote by $G'$ the essential intersection of $G$ and $dG$.

Then, by the very definition, $(H - G')^-$ is the essential union of $(H - G)^-$ and $(H - dG)^-$. Since $W$ is a good submanifold, $G$ has the square- and annulus-property, and it follows that $G'$ has also this property.

$U(F)$ is a product I-bundle. Setting $X = U(F)$ and $h = h|U(F)$, we see that we may apply 26.3. Hence it follows the existence of an admissibly isotopy $\varphi_t$, $t \in I$, of $h|H$, with $\varphi_t(H) = H$ and $\varphi_t(G) = G$, for all $t \in I$, such that $\varphi_1|(H - G')^- = id|(H - G')^-$.

Let $G_1$ be a component of $G$. Then, to prove 27.5, it remains to show that $G_1$ contains precisely one component $G_1'$ of $G'$ and that $G_1$ can be admissibly contracted in $H$ to $G_1'$.

Case 1. $G_1$ is an inner square or annulus in $H$.

It follows from the existence of the isotopy $\varphi_t$ that $G_1$ contains at least one component $G_1'$ of the essential intersection $G'$. For otherwise, removing trivial components from $(W - U(G_1))^-$ (if necessary) we obtain a good submanifold $W'$ such that $G$ cannot be admissibly isotoped into $W' \cap H$ (recall that $G$ has the square- and annulus-property), where $U(G_1)$ is a regular neighborhood of $G_1$ in $(N_1, \underline{\underline{n}}_1)$. This, however, contradicts our choice of $W$. Since $G'$ has the square- and annulus-property, $G_1$ contains at most one component of $G'$. Thus, of course, $G_1$ can be admissibly contracted in $H$ to $G_1'$.

Case 2. $G_1$ is not an inner square or annulus in $H$.

Recall that $G_1$ is a component of $H \cap W$. Let $X$ be the component of $W$ which contains $G_1$. Since we are in Case 2 and since $W$ is an essential F-manifold, it follows that $X$ is an I-bundle and that $G_1$ is one lid of $X$.

Let $p\colon X \to B$ be the projection, and let $G_1^+ = (\partial X - p^{-1}\partial B)^-$. Then $G_1$ is a component of $G_1^+$. Denote by $e\colon G_1^+ \to G_1^+$ the involution given by the reflections in the I-fibres of $X$. As boundary-pattern of $G_1^+$, we fix the boundary-pattern induced by $\underline{\underline{n}}_0$, together with the set of components of $(\partial G_1^+ - \partial H)^-$. Then $e$ is an admissible homeomorphism of $G_1^+$.

Define $G_1'' = G' \cap G_1^+$. Since $G'$ is the essential intersection of $G$ and $dG$, $G_1''$ is an essential surface in $H$. Since $G$ is in a very good position to $dG$, it follows that $G_1''$ is even an essential surface in $G_1^+$. Moreover, we may suppose that $W$ is admissibly isotoped so that $G_1''$ is in a very good position with respect to $e(G_1'')$.

Since $W$ is a good submanifold, we may suppose that $h$ is admissibly isotoped so that $h(W) = W$ and $h|(H - W)^- = id$. In particular, $h|X$ is an admissible homeomorphism of $X$. Setting $h = h|X$ and $G = G_1''$, we claim that 26.3 may be applied. For this it remains to show that $h|X$ can be admissibly isotoped in $X$ so that afterwards $h|(G_1^+ - G_1'')^- = id$. But this follows immediately from the existence of the admissible isotopy $\varphi_t$ of $h|H$ defined in the beginning of 27.5.

Now, by 26.3, $h|X$ can be admissibly isotoped in $X$ so that afterwards $h|p^{-1}R = id$, where $R$ is the essential union of $p(G_1^+ - G_1'')^-$ and $p(G_1^+ - eG_1'')^-$. In general, however, this isotopy cannot be chosen to be constant on $(\partial X - \partial N_1)^-$. Therefore we also fix a regular neighborhood $U$ of $(\partial X - \partial N_1)^-$ in $X$, and we define

$$W' = (W - X) \cup p^{-1}R \cup U.$$

Then it is easily checked that $W'$ is an essential F-manifold in $(N_1, \underset{=}{n}_1)$ with property (ii). Without loss of generality, $W'$ also has property (i), i.e. $W'$ is a good submanifold. For, if this is not the case, we simply have to add the components of $(X - W')^-$ to $W'$ which are I-bundles over the square or annulus (recall that $W$ has property (i)).

By our choice of $W$, the essential surface $H \cap W$ can be admissibly isotoped in $H$ into $W' \cap H$. In particular, $H \cap X$ can be admissibly isotoped into $H \cap p^{-1}R$. By definition of $R$, this implies that $G_1$ can be admissibly contracted to $G_1'$ (recall that $W'$ has property (i)).

This completes the proof of 27.5.

Since, by 27.5, we may suppose that $W \cap H = d(W \cap H)$, there is a system $Z$ of I-bundles in $U(F)$ with $Z \cap H = W \cap H$. The submanifold

$$W^+ = W \cup Z$$

consists of essential I-bundles, Seifert fibre spaces, and Stallings manifolds in $(N_0, \underline{\underline{n}}_0)$.

Since $N_0 = M_{2i}$ is a simple 3-manifold, the characteristic submanifold $V_0$ of $(N_0, \underline{\underline{n}}_0)$ is trivial. Hence also $W^+$ is trivial, i.e. $W^+$ is contained in a regular neighborhood of some sides of $(N_0, \underline{\underline{n}}_0)$ (note that, by the suppositions of 27.2, $N_0$ is not a Stallings manifold and that, by 10.7, $(\partial W^+ - \partial N_0)^-$ can be admissibly isotoped into $V_0$). In particular, $H \cap W$ is contained in a regular neighborhood of some sides of $H$. Hence it follows from property (ii) of $W$, that $h|H$ can be admissibly isotoped in $H$ into the identity. This completes the proof of 27.2. q.e.d.

<u>27.6 Corollary</u>. <u>Let</u> $(M, \underline{\underline{m}})$ <u>be a sufficiently large, irreducible</u> 3-<u>manifold with complete and useful boundary-pattern</u>.

<u>Then the admissible Dehn twists of</u> $(M, \underline{\underline{m}})$ <u>generate a subgroup of finite index in</u> $H(M, \underline{\underline{m}})$.

<u>Proof</u>. By 10.9, the characteristic submanifold $V$ of $(M, \underline{\underline{m}})$ is unique, up to admissible ambient isotopy. Hence it follows that every admissible homeomorphism $h$ of $(M, \underline{\underline{m}})$ can be admissibly isotoped so that afterwards $h(V) = V$.

Now, $(M - V)^-$ is a union of simple 3-manifolds, together with components which are product I-bundles over the square or annulus. Thus, by 27.1, there are finitely many homeomorphisms $h_1, \ldots, h_m$ of $(M, \underline{\underline{m}})$ with $h_i(V) = V$, for all $1 \leq i \leq m$, such that $h$ can be admissibly isotoped so that afterwards

$$h|(M - V)^- = h_j|(M - V)^-, \quad \text{for some} \quad 1 \leq j \leq m.$$

$h \cdot h_j^{-1}$ is a homeomorphism of $H_{\overline{M-V}}(M, \underline{\underline{m}})$ , i.e. the identity on $(M - V)^-$. Recall that $V$ consists of essential I-bundles and

Seifert fibre spaces. The mapping class group of a surface (orientable or not) contains a subgroup of finite index generated by Dehn twists [De 1], [Li 1], [Li 2,4]. Hence this is also true for I-bundles since every homeomorphism of an I-bundle can be admissibly isotoped into a fibre-preserving one (see 5.9). Hence and by 25.10, it follows that the Dehn twists of $H_{\overline{M-V}}(M,\underline{\underline{m}})$ generate a subgroup of finite index in $H_{\overline{M-V}}(M,\underline{\underline{m}})$. Since every conjugate of a Dehn twist is again a Dehn twist, this subgroup is even a normal subgroup. Hence there are finitely many admissible homeomorphism $g_1,\dots,g_n$ in $(M,\underline{\underline{m}})$ so that every admissible homeomorphism in $(M,\underline{\underline{m}})$ which is the identity on $(M - V)^-$ is admissibly isotopic to $\sigma g_k$, for some $1 \leq k \leq n$ and some product $\sigma$ of admissible Dehn twists.

In particular, $h \cdot h_j^{-1}$ is admissibly isotopic to $\sigma g_k$, for some $1 \leq k \leq n$ and some product $\sigma$ of admissible Dehn twists. Then $h = \sigma \cdot g_k \cdot h_j$, up to admissible isotopy. Of course, there are only finitely many products $g_k \circ h_j$, $1 \leq j \leq m$ and $1 \leq k \leq n$, and this proves that the admissible Dehn twists of $(M,\underline{\underline{m}})$ generate a subgroup of finite index in $H(M,\underline{\underline{m}})$. q.e.d.

As an application of the above result, we consider the problem of extending a homeomorphism of a surface $F$ to a 3-manifold $M$ with $\partial M = F$. Since in the view of 27.1 the mapping class group of 3-manifolds is not very large, it seems to be unlikely that every homeomorphism of a surface can be extended. Indeed we shall construct below, for any surface different from $S^2$, homeomorphisms which cannot be extended. For a more algebraic approach to the extension problem see [Ne 1]. There it is also shown that the cobordism group of homeomorphisms of surfaces is very complicated (at least not finitely generated). But the precise structure of this group is not known. This contrasts with the cobordism group of 3-manifold homeomorphisms which is known to be trivial [Me 1].

Our examples are based on the following observations.

The first one is a consequence of 27.6. To describe it, let $M$ be any irreducible 3-manifold whose boundary consists of tori $T_1,\dots,T_n$, $n \geq 1$, and let $g$ be a homeomorphism of $M$ with

$g(T_i) = T_i$. Then the following holds:

<u>27.7 Lemma</u>. <u>Suppose</u> $g|T_1$ <u>is not isotopic to a periodic homeomorphism and that there is no essential curve</u> $k$ <u>in</u> $T_1$ <u>which is isotopic to its image under</u> $g|T_1$. <u>Then</u> $M$ <u>is torus</u> $\times$ I.

<u>Proof</u>. $M$ cannot be a solid torus. Otherwise take a meridian disc $D$ in $M$. Certainly the intersection number of $\partial D$ and $g(\partial D)$ is zero. This means, since $\partial M$ consists of tori, that $\partial D$ is isotopic to $g(\partial D)$.

Since $M$ is irreducible and since $\partial M$ consists of tori, this implies that $M$ is boundary-irreducible. Hence, by 9.4, the characteristic submanifold $V$ of $M$ exists. Furthermore, by 10.9, we may suppose that $g$ is isotoped so that $g(V) = V$. Since $\partial M$ consists of tori, $\partial M$ is contained in $V$ (see 10.6), and let $X_1$ be the component of $V$ which contains $T_1$.

$X_1$ cannot be a regular neighborhood of $T_1$. Otherwise the component $Y$ of $(M - V)^-$ which meets $X_1$ is a simple 3-manifold (see 10.4) and so, by 27.1, some power of $g|Y$ is isotopic to the identity. It is well-known (see [Ni 4]) that then $g|T_1$ is isotopic to a periodic map which contradicts our suppositions on $g|T_1$.

$X_1$ cannot be a Seifert fibre space, since then $g|X_1: X_1 \to X_1$ can be admissibly isotoped into a fibre preserving map (see 5.9) which contradicts our supposition that $g|T_1$ has no invariant curve.

So $X_1$ is an I-bundle, and so $M$. But it cannot be a twisted I-bundle, i.e. an I-bundle over the Klein bottle, for $g$ can be isotoped into a fibre preserving homeomorphism and the mapping class group of the Klein bottle is finite (see [Li 2]).

Thus $M$ is a product I-bundle. q.e.d.

The second observation is concerned with the question whether an extendable homeomorphism can be extended to an irreducible 3-manifold.

<u>27.8 Lemma</u>. <u>If a homeomorphism of a closed (not necessarily connected) surface can be extended to a</u> 3-<u>manifold, it can be extended to an</u>

*irreducible* 3-*manifold* (*not necessarily connected*).

*Proof*. Suppose the homeomorphism in question extends to a homeomorphism $h: M \to M$ of a 3-manifold $M$. If $M$ is irreducible, we are done. In the other case recall from [Kn 1] the existence of a complete system $S$ of pairwise disjoint 2-spheres in $M$. This means that any 2-sphere in $M$, disjoint from $S$, bounds, together with some spheres from $S$, a 3-ball with or without holes.

$T = h^{-1}S$ is again a system of 2-spheres. Without loss of generality we may suppose that $h$ is isotoped so that $T$ is transversal to $S$. In particular, $T - \mathring{U}(S)$ and $S - \mathring{U}(T)$ are surfaces, where $U(S)$ and $U(T)$ are regular neighborhoods in $M$.

Denote by $M_1$ and $N$ the union of all the components of $M - (\mathring{U}(S) \cup \mathring{U}(T))$ resp. $M - \mathring{U}(S)$ which meet $\partial M$. Every component of $\partial N$ different from $\partial M$, is a 2-sphere. Define $N^+$ to be the manifold obtained from $N$ be attaching 3-balls to the sphere-components of $\partial N$. Then $N^+$ is irreducible.

Of course, $h(M_1)$ is contained in $N$, i.e. $h|M_1$ is an embedding into $N$, and so into $M^+$. 27.8 is proved, if we can show that $h|M_1$ extends to a homeomorphism of $N^+$.

Let $G = \partial U(T) \cap M_1$. By definition, $\partial G$ lies in $\partial N - \partial M = \partial(N^+ - N)^-$. Since $(N^+ - N)^-$ consists of 3-balls, we find a system $D$ of (disjoint) discs in $(N^+ - N)^-$ with $D \cap \partial(N^+ - N)^- = \partial G$. Define $G^+ = G \cup D$. As a subsurface of 2-spheres of $\partial U(T)$, $G$ is a planar surface. Hence $G^+$ is a system of 2-spheres. By our choice of $S$ and $D$, each of these 2-spheres bounds a 3-ball in $N^+$. By our choice of $G$, all these 3-balls are disjoint. Denote $M_1^+ = M_1 \cup U(D)$, where $U(D)$ is a regular neighborhood in $M$. Then, by what we have seen so far, $(N^+ - M_1^+)^-$ is a system of 3-balls.

Since $\partial G \subset \partial U(T) \cap M_1$, it follows that also $h(\partial G)$ is contained in $\partial N - \partial M = \partial(N^+ - N)^-$. As above there is a system $D'$ of discs in $(N^+ - N)^-$ with $D' \cap \partial(N^+ - N)^- = h(\partial G)$. By the cone construction, $h$ can be extended to an embedding $h^+: M_1^+ \to N^+$ with $h^+(D) = D'$.

Now $h^+(\partial M_1^+)$ is a system of 2-spheres in the ireeducible 3-manifold $N^+$, and so it bounds a system of 3-balls in $N^+$. Hence using the cone construction once more, $h^+$ can be extended to a homeomorphism of $N^+$. q.e.d.

We now come to the description of our non-extendable homeomorphisms.

First of all, there are infinitely many, pairwise non-conjugate homeomorphisms $h$ of the torus which have no invariant, essential curve, i.e. no essential curve which is isotopic to its image under $h$. To give examples, consider the homeomorphism $h_n$, $n \geq 1$, given by the matrix $\begin{pmatrix} 1 & n \\ 1 & n+1 \end{pmatrix}$. Then $h_n$ has no invariant curve since the corresponding matrix has no integer eigenvalues. One also proves easily that $n = m$ if, for all essential curves, the intersection numbers (see [ST 1]) with their images under $h_n$, resp. $h_m$, are the same. So $h_n$ cannot be conjugate to $h_m$, if $n \neq m$. Furthermore, there is no integer $g$ such that $h_n^g$ is isotopic to id.

Now take $n$ copies, $n \geq 1$, of torus $\times$ I and specify one disc in each torus $\times$ 1. Attach these $n$ discs to $n$ discs in the boundary of a 3-ball. In this way we obtain a 3-manifold $N$ whose boundary consists of $n$ tori $T_1, \ldots, T_n$ and a surface $F$ with $n$ handles. For every $1 \leq i \leq n$, fix a homeomorphism $h_i: T_i \to T_i$ which has no invariant, essential curve, such that $h_i$ is not conjugate to $h_j$, if $i \neq j$, and such that $h_i$ is not isotopic to a periodic homeomorphism.

Of course, the homeomorphism $h_1 \cup \ldots \cup h_n$ can be extended to a homeomorphism $g$ of $N$. In fact, there are, for $n \geq 2$, infinitely many isotopy classes of such extensions. By the following result this gives us, for every surface with genus $\geq 1$, many homeomorphisms which are not null-cobordant.

<u>27.9 Corollary</u>. <u>If the genus of</u> $F$ <u>is non-zero</u>, $g|F$ <u>cannot be extended to any</u> 3-<u>manifold</u> $M$ <u>with</u> $F = \partial M$.

<u>Proof</u>. Assume the contrary, and attach $M$ to the "other side" of $F$. Then $Q = M \cup N$ defines a 3-manifold to which $h_1 \cup \ldots \cup h_n$ can be extended. By 27.8, we may suppose that $Q$ is irreducible. Hence it follows from 27.7 that at least two of the homeomorphisms

$h_1,\ldots,h_n$ are conjugate. This contradicts our choice of $h_1,\ldots,h_n$.

q.e.d.

Altogether we have,

27.10 Corollary. *For every surface F with genus $\geq 1$, there are infinitely many homeomorphisms $h: F \to F$ which cannot be extended to any 3-manifold M with $F = \partial M$.*

*Remark*. Observe that our constructions lead to surface-homeomorphisms which are not periodic. J. Birman has informed us that she has examples of periodic surface-homeomorphisms which cannot be extended to 3-manifolds.

## Chapter X: Dehn flips of 3-manifolds.

Given a solid torus $W$ in a 3-manifold $M$ such that $(\partial W - \partial M)^-$ consists of essential annuli, the following "local" procedure is conceivable: cut out $W$ and glue it differently back again. A little bit more general is the following: we say that a 3-manifold $M'$ is obtained from $M$ by a Dehn flip along $W$ if there is a solid torus $W'$ in $M'$ such that (1) $(\partial W' - \partial M)^-$ consists of essential annuli, (2) there is a homeomorphism $h: (M - W)^- \to (M' - W')^-$ with $h(\partial W - \partial M)^- = (\partial W' - \partial M')^-$, and there is (3) a homotopy equivalence $f: W \to W'$ with $f(\partial W - \partial M)^- = (\partial W' - \partial M')^-$.

Some of these Dehn flips lead to a homotopy equivalent 3-manifolds. E.g., this is true for any Dehn flip along components of $U(\partial X - \partial M)^-$, provided $X$ is an essential Seifert fibre space in the 3-manifold $M$ with non-orientable orbit (to see this use the example in the remark 4 following the proof of 28.4). On the other hand, we will see that, for a given Haken 3-manifold $M$ which is boundary-irreducible, there is a system of disjoint solid tori so that Dehn flips along these solid tori generate the whole homotopy type of $M$ (see §29). In particular, there are only finitely many Haken 3-manifolds homotopy equivalent to $M$.

To prove this we need some facts concerning homotopy equivalences of I-bundles and Seifert fibre spaces. These will be established in §28, and these have also some interest on their own.

### §28. Geometric obstructions for homotopy equivalences

In this paragraph we shall describe a certain geometric obstruction for a given homotopy equivalence between certain 3-manifolds to be a homeomorphism. To construct this obstruction and to prove its existence (see 28.5) we first have to consider essential maps between Seifert fibre spaces. It turns out that such maps are in general homotopic to fibre preserving maps (see 28.4). This result is a generalization of 5.9. To prove it we need the following two technical lemmas.

<u>28.1 Lemma. Let $(M_1, \underline{\underline{m}}_1)$ and $(M_2, \underline{\underline{m}}_2)$ be two solid tori with fixed admissible Seifert fibrations. Suppose that $(M_1, \underline{\underline{m}}_1)$ has at least one free side. Let f: $(M_1, \underline{\underline{m}}_1) \to (M_2, \underline{\underline{m}}_2)$ be an admissible map, and suppose that the restriction of f to the free sides of $(M_1, \underline{\underline{m}}_1)$ is an essential map into $(M_2, \underline{\underline{m}}_2)$.</u>

<u>Then f can be admissibly deformed into a fibre preserving map g.</u>

<u>In addition: If $(M_1, \underline{\underline{m}}_1)$ is a product $S^1$-bundle or if card$(\underline{\underline{m}}_1) = 1$, then g may be chosen so that it maps the exceptional fibre of $(M_1, \underline{\underline{m}}_1)$ onto the exceptional fibre of $(M_2, \underline{\underline{m}}_2)$ and the ordinary fibres onto the ordinary fibres.</u>

<u>Proof</u>. Let D be the unit disc in $\mathbb{R}^2$, and fix polar coordinates $(a,\alpha)$ in D. The arcs $(a,\alpha) \times I$ define a fibration of $D \times I \subset \mathbb{R}^2 \times \mathbb{R}$-- the standard fibration.

Now let $(m,n)$ $1 \leq m \leq n$, be a given pair of coprime integers. This pair gives rise to the following definitions, with $t \in I$ :

$$\varphi_t(\alpha) = \alpha + t\,\frac{m}{n}\cdot\pi, \quad \varphi_t(a,\alpha) = (a,\varphi_t(\alpha)),$$

and

$$\varphi(a,\alpha,t) = (a,\varphi_t(\alpha),t).$$

In this way $\varphi$ defines a homeomorphism of $D \times I$ with $\varphi(D \times i) = D \times i$, $i = 0,1$. By abuse of language, we also denote by $\varphi$ the image under $\varphi$ of the standard fibration. Attaching the two lids of $D \times I$ under the identity, we obtain a solid torus and furthermore, from the fibration $\varphi$, a Seifert fibration of the solid torus of type $(m,n)$. It is well-known that every Seifert fibration of a solid torus of type $(m,n)$ is isotopic to the preceding one.

Observe that in the above description two points $x_1 = (a_1,\alpha_1,t_1)$ and $x_2 = (a_2,\alpha_2,t_2)$ of $D \times I$ lie on the same fibre of $\varphi$ if and only if

$$a_1 = a_2 \text{ and } \alpha_1 = \varphi_{t_1}(\alpha),\ \alpha_2 = \varphi_{t_2}(\alpha), \text{ for some } \alpha. \qquad (*)$$

Finally call a map f: $D \to D$ <u>straightened</u> if

$f(a,\alpha) = (af_1(1,\alpha), f_2(1,a))$, where $f_1, f_2$ denote the coordinate maps of $f$.

Now fix two fibrations $\varphi$, $\psi$ of $D \times I$ as described above. Furthermore let $f\colon D \times I \to D \times I$ be a map with $f(D \times i) \subset D \times i$, $i = 0,1$, and such that

1) $f|D \times 0 = f|D \times 1$, $f|D \times 0$ is straightened, and
2) $f|\partial D \times I$ maps $\varphi$-fibres to $\psi$-fibres.

<u>28.2 Assertion</u>. <u>$f$ can be deformed (rel $\partial(D \times I)$) into a map which maps the $\varphi$-fibration into the $\psi$-fibration</u>.

To see this observe that, by the asphericity of $D \times I$, it suffices to prove that $f|\partial(D \times I)$ extends to such a fibre preserving map, say $\hat{f}$. For this define

$$\hat{f}(a,\alpha,t) = \begin{pmatrix} a & f_1(1,\alpha,t) \\ & f_2(1,\alpha,t) \\ & f_3(1,\alpha,t) \end{pmatrix},$$

where $f_1, f_2, f_3$ are the coordinate maps of $f$. By our suppositions of $f$, it follows at once that $\hat{f}$ is indeed an extension of $\varphi|\partial(D \times I)$. To show that $\hat{f}$ is fibre preserving, let $x_1 = (a_1,\alpha_1,t_1)$ and $x_2 = (a_2,\alpha_2,t_2)$ be two points lying on the same $\varphi$-fibre. This means that $a_1 = a_2$ and $\alpha_1 = \varphi_{t_1}(\alpha)$, $\alpha_2 = \varphi_{t_2}(\alpha)$, for some $\alpha$. Hence, by (*) $(1,\alpha_1,t_1)$ and $(1,\alpha_2,t_2)$ lie also on the same $\varphi$-fibre, and so $(b_1,\beta_1,s_1) = f(1,\alpha_1,t_1)$ and $(b_2,\beta_2,s_2) = f(1,\alpha_2,t_2)$ lie on the same $\psi$-fibre since $f|\partial D \times I$ is fibre preserving. Therefore by (*) again, $(ab_1,\beta_1,s_1) = \hat{f}(x_1)$ and $(ab_2,\beta_2,s_2) = \hat{f}(x_2)$ lie on the same $\psi$-fibre. This completes the proof of the assertion.

To prove the lemma fix an essential horizontal disc $D$ in $(M_2,\bar{\underline{m}}_2)$. Applying 4.4 and 5.6, we see that $f$ can be admissibly deformed so that $f^{-1}D$ consists of such discs in $(M_1,\bar{\underline{m}}_1)$. Without loss of generality, we may suppose that $f^{-1}D$ is connected (otherwise choose an admissible q-sheeted covering map $p\colon (N_2,\underline{\underline{n}}_2) \to (M_2,\underline{\underline{m}}_2)$, where $q$ denotes the circulation number of the image under $f$ of the core of $M_1$, and consider a lifting $\tilde{f}: M_1 \to N_2$ of $f$ instead of $f$). Let $C$ be a free side of $(M_1,\underline{\underline{m}}_1)$ (this exists by suppositions) and let $c$ be an arc $C \cap f^{-1}D$. $f|c$ can be admissibly deformed (rel $c$) in $D$ near $\partial D$, and this deformation can be extended to an admissible homotopy of $f$ which pulls $f|C$ near $\partial M_2$, and then into a vertical map. This is in general not true for admissible singular annuli (observe that there are non-homotopic curves in the boundary torus which are homotopic in the solid torus; so this homotopy cannot be deformed into the boundary). Since $M_2$ is aspherical, it holds, however, for singular annuli whose sides are disjoint. In particular, it is true for essential singular annuli, for they can be essential only if the completed boundary pattern of $(M_2,\underline{\underline{m}}_2)$ has at least two sides. Hence, without loss of generality, $f|\partial M_1$ is fibre preserving (note that $\bar{\underline{\underline{m}}}_1$ consists of annuli). Since $D$ is aspherical, $f|f^{-1}D: f^{-1}D \to D$ is homotopic (rel boundary) to a straightened map. Thus, cutting $M_1, M_2$ along $f^{-1}D$, resp. $D$, we get a situation as described in 28.2 and the lemma follows from 28.2.

For the additional remark, suppose first that $\text{card}(\underline{\underline{m}}_1) = 1$ and let $C$ be the free side of $(M_1,\underline{\underline{m}}_1)$. Then $f|C$ is an admissible singular annulus in $(M_2,\underline{\underline{m}}_2)$ whose sides lie both in the same bound side of $(M_2,\underline{\underline{m}}_2)$. Moreover, by the suppositions of 28.1, $f|C$ is essential in $(M_2,\underline{\underline{m}}_2)$. Hence $(M_2,\underline{\underline{m}}_2)$ cannot be a product $S^1$-bundle, and so it has an exceptional fibre. Then checking the above constructions, the additional remark follows. Now suppose that $(M_1,\underline{\underline{m}}_1)$ is a product $S^1$-bundle. Let $C_1,\dots,C_m$, $m \geq 1$, be all the free sides of $(M_1,\underline{\underline{m}}_1)$. Pushing $f|C_i$ admissibly near $\partial M_2$, we see that there is an admissible homotopy $f_t$, $t \in I$, of $f$ so that the restriction of $f_1$ to a regular neighborhood $U$ of $(\partial M_1 - C_m)^-$ is fibre preserving. Since the admissible fibration of $(M_1,\underline{\underline{m}}_1)$ is supposed to be a product fibration, there is a fibre preserving, admissible isotopy $\alpha_t$, $t \in I$, with $\alpha_0 = \text{id}$ and $\alpha_1(M_1) \subset U$. Then $f_t \circ \alpha_t$ is the required admissible homotopy. q.e.d.

<u>28.3 Lemma.</u> <u>Let $(M_1,\underline{\underline{m}}_1)$ be a solid torus with fixed admissible fibration as Seifert fibre space. Suppose that $(M_1,\underline{\underline{m}}_1)$ has at least</u>

*one free side. Let* $(M_2,\underline{\underline{m}}_2)$ *be a Seifert fibre space with fixed admissible fibration and non-empty boundary. Suppose that* $(M_2,\underline{\underline{m}}_2)$ *is not one of the exceptions of 5.10.2. Let* $f\colon (M_1,\underline{\underline{m}}_1) \to (M_2,\underline{\underline{m}}_2)$ *be an admissible map, and suppose that the restriction of* $f$ *to the free sides of* $(M_1,\underline{\underline{m}}_1)$ *are essential in* $(M_2,\underline{\underline{m}}_2)$.

*Then* $f$ *can be admissibly deformed into a fibre preserving map* $g$.

*In addition: If* $(M_1,\underline{\underline{m}}_1)$ *is a product* $S^1$*-bundle or if* $\mathrm{card}(\underline{\underline{m}}_1) = 1$, *then* $g$ *may be chosen so that it maps the exceptional fibre of* $(M_1,\underline{\underline{m}}_1)$ *onto some exceptional fibre of* $(M_2,\underline{\underline{m}}_2)$, *and the ordinary fibres onto the ordinary ones.*

*Proof*. Let $A$ be a system of pairwise disjoint, essential and vertical annuli in $(M_2,\bar{\underline{\underline{m}}}_2)$, which splits $(M_2,\bar{\underline{\underline{m}}}_2)$ into a system of solid tori. Such a system exists, by 5.4, for $\partial M \neq \emptyset$. $(M_1,\underline{\underline{m}}_1)$ cannot be one of the exceptions of 5.1.2, for at least one side of $(M_1,\underline{\underline{m}}_1)$ is a free side and the restriction of $f$ to the free sides is an essential map into $(M_2,\underline{\underline{m}}_2)$. Thus, by 5.2, the complete boundary-pattern of $(M_1,\underline{\underline{m}}_1)$ is useful. Hence, by 4.4, $f$ can be admissibly deformed so that $f^{-1}A$ is an essential surface in $(M_1,\bar{\underline{\underline{m}}}_1)$, and that no component of $f^{-1}A$ is a 2-sphere or an admissible i-faced disc, $1 \leq i \leq 3$, in $(M_1,\bar{\underline{\underline{m}}}_1)$. Suppose $f$ is admissibly deformed so that the above holds and that, in addition, the number of components of $\cup C_i \cap f^{-1}A$ is minimal (here $\cup C_i$ denotes the union of all the free sides of $(M_1,\underline{\underline{m}}_1)$).

Denote by $(\tilde{M}_1,\tilde{\underline{\underline{m}}}_1)$ and $(\tilde{M}_2,\tilde{\underline{\underline{m}}}_2)$ the manifolds obtained from $(M_1,\underline{\underline{m}}_1)$ and $(M_2,\underline{\underline{m}}_2)$ by splitting at $f^{-1}A$ and $A$, respectively. Then $\tilde{f} = f|\tilde{M}_1\colon (\tilde{M}_1,\tilde{\underline{\underline{m}}}_1) \to (\tilde{M}_2,\tilde{\underline{\underline{m}}}_2)$ is an admissible map. By our minimality condition on $\cup C_i \cap f^{-1}A$ and the surgery arguments of 4.4, it follows that the restriction of $\tilde{f}$ to any free side of $(\tilde{M}_1,\tilde{\underline{\underline{m}}}_1)$ is an essential map into $(\tilde{M}_2,\tilde{\underline{\underline{m}}}_2)$. This implies, in particular, that $f^{-1}A$ cannot be horizontal, since, by 5.10 and our suppositions on $(M_2,\underline{\underline{m}}_2)$, the restriction of $f$ to any free side of $(M_1,\underline{\underline{m}}_1)$ can be admissibly deformed into a vertical map. Thus, by 5.6, we may suppose that $f^{-1}A$ is vertical in $(M_1,\bar{\underline{\underline{m}}}_1)$. Then it follows, by comparing fundamental groups, that $(\tilde{M}_1,\tilde{\underline{\underline{m}}}_1)$ as well as $(\tilde{M}_2,\tilde{\underline{\underline{m}}}_2)$ are systems of solid tori, and these carry admissible fibrations as Seifert fibre spaces induced from that of $(M_1,\underline{\underline{m}}_1)$ and $(M_2,\underline{\underline{m}}_2)$, respectively. Thus 28.3 follows from 28.1.

For the additional remark, note that the components of $f^{-1}A$ are admissibly parallel in $(M_1,\bar{\underline{\underline{m}}}_1)$, if $\mathrm{card}(\underline{\underline{m}}_1) = 1$. To see this, consider the fibre projection of $f^{-1}A$ into the orbit surface of $(M_1,\underline{\underline{m}}_1)$. q.e.d.

The next proposition is an extension of 5.9 to essential maps, and so also to admissible homotopy equivalences.

*28.4 Proposition. Let* $(M_i,\underline{\underline{m}}_i)$, $i = 1,2$, *be an I-bundle or a Seifert fibre space with fixed admissible fibration. Suppose that* $M_1$ *is neither a ball nor a solid torus, if* $(M_2,\underline{\underline{m}}_2)$ *is an I-bundle, and that* $(M_i,\underline{\underline{m}}_i)$, $i = 1,2$, *is not one of the following exceptions:*

1. $M_i$ *is one of* 5.1.1-5.1.5.
2. $(M_i,\underline{\underline{m}}_i)$ *admits an admissible fibration as I-bundle over the square, annulus, Möbius band, torus, or Klein bottle.*
3. $M_i$ *is one of the closed 3-manifolds which can be obtained by glueing two I-bundles over the torus or Klein bottle together along their boundaries.*

*Then every essential map* $f\colon (M_1,\underline{\underline{m}}_1) \to (M_2,\underline{\underline{m}}_2)$ *can be admissibly deformed into a fibre preserving map.*

*In addition: If* $(M_1,\underline{\underline{m}}_1)$ *has at least one free side, then* $f$ *need not be essential, but it suffices to suppose that* $f$ *induces a monomorphism on the fundamental groups and that the restriction of* $f$ *to any free side is essential.*

*Remark 1.* In general the homotopy of $f$ cannot be chosen to be constant on the free sides, even not if the restriction of $f$ to the free sides is already a fibre preserving map.

2. For the case that $(M_i,\bar{\underline{\underline{m}}}_i)$ is an I-bundle, but not $(M_i,\underline{\underline{m}}_i)$, i.e. for the case that at least one of the lids is free, see 5.5, 3.4 and 5.9.

Proof. $\bar{\underline{\underline{m}}}_i$, $i = 1,2$, is a useful boundary-pattern of $M_i$ (see 5.2). Moreover, $M_i$ is sufficiently large (see 5.4). Hence, by [Wa 4] and 3.4, f can be admissibly deformed into a covering map, provided $\underline{\underline{m}}_1 = \bar{\underline{\underline{m}}}_1$. After such a deformation, we may use f to lift the admissible fibration of $(M_2, \underline{\underline{m}}_2)$ to another admissible fibration of $(M_1, \underline{\underline{m}}_1)$. But, by 5.9, it follows the existence of an admissible ambient isotopy $\alpha_t$, $t \in I$, of $(M_1, \underline{\underline{m}}_1)$ with $\alpha_0 = \mathrm{id}$ and such that $\alpha_1$ maps the fixed admissible fibration of $(M_1, \underline{\underline{m}}_1)$ to the new one. This means that the admissible homotopy $f \cdot \alpha_t$, $t \in I$, moves f into a fibre preserving map, and we are done.

Thus we may suppose that $(M_1, \underline{\underline{m}}_1)$ has at least one free side, and we denote by $C_1, \ldots, C_m$, $m \geq 1$, all the free sides of $(M_1, \underline{\underline{m}}_1)$. In this case, we will not use the fact that f itself is essential, but just that $f_*: \pi_1 M_1 \to \pi_1 M_2$ is a monomorphism and that $f|C_i$, $1 \leq i \leq m$, is essential in $(M_2, \underline{\underline{m}}_2)$.

Although we supposed that $M_1$ is not a ball if $M_2$ is an I-bundle, the assertion of 28.4 is true in the following case, which will be needed below.

Case 1. $(M_2, \underline{\underline{m}}_2)$ is an I-bundle over the disc, and at least one side of $\underline{\underline{m}}_1$ different from a lid will be mapped under f into a side of $\underline{\underline{m}}_2$ different from a lid.

Since $f_*: \pi_1 M_1 \to \pi_1 M_2$ is a monomorphism, it follows that $(M_1, \underline{\underline{m}}_1)$ is also an I-bundle over a disc D. By 5.10, $f|C_i$, $1 \leq i \leq m$, can be admissibly deformed into a vertical map. We claim that $f|C_i$ can in fact be admissibly deformed into a fibre preserving map. This follows since, by suppositions on $M_1$, there are at least two neighboring sides of $(M_1, \bar{\underline{\underline{m}}}_1)$ which are not mapped into a lid of $M_2$, and so the common arc of these two discs is mapped under f into a fibre of $M_2$ ($M_1$ is not the I-bundle over the square). Hence we may suppose that f is admissibly deformed so that $f|\partial D \times I$ is fibre preserving. Then, of course, $f|\partial D \times I$ extends to a fibre preserving map f'. It follows, from the asphericity of $M_2$ and the lids of $(M_2, \underline{\underline{m}}_2)$, that f is admissibly homotopic (rel $\partial D \times I$) to f'.

Case 2. $(M_2,\underline{\underline{m}}_2)$ *is a solid torus and admissibly fibered as Seifert fibre space.*

Since $f_*$ is a monomorphism, it follows that $M_1$ is a solid torus. $(M_2,\underline{\underline{m}}_2)$ cannot be an I-bundle since $(M_2,\underline{\underline{m}}_2)$ admits no fibration as I-bundle over the annulus or Möbius band. Hence to prove 28.4 in Case 2, it suffices to show that $(M_1,\underline{\underline{m}}_1)$ is admissibly fibered as Seifert fibre space, and then to apply 28.1. Assume the converse. Then at least one free side of $(M_1,\underline{\underline{m}}_1)$, say $C_1$, must be a square, for, by 2 of 28.4, $(M_1,\underline{\underline{m}}_1)$ cannot be admissibly fibered as the I-bundle over the annulus or Möbius band. By 5.10, $f|C_1$ can be admissibly deformed into a vertical map, which is a contradiction since we are in Case 2.

Case 3. $(M_2,\underline{\underline{m}}_2)$ *is either an I-bundle whose orbit surface has non-empty boundary, or Seifert fibre space with non-empty boundary.*

If $(M_2,\underline{\underline{m}}_2)$ is an I-bundle, we may fix a system $A$ of pairwise disjoint, vertical and essential squares which split $M_2$ into a system of balls. If $(M_2,\underline{\underline{m}}_2)$ is a Seifert fibre space, we find such a system of annuli which split $M_2$ into solid tori (see 5.4). Let $A$ be chosen such that the above holds and that, in addition, the number of components of $A$ is as small as possible. By 4.4, $f$ can be admissibly deformed so that $f^{-1}A$ is essential in $(M_1,\bar{\underline{\underline{m}}}_1)$ and that no component of $f^{-1}A$ is a 2-sphere or an admissible i-faced disc, $1 \leq i \leq 3$, in $(M_1,\bar{\underline{\underline{m}}}_1)$. Suppose that $f$ is admissibly deformed so that the above holds, and that, in addition, the number of components of $\cup C_i \cap f^{-1}A$ is minimal. Let $(\tilde{M}_1,\tilde{\underline{\underline{m}}}_1)$ and $(\tilde{M}_2,\tilde{\underline{\underline{m}}}_2)$ be the manifolds obtained from $(M_1,\underline{\underline{m}}_1)$ and $(M_2,\underline{\underline{m}}_2)$ by splitting at $f^{-1}A$ and $A$, respectively, and denote by $\tilde{C}_i$, $1 \leq i \leq m$, the surface obtained from $C_i$ by splitting at $f^{-1}A$. Then $\tilde{f} = f|\tilde{M}_1\colon (\tilde{M}_1,\tilde{\underline{\underline{m}}}_1) \to (\tilde{M}_2,\tilde{\underline{\underline{m}}}_2)$ is an admissible map. Since $f^{-1}A$ is essential, it follows that the restriction of $\tilde{f}$ to any component of $\tilde{M}_1$ induces a monomorphism on the fundamental groups. Now, recall that $f|C_i$, $1 \leq i \leq m$, is essential in $(M_2,\underline{\underline{m}}_2)$ and that $\cup C_i \cap f^{-1}A$ is minimal. Moreover, by 5.6, we may suppose that $f^{-1}A$ is either horizontal or vertical. Hence, using the surgery arguments of 4.4, it is easily checked that $\tilde{f}|\tilde{C}_i$, $1 \le i \le m$, is an essential map into $(\tilde{M}_2, \tilde{\underline{\underline{m}}}_2)$.

We assert that $f^{-1}A$ cannot be horizontal. Assume the converse. Note that $f^{-1}A$ (without boundary-pattern) consists of discs or annuli since $f_*$ is a monomorphism and since $A$ consists of squares or annuli. By 2 of 28.4, $(M_1, \underline{\underline{m}}_1)$ cannot admit an admissible fibration as I-bundle over the torus or Klein bottle. Hence it follows that at least one free side of $(M_1, \underline{\underline{m}}_1)$, say $C_1$, must be a square or annulus. Now, $C_1 \cap f^{-1}A$ is non-empty since $f^{-1}A$ is horizontal, and it consists of arcs which are essential in $C_1$. Let $k$ be one of such arcs, and let $A_1$ be the component of $A$ which contains $f(k)$. Then $f|k$ joins two disjoint sides of $A_1$, for $f|C_1$ is essential in $(M_2, \underline{\underline{m}}_2)$. This implies that $A_1$ cannot be an annulus, for otherwise $f|C_1$ cannot be admissibly deformed into a vertical map (recall our choice of $A$ and note that $\tilde{f}|\tilde{C}_1$ is essential in $(\tilde{M}_2, \tilde{\underline{\underline{m}}}_2)$) which contradicts 5.10. Hence $A_1$ must be a square, i.e. $(M_2, \underline{\underline{m}}_2)$ must be an I-bundle, by our choice of $A$. Then $f^{-1}A$ consists of discs, and so $M_1$ must be a ball or a solid torus, since $f^{-1}A$ is horizontal. But this contradicts the suppositions of 28.4.

Thus, by 5.6, we may suppose that $f^{-1}A$ is vertical. Then $(\tilde{M}_1, \tilde{\underline{\underline{m}}}_1)$ and $(\tilde{M}_2, \tilde{\underline{\underline{m}}}_2)$ carry admissible fibrations induced by that of $(M_1, \underline{\underline{m}}_1)$ and $(M_2, \underline{\underline{m}}_2)$, respectively. Let $(N_1, \underline{\underline{n}}_1)$ be any component of $(\tilde{M}_1, \tilde{\underline{\underline{m}}}_1)$ and let $(N_1', \underline{\underline{n}}_1')$ be the component of $(\tilde{M}_2, \tilde{\underline{\underline{m}}}_2)$ containing $\tilde{f}(N_1)$. To prove 28.4 in Case 3, we still have to show that $\tilde{f}|N_1\colon (N_1, \underline{\underline{n}}_1) \to (N_1', \underline{\underline{n}}_1')$ can be admissibly deformed into a fibre preserving map.

By our choice of $A$, $N_1'$ is either a ball or a solid torus, according to whether $(M_2, \underline{\underline{m}}_2)$ is an I-bundle or a Seifert fibre space. Hence, applying Case 1 and Case 2, we are done if $(N_1', \underline{\underline{n}}_1')$ does not admit an admissible fibration as I-bundle over the square, annulus, or Möbius band. By our minimality condition on $A$ and since $(M_2, \underline{\underline{m}}_2)$ is not 2 of 28.4, $(N_1', \underline{\underline{n}}_1')$ cannot admit an admissible fibration as I-bundle over the square or annulus. Thus we suppose it admits such a fibration over the Möbius band. Then, in particular, $(M_2, \underline{\underline{m}}_2)$ is a Seifert fibre space. Now, we assert that $(N_1, \underline{\underline{n}}_1)$

cannot be an I-bundle. For otherwise $(M_1,\underline{\underline{m}}_1)$ is an I-bundle, and the lids of $(M_1,\underline{\underline{m}}_1)$ are mapped under $f$ into bound sides of $(M_2,\underline{\underline{m}}_2)$. This implies that the lids (without boundary-pattern) of $(M_1,\underline{\underline{m}}_1)$ consist of annuli, or discs. Since $(M_1,\underline{\underline{m}}_1)$ is not 2 of 28.4, it follows that at least one free side of $(M_1,\underline{\underline{m}}_1)$, say $C_1$, must be a square. But $f|C_1$ is an essential map and so, by 5.10, $f|C_1$ can be admissibly deformed in $(M_2,\underline{\underline{m}}_2)$ into a vertical map. This contradicts the fact that $(M_2,\underline{\underline{m}}_2)$ is a Seifert fibre space. Thus $(N_1,\underline{\underline{n}}_1)$ cannot be an I-bundle, and so, by 28.1, $\tilde{f}|N_1\colon (N_1,\underline{\underline{n}}_1) \to (N_1',\underline{\underline{n}}_1')$ can be admissibly deformed into a fibre preserving map. This completes the proof in Case 3.

Case 4. Case 1, Case 2, and Case 3 do not hold.

Applying 5.4, we find either an essential and vertical annulus, or torus $A$ in $(M_2,\underline{\underline{m}}_2)$, according to whether $(M_2,\underline{\underline{m}}_2)$ is an I-bundle or a Seifert fibre space. Let $(\tilde{M}_1,\tilde{\underline{\underline{m}}}_1)$ and $(\tilde{M}_2,\tilde{\underline{\underline{m}}}_2)$ have the usual meaning. As in Case 3, deform $f$ admissibly so that $f^{-1}A$ is an essential surface in $(M_1,\bar{\underline{\underline{m}}}_1)$ such that no component of $f^{-1}A$ is a 2-sphere or an admissible i-faced disc, $1 \leq i \leq 3$, in $(M_1,\bar{\underline{\underline{m}}}_1)$, and that, in addition, the number of components of $\cup C_i \cap f^{-1}A$ is as small as possible.

We assert that $f^{-1}A$ cannot be horizontal. This is clear if $A$ is a torus. For then, $f^{-1}A$ consists either of annuli, or of tori. In the first case, $(M_1,\underline{\underline{m}}_1)$ is one of the exceptions of 2 of 28.4, and in the second case it has no free sides which contradicts our suppositions on $(M_1,\underline{\underline{m}}_1)$. If, on the other hand, $A$ is an annulus, then, by our choice of $A$, $(M_2,\underline{\underline{m}}_2)$ is an I-bundle. Moreover, $f^{-1}A$ (without boundary-pattern) consists either of discs or of annuli. In the first case, $M_1$ must be a ball or a solid torus which gives a contradiction to our suppositions of 28.4. In the second case, $(M_1,\bar{\emptyset})$ admits an admissible fibration as I-bundle over the torus or Klein bottle. Fix a torus $T$ near $\partial M_1$, and let $\tilde{T}$ be the surface obtained from $T$ by splitting at $f^{-1}A$. Then $f|\tilde{T}$ consists of essential singular annuli in $(\tilde{M}_2,\tilde{\underline{\underline{m}}}_2)$ (this follows, as in Case 3, from the facts that $f_*$ is a monomorphism and that

$\cup C_i \cap f^{-1}A$ is minimal). On the other hand, $f|\tilde{T}$ cannot be admissibly deformed in $(\tilde{M}_2,\tilde{\underline{m}}_2)$ into a vertical map since A is a vertical annulus in the I-bundle $(M_2,\underline{m}_2)$. Thus, by 5.10 $(\tilde{M}_2,\tilde{\underline{m}}_2)$ consists of I-bundles over the annulus or Möbius band, and so $(M_2,\underline{m}_2)$ must be the I-bundle over the torus or Klein bottle. But this contradicts the suppositions of 28.4.

Hence, by 5.6, we may suppose that $f^{-1}A$ is vertical. Then $(\tilde{M}_1,\tilde{\underline{m}}_1)$ and $(\tilde{M}_2,\tilde{\underline{m}}_2)$ carry admissible fibrations induced by that of $(M_1,\underline{m}_1)$ and $(M_2,\underline{m}_2)$, respectively. Let $(N_1,\underline{n}_1)$ be a component of $(\tilde{M}_1,\tilde{\underline{m}}_1)$ and $(N_1',\underline{n}_1')$ be the component of $(\tilde{M}_2,\tilde{\underline{m}}_2)$ containing $f(N_1)$. Again we still have to show that $f|N_1\colon (N_1,\underline{n}_1) \to (N_1',\underline{n}_1')$ can be admissibly deformed into a fibre preserving map. This follows, by Case 3, provided $(N_1',\underline{n}_1')$ does not admit an admissible fibration as I-bundle over the annulus, torus, Möbius band, or Klein bottle. The first two cases cannot occur, for A is connected and $(M_2,\underline{m}_2)$ is not one of the exceptions 2 or 3 of 28.4. In the other cases, A is separating. Let $(N_2',\underline{n}_2')$ be the component of $(\tilde{M}_2,\tilde{\underline{m}}_2)$ different from $(N_1',\underline{n}_1')$. Then $(N_2',\underline{n}_2')$ cannot admit an admissible fibration as I-bundle over the Möbius band or Klein bottle, for $(M_2,\underline{m}_2)$ is not the exception 2 or 3 of 28.4. Denote by $(N_2,\underline{n}_2)$ the union of all the components of $(\tilde{M}_1,\tilde{\underline{m}}_1)$ which are mapped under f into $(N_2',\underline{n}_2')$. Then, by Case 3, $f|N_2\colon (N_2,\underline{n}_2) \to (N_2',\underline{n}_2')$ can be admissibly deformed into a fibre preserving map. Hence we may suppose that $f|\partial N_1\colon \partial N_1 \to \partial N_1'$ is fibre preserving. This implies that $f|N_1\colon N_1 \to N_1'$ can be deformed (rel $\partial N_1$) into a fibre preserving map. To see this observe that a Klein bottle has only two 2-sided curves [Li 2], so $N_1'$ only has two essential annuli, and so at most two Seifert fibrations, up to isotopy (the annuli are vertical with respect to the Seifert fibrations, respectively). So, taking any essential vertical annulus B in $N_1$, and considering $f|B$, our claim follows from 5.11. q.e.d.

<u>Remark 3</u>. Recall from [Wa 1] (see 5.9) that admissible homeomorphisms of Seifert fibre spaces are admissibly isotopic to fibre preserving homeomorphisms, under the conditions of 28.4. Hence, from this point of view, homotopy equivalences behave like homeomorphisms. On the other hand, observe (use a hierarchy) that also every homeomorphism

of the orbit surface of a sufficiently large Seifert fibre space lifts to a fibre preserving homeomorphism of the whole Seifert fibre space. This in turn is in general not true for homotopy equivalences. Here is an example. Let $F$ be the non-orientable surface obtained by attaching two Möbius bands along one arc in their boundaries. If we fix a base point $x$ of $F$ in the common arc, the cores of the Möbius bands define a canonical base $\alpha$, $\beta$ for $\pi_1(F,x)$. Define $\varphi: \pi_1 F \to \pi_1 F$ to be the isomorphism mapping $\alpha$ to $\alpha$ and $\beta$ to $\alpha\cdot\beta$. Of course, $\varphi$ is induced by a homotopy equivalence $\bar{f}: F \to F$. However, $\bar{f}$ cannot be lifted to the $S^1$-bundle over $F$. To see this recall that $\varphi$ maps the one-sided curve $\beta$ to the 2-sided curve $\alpha\cdot\beta$ and observe that there is no essential map from the Klein bottle (above $\beta$) to the torus (above $\alpha\cdot\beta$).

Remark 4. The third remark reflects a special phenomenon for Seifert fibre spaces over non-orientable orbits. Another one (but closely connected to the first) is the following. If $A$ is an annulus from the boundary-pattern of a Seifert fibre space, $M$, as above, then there is an admissible homotopy equivalence $f: M \to M$ with $f(A) = A$ such that the restriction $f|s: s \to s$ to the core $s$ of $A$ is orientation-reversing (it is easy to see that this cannot happen for Seifert fibre spaces over orientable orbits). In order to construct such homotopy equivalences, it suffices to establish them for the $S^1$-bundle $M$ over the Möbius band. But the latter follows from the observation that an oriented fibre in the interior of $M$ is both homotopic to $s$ and its inverse.

Remark 5. Let $M_1, M_2$ be two Seifert fibre spaces, but not the exceptions of 28.4. Denote by $M_1^*$ and $M_2^*$ the $S^1$-bundles obtained from $M_1$ resp. $M_2$ by drilling out the exceptional fibres. Then, by 28.4, any homotopy equivalence $f: M_1 \to M_2$ can be deformed so that afterwards the restriction to $M_1^*$ defines an admissible fibre preserving map $f^*: M_1^* \to M_2^*$. This map is in general not a homotopy equivalence itself. The true reason for this is that one finds isomorphisms of Fuchsian groups which are not isomorphisms of the corresponding free groups. E.g. the Fuchsian group

$G = \{a_1, a_2, x | x^3 = 1\}$ admits an isomorphism $\varphi\colon G \to G$ defined by $a_1 \to a_1 a_2 x^2 a_1$, $a_2 \to a_2 x^{-1} a_1$, $x \to x$ (an inverse can be defined by $a_1 \to a_1 a_2^{-1}$, $a_2 \to a_2 a_2 a_1^{-1} x$, $x \to x$). But $\varphi$ is not an isomorphism of the free group $\{a_1, a_2, x | -\}$, for $\varphi a_1, \varphi a_2, \varphi x$ is not a base since neither the first nor the second Nielsen process is possible (see [ZVC 1] for the definition of Nielsen processes).

We now come to the definition of the obstruction submanifold for homotopy equivalences of 3-manifolds. For this let $f\colon (M_1, \underline{\underline{m}}_1) \to (M_2, \underline{\underline{m}}_2)$ be an admissible homotopy equivalence between Haken 3-manifolds whose completed boundary-patterns are useful. Then an <u>obstruction submanifold</u> $O_f$ <u>of</u> $f$ is defined to be an essential F-manifold in $(M_2, \bar{\underline{\underline{m}}}_2)$ satisfying:

1. $f$ can be admissibly deformed so that afterwards $f^{-1} O_f$ is an essential F-manifold and that $f | f^{-1} O_f\colon f^{-1} O_f \to O_f$ is an admissible homotopy equivalence, and $f | (M_1 - f^{-1} O_f)^-\colon (M_1 - f^{-1} O_f)^- \to (M_2 - O_f)^-$ is an admissible homeomorphism (with respect to the proper boundary-patterns),
2. every essential F-manifold in $(M_2, \bar{\underline{\underline{m}}}_2)$ satisfying 1 can be admissibly isotoped in $(M_2, \bar{\underline{\underline{m}}}_2)$ into $O_f$,
3. $O_f$ minus a component of $O_f$ does not satisfy 1.

Observe that, by 2 and 3, an obstruction submanifold of $f$ is unique, up to admissible ambient isotopy. Furthermore, the obstruction submanifold of admissible homotopy equivalences between Haken 3-manifolds with complete and useful boundary-patterns has to be empty (see [Wa 4]).

Our aim is to prove the existence of obstruction submanifolds in 3-manifolds without exceptional curves. To do this we have first to make precise what we mean by an "exceptional curve."

For this let $(M, \underline{\underline{m}})$ be a Haken 3-manifold whose complete boundary-pattern is useful. Let $V$ be the characteristic submanifold of $(M, \underline{\underline{m}})$ and fix an admissible fibration of $V$. Then a curve $k$ in $M$ is called an <u>exceptional curve of</u> $(M, \underline{\underline{m}})$ if $k$ is an exceptional fibre of $V$. It is to be understood that the fibration of $V$ is chosen so that the number of exceptional curves of $(M, \underline{\underline{m}})$ is as small as possible (observe that here is a choice involved, for the $S^1$-bundle over the Möbius band also admits a fibration as Seifert fibre space with two exceptional fibres). Note that the exceptional curves of $(M,\underline{\underline{m}})$ are well-defined, up to admissible ambient isotopy, since, by 10.9, $V$ is unique, and since, by 5.9 and our suppositions on the fibrations of $V$, the fibration of $V$ is unique, up to ambient isotopy. Furthermore, note that the exceptional curves of $(M,\underline{\underline{m}})$ and $(M,\bar{\underline{\underline{m}}})$ might be different, e.g. it might happen that $(M,\bar{\emptyset})$ has exceptional curves while $(M,\emptyset)$ has none at all.

With the above notation we can prove

<u>28.5 Proposition</u>. <u>Let</u> $(M_1,\underline{\underline{m}}_1)$ <u>and</u> $(M_2,\underline{\underline{m}}_2)$ <u>be two irreducible</u> 3-<u>manifolds whose completed boundary-patterns are useful and non-empty</u>. <u>Suppose</u> $(M_i,\bar{\underline{\underline{m}}}_i)$ <u>contains no exceptional curves and no Klein bottles</u>, <u>or Möbius bands</u>, $i = 1,2$.

<u>Then for every admissible homotopy equivalence</u> $f\colon (M_1,\underline{\underline{m}}_1) \to (M_2,\underline{\underline{m}}_2)$ <u>there exists an obstruction submanifold</u>, <u>and this is unique up to ambient admissible isotopy</u>.

<u>Remark</u>. It will be apparent from the proof that the assumption that $(M_i,\bar{\underline{\underline{m}}}_i)$ contains no exceptional curve can be weakened to the assumption that no essential annulus in $(M_i,\bar{\underline{\underline{m}}}_i)$ separates a solid torus.

<u>Proof.</u> The uniqueness being straightforward, it remains to show the existence of an obstruction submanifold for $f$. We split this proof into two cases.

<u>Case 1</u>. $(M_2,\underline{\underline{m}}_2)$ <u>is an</u> I- <u>or</u> $S^1$-<u>bundle over a surface</u>.

Applying 24.2 to $f$ or its admissible homotopy inverse, we see that $M_1$ and $M_2$ are both either I- or $S^1$-bundles. We may suppose that $(M_2,\underline{\underline{m}}_2)$ is not one of the exceptions of 28.4, for otherwise it follows from 3.4, 5.5, and [Wa 4] that $f$ is admissibly homotopic to a homeomorphism, and so the obstruction submanifold is empty. Denote by $p_i\colon M_i \to F_i$ and $s_i\colon F_i \to M_i$, $i = 1,2$, the fibre projection

and a fixed section (the latter exists since $\partial M_i \neq \emptyset$). Observe that the map $g = p_2 \circ f \circ s_1 \colon F_1 \to F_2$ is an admissible homotopy equivalence, and that $F_1$ and $F_2$ are orientable (by supposition, $M_i$ contains no Klein bottle). Therefore, by 30.15, there exists an obstruction surface $C_2, C_2 \subset F_2$, for $g$. Since $f$ is fibre preserving (28.4) and $X_2$ is a bundle, every admissible homotopy of $g$ can be lifted to an admissible homotopy of $f$. So it follows that the subbundle in $M_2$ over $C_2$ is indeed an obstruction submanifold for $f$.

<u>Case 2</u>. $(M_2, \underline{\underline{m}}_2)$ <u>is not an</u> I- <u>or</u> $S^1$-<u>bundle over a surface</u>.

Let $V_i'$, $i = 1,2$, be the union of all the components of the characteristic submanifold of $M_i$ which meet the boundary. By 24.2 and 3.4 we may suppose that $f$ splits into an admissible homotopy equivalence $V_1' \to V_2'$ and an admissible homeomorphism $\overline{M_1 - V_1'} \to \overline{M_2 - V_2'}$. Furthermore, $f|V_1'$ maps $S^1$-bundles to $S^1$-bundles and I-bundles to I-bundles, and without loss of generality it maps the lids of the I-bundles to lids (see 5.5). Then it follows easily from Case 1 that there is an obstruction submanifold $W_2$, $W_2 \subset V_2'$, for the map $f|V_1' \colon V_1' \to V_2'$. Hence the union of $W_2$ with some components of the regular neighborhood $U_2$ of $(\partial V_2 - \partial M_2)^-$ is an essential F-manifold with property 1 of an obstruction submanifold. Let $O_f$ be defined to be the essential F-manifold with property 1 and given as the union of $W_2$ with a minimal number of components of $U_2$.

We claim that $O_f$ is the required obstruction submanifold for $f$. To prove this let $O'$ be any other essential F-manifold with property 1. Then we have to show that $O_f$ can be admissibly isotoped into $O'$.

By 10.8, we may suppose that $O'$ is contained in $V_2'$ (as a vertical submanifold). Let $O''$ be $O'$ minus all the components of $O'$ which can be admissibly isotoped into the regular neighborhood of $(\partial V_2' - \partial M_2)^-$. By construction of $O_f$, it remains to show that $f$ can be admissibly deformed so that afterwards

(i) $f$ splits into two admissible homotopy equivalences $V_1' \to V_2'$ and $\overline{M_1 - V_1'} \to \overline{M_2 - V_2'}$.

(ii) $O''$ has property 1 with respect to $f|V_1' \colon V_1' \to V_2'$.

By our choice of $O'$, $f$ can be admissibly deformed so that afterwards it splits into an admissible homotopy equivalence $f^{-1}O' \to O'$ between essential F-manifolds and an admissible homeomorphism $(M_1 - f^{-1}O')^{-} \to (M_2 - O')^{-}$. It follows from the normalization that <u>both</u> $f^{-1}O'$ and $f^{-1}V_2'$ are essential F-manifolds. Hence, by 10.8, $V_1'$ can be admissibly isotoped so that afterwards $V_1'$ contains both $f^{-1}O'$ and $f^{-1}V_2'$. Then it is easily seen that $f(V_1')$ is an essential F-manifold in $M_2$ containing $V_2'$. This means that $f(V_1')$ is a regular neighborhood of $V_2'$ (see 10.6.1). So an admissible ambient isotopy of $(M_2, \bar{\underline{\underline{m}}}_2)$, which is constant on $O'$, changes $f$ so that it splits into essential maps $f|V_1$: $V_1 \to V_2$ and $f|(M_1 - V_1)^{-}$: $(M_1 - V_1)^{-} \to (M_2 - V_2)^{-}$, and that $f|(M_1 - f^{-1}O')^{-}$: $(M_1 - f^{-1}O')^{-} \to (M_2 - O')^{-}$ is still an admissible homeomorphism. Now, by 18.3, any admissible homotopy inverse of $f$ can also be split (up to admissible homotopy) into essential maps $V_2 \to V_1$ and $(M_2 - V_2)^{-} \to (M_1 - V_1)^{-}$. So it follows from 18.2 that $f|V_1'$ and $f|(M_1 - V_1')^{-}$ are in fact admissible homotopy equivalences, i.e. $f$ satisfies (i) and (ii) above. q.e.d.

## §29. On the homotopy type of 3-manifolds and the isomorphism problem for 3-manifold groups

By a sufficiently large 3-manifold-group we understand a group which is known to be the fundamental group of a Haken 3-manifold which is boundary-incompressible. In this paragraph we are going to reduce the isomorphism problem for sufficiently large 3-manifold-groups to the homeomorphism problem for (boundary-incompressible) Haken 3-manifolds. Recall from [Ha 2] and [He 1] that the latter problem is solvable. Hence, in particular, the isomorphism problem for knot groups and non-splittable link groups are solvable.

The crucial step in our reduction can be formulated as follows:

29.1 Theorem. *Let* $(M,\emptyset)$ *be a Haken 3-manifold whose completed boundary-pattern is useful. Then one can construct, in a finite number of steps, a system* $W$ *of solid tori in* $M$ *with:*

1. $(\partial W - \partial M)^{-}$ *consists of essential annuli in* $(M,\bar{\emptyset})$, *and*
2. *any boundary-incompressible Haken 3-manifold homotopy equivalent to* $M$ *can be obtained from* $M$ *by a number of Dehn flips along components of* $W$.

Proof. The main ingredients of the proof are 24.2, together with the result of [Ha 1] that, if there is an essential surface in $(M,\underline{\underline{m}})$ with given admissible homeomorphism type and given boundary, then such a surface can actually be constructed in a finite number of steps (using a triangulation of $M$).

Furthermore note the following simple fact (whose converse is also true, by 28.1).

29.2 Assertion. *Let* $(W_1,\underline{\underline{w}}_1)$ *and* $(W_2,\underline{\underline{w}}_2)$ *be two solid tori whose boundary-patterns are useful and consist of disjoint annuli. Define* $m_i$, $i = 1,2$, *the circulation number of one side of* $W_i$ *with respect to* $W_i$. *Then* $(W_1,\underline{\underline{w}}_1)$ *and* $(W_2,\underline{\underline{w}}_2)$ *are admissibly homotopy equivalent if* card $\underline{\underline{w}}_1$ = card $\underline{\underline{w}}_2$ *and* $m_1 = m_2$.

It suffices to prove this for solid tori $(W_1,\underline{\underline{w}}_1)$ and $(W_2,\underline{\underline{w}}_2)$ whose boundary-patterns consist of precisely one annulus $A_i$, $i = 1,2$. Fix a meridian disc $D_i$ in $W_i$ (i.e. an essential disc). Since, by supposition, the circulation numbers of $A_1$ and $A_2$ are equal, there is a homeomorphism $\bar{f}: A_1 \to A_2$ with $\bar{f}(A_1 \cap D_1) = A_2 \cap D_2$. Since $D_2$ and $W_2$ are aspherical, there is no obstruction to extending $\bar{f}$ to an admissible map $f: W_1 \to W_2$ with $f^{-1}D_2 = D_1$. In the same way, $\bar{f}^{-1}$ can be extended to an admissible map $g: W_2 \to W_1$ with $g^{-1}D_1 = D_2$. Now observe that every admissible map of a disc which is the identity on the sides is admissibly homotopic to the identity. Hence there is no obstruction to deforming $f \circ g$ and $g \circ f$ admissibly to the identity. This proves the assertion.

For convenience we split the remainder of the proof into three cases.

<u>Case 1. $(M,\underline{\underline{m}})$ is a Seifert fibre space over the disc whose boundary-pattern is useful and consists of disjoint annuli.</u>

For any given Seifert fibre space $(M_i,\underline{\underline{m}}_i)$, $i \geq 1$, as considered in this case, we are going to construct a system $W_i$ of essential solid tori in $(M_i,\underline{\underline{m}}_i)$ with $W_i \cap \partial M_i$ contained in free sides of $(M_i,\underline{\underline{m}}_i)$. These systems will have the following properties: if $M_1$ is admissibly homotopy equivalent to $M_2$, any homeomorphism $\cup\underline{\underline{m}}_1 \to \cup\underline{\underline{m}}_2$ can be extended to an admissible homeomorphism $h: (M_1 - W_1)^- \to (M_2 - W_2)^-$ such that for any component $X_1$ of $W_1$ there is a component $X_2$ of $W_2$ admissibly homotopy equivalent to $X_1$ with $h(\partial X_1 - \partial M_1)^- = (\partial X_2 - \partial M_2)^-$.

Let $C_i$ be any free side of $M_i$ (this exists for otherwise the theorem is trivial, by 3.4). Construct a complete system of essential annuli in $M_i$ with one side in $C_i$ (apply the forementioned result of [Ha 1]). Furthermore, check which annuli of $C_i$ separate a solid torus from $M_i$ which meets $\partial M_i$ only in $C_i$. (This can be checked in a finite number of steps, by constructing essential discs). Let $A_i$ be the subsystem of all the latter annuli of $C_i$. Then each component of $A_i$ separates a solid torus which contains precisely one exceptional fibre of $M_i$. Indeed, by 5.4, the components of $A_i$ are in one-to-one correspondence to the exceptional fibres of $M_i$. Define $W_i = (M_i - U(\cup \underline{\underline{m}}_i \cup A_i))^-$, where $U(\cup \underline{\underline{m}}_i \cup A_i)$ is a regular neighborhood. Of course, $W_i$ is a system of solid tori since we are in Case 1, and we claim that the $W_i$'s satisfy the above property.

Let $M_1$ and $M_2$ be two admissibly homotopy equivalent Seifert fibre spaces (as considered under Case 1). Then, by 28.4, there is a fibre preserving admissible homotopy equivalence $f: M_1 \to M_2$ which maps exceptional fibres to exceptional fibres (if $M_2$ is the $S^1$-bundle over the annulus or Möbius band, then, by 5.5 and 3.4, $f$ can be deformed into a homeomorphism and the theorem is trivial). Furthermore, there is an admissible homotopy inverse of $f$ with the same properties (see 28.4). Observe that an exceptional fibre is neither homotopic to a different exceptional fibre nor homotopic to a non-trivial multiple of itself (the homotopy cannot go across $A_i$). In particular, it follows that $f$ defines a bijection on the set of exceptional fibres, i.e. the number of components of $A_1$ and $A_2$ are the same. Recalling 29.2, we therefore easily see that $W_1$ and $W_2$ satisfy the required properties.

<u>Case 2. $(M,\underline{\underline{m}})$ is any Seifert fibre space whose boundary-pattern is useful and consists of disjoint annuli.</u>

We are going to construct a system $W$ of essential solid tori in $(M,\underline{\underline{m}})$ with $W \cap \partial M$ contained in free sides of $(M,\underline{\underline{m}})$. This system will have the following properties with respect to any admissible homotopy equivalence $f: (M',\underline{\underline{m}}') \to (M,\underline{\underline{m}})$ whose restriction $f|\cup\underline{\underline{m}}': \cup\underline{\underline{m}}' \to \cup\underline{\underline{m}}$ is a homeomorphism: there is a system $W'$ of essential solid tori in $M'$ so that $f|\cup\underline{\underline{m}}'$ can be extended to an admissible homeomorphism $h: (M' - W')^- \to (M - W)^-$ such that for any component $X'$ of $W'$ there is a component $X$ of $W$ admissibly homotopy equivalent to $X'$ with $h(\partial X' - \partial M')^- = (\partial X - \partial M)^-$.

Construct a maximal, non-separating system $A$ of vertical annuli in $(M\ \underline{\underline{m}})$ which have at least one free side. Let $U$ be a regular neighborhood of $\cup\underline{\underline{m}} \cup A$. Then the manifold obtained from $M$ by splitting along $U$ is a Seifert fibre space as in Case 1. Hence we may construct a system W of essential solid tori in $(M - U)^-$ as in Case 1. We claim that W has the required properties.

Let f: M' → M be an admissible homotopy equivalence. Then, by 24.2, M' is again a Seifert fibre space. The orbit B' of M' is homotopy equivalent to the orbit B of M (this follows e.g. from [Wa 3], or more geometrically as follows: Denote by $B_0'$, $B_0$ the orbits minus exceptional points. Since, by 28.4, f can be chosen to be fibre preserving, the composition of a section s: $B_0' \to M'$ with f and the projection p: M → B can be extended to an essential map B' → B. The induced map $\pi_1 B' \to \pi_1 B$ is in fact an isomorphism since we find an inverse, using a homotopy inverse of f). Hence we find a maximal non-spearating system A' of vertical annuli in $(M', \underline{\underline{m}}')$ which have at least one free side, such that $f(A' \cap \cup \underline{\underline{m}}') = A \cap \cup \underline{\underline{m}}$ and that the number of components is equal to that of A. Of course, there is an extension of $f|\cup\underline{\underline{m}}'$ to a homeomorphism g: U' → U, where U' is a regular neighborhood of $\cup\underline{\underline{m}}' \cup A'$. The manifold obtained from M' by aplitting along U' is a Seifert fibre space as in Case 1, and we may construct a system W' of essential solid tori in $(M' - U')^-$ as in Case 1. By construction, W' and W are essential in M' and M, resp., and, as in Case 1, it follows that W has the required properties.

<u>Case 3.</u> $(M,\emptyset)$ <u>is any boundary-incompressible Haken</u> 3-<u>manifold</u>.

Let V be the characteristic submanifold of $(M,\bar{\emptyset})$. Denote by X the union of all components of V which are Seifert fibre spaces meeting ∂M, and by Y the union of all components of V which are I-bundles. For every component of X we can construct a system of essential solid tori (which meet ∂X in free sides) as in Case 2. Let W be the union of all these solid tori with a regular neighborhood of $(\partial Y - \partial M)^-$. W is a system of essential solid tori in $(M,\bar{\emptyset})$ (see 4.6), i.e. it satisfies 1 of 29.1. That it also satisfies 2 of 29.1 can be seen by a straightforward application of 24.2, 5.5, and 3.4. Hence we still have to show that W can be <u>constructed</u> in a finite number of steps.

To prove the latter it remains to construct the system $\bar{S}$

of essential annuli and tori which splits $M$ into a manifold $\tilde{M}$ whose components are either components of $X \cup Y$ or of $(M - X \cup Y)^{-}$. Then $X$ can be characterized as the union of all components of $\tilde{M}$ meeting $\partial M$ whose boundary consists of tori. In particular, it can be checked in a finite number of steps which components of $\tilde{M}$ belong to $X$. Then $W$ can be constructed using arguments given in Case 2.

To construct $\bar{S}$ we first give another characterization of $\bar{S}$. For this let $T$ be any essential annulus or torus in $(M,\bar{\emptyset})$, and let $(M^*,\underline{\underline{m}}^*)$ be the 3-manifold obtained from $(M,\bar{\emptyset})$ by splitting along $T$. We call $T$ a <u>bad</u> annulus or torus if the following holds:

1. there is an admissible annulus in $M^*$ which joins $\partial M$ with a copy of $T$, and
2. there is no essential annulus in $M$ so that $A \cap T \neq \emptyset$ and $A \cap M^*$ is essential in $M^*$.

Since $V$ is a complete and full F-manifold, every component of $(\partial(X \cup Y) - \partial M)^{-}$ is a bad annulus or torus and, using 10.7, we see that bad annuli and tori can be admissibly isotoped into $(\partial(X \cup Y) - \partial M)^{-}$. Hence the complete system $S$ of all bad annuli and tori is admissibly isotopic to the system $\bar{S}$ defined above. The advantage of our characterization of $S$ is, that we now can easily see how to construct $S = \bar{S}$. For this construct a complete system $C$ of essential annuli and tori. Using the characteristic submanifold $V$ one sees that $C$ contains $S$, up to admissible isotopy, and that the manifold obtained from $M$ by splitting along $C$ contains only finitely many (not necessarily disjoint) essential annuli, up to admissibly isotopy (see 5.7). By [Ha 1] again, all these annuli can be constructed. Hence one can decide, in a finite number of steps, which components of $C$ are bad annuli or tori.

q.e.d.

In general, not every homeomorphism of the boundary-pattern extends to the 3-manifold, even not for Seifert fibre spaces. But observe that this is true for $S^1$-bundles over the Möbius band whose boundary-pattern consists of two disjoint annuli (to see this consider the third remark after the proof of 28.4). Using this, together with the characteristic submanifold, it is not difficult to show the following: all Dehn flips along a given bad annulus $A$ of a Haken 3-manifold $(M,\emptyset)$ lead to homotopy equivalent 3-manifolds if and only if there is an annulus $B$ in $(M - U(A))^-$ whose boundary is contained in one component of $(\partial U(A) - \partial M)^-$ and such that $U(A) \cup U(B)$ is the $S^1$-bundle over the Möbius band. We leave it to the reader to put this fact together with arguments of the proof of 29.1, in order to establish a procedure which decides in a finite number of steps whether or not Dehn flips lead to homotopy equivalent 3-manifolds, i.e. to give a rigorous proof of the second part of the following corollary of 29.1.

29.3 Corollary. *There are only finitely many boundary-incompressible Haken 3-manifolds homotopy equivalent to a given Haken 3-manifold, and the set of all these 3-manifolds can be constructed in a finite number of steps.*

This in turn implies the following:

29.4 Corollary. *The isomorphism problem for sufficiently large 3-manifold groups can be solved, if the homeomorphism problem for Haken 3-manifolds is solvable.*

Remark. Recall from [Ha 2] and [He 1] that the supposition of the corollary is true.

Proof of 29.4. The isomorphism problem for sufficiently large 3-manifold groups asks for an algorithm which decides in a finite number of steps whether or not two given fundamental groups $\pi_1 M_1$ and $\pi_1 M_2$ of boundary-incompressible Haken 3-manifolds $M_1, M_2$ are isomorphic. Since Haken 3-manifolds are aspherical, this is the same as asking for an algorithm which decides in a finite number of steps whether or not two given boundary-incompressible Haken 3-manifolds $M_1, M_2$ are homotopy equivalent. Our algorithm for the latter question consists simply in constructing the set $\underline{\underline{M}}_1$ of all boundary-incompressible

Haken 3-manifolds homotopy equivalent to $M_1$ (this can be done, by 29.3) and checking whether or not $M_2$ is an element of $\underline{\underline{M}}_1$. q.e.d.

## Part VI. APPENDIX

### Chapter XI. Homotopy equivalences of surfaces and I-bundles.

The object of this chapter is to give rigorous proofs for the results on homotopy equivalences of surfaces which were needed in this book (and which are also of some interest in their own right).

Let $(F,\underline{\underline{f}})$ and $(F',\underline{\underline{f}}')$ be two orientable surfaces. Let $G$ and $G'$ be essential surfaces in $(F,\underline{\underline{f}})$ and $(F',\underline{\underline{f}}')$, respectively, and denote by $(\tilde{F},\tilde{f})$ and $(\tilde{F}',\tilde{f}')$ the surfaces obtained from $(F,\underline{\underline{f}})$ and $(F',\underline{\underline{f}}')$ by splitting at $(\partial G - \partial F)^-$ and $(\partial G' - \partial F')^-$, respectively. Denote by $\underline{\underline{g}}$, $\underline{\underline{g}}'$ the boundary-patterns of $G$, $G'$ induced by $\tilde{\underline{\underline{f}}}$, $\tilde{\underline{\underline{f}}}'$, resp., i.e. the proper boundary-patterns.

Keeping the above notations in mind, an admissible map $f\colon (F,\underline{\underline{f}}) \to (F,\underline{\underline{f}}')$ is called an <u>admissible</u> $(G,G')$-<u>homeomorphism</u> if

1. $f|G\colon (G,\underline{\underline{g}}) \to (G',\underline{\underline{g}}')$ is an admissible homeomorphism, and
2. $f|\tilde{F}\colon (\tilde{F},\tilde{\underline{\underline{f}}}) \to (\tilde{F}',\tilde{\underline{\underline{f}}}')$ is an admissible map.

Moreover, an admissible homotopy $f_t\colon (F,\underline{\underline{f}}) \to (F',\underline{\underline{f}}')$, $t \in I$ is called an <u>admissible</u> $(G,G')$-<u>homotopy</u> if

1. $f_0$ and $f_1$ are admissible $(G,G')$-homeomorphisms, and
2. $f_t^{-1}(\partial G' - \partial F')^- = (\partial G - \partial F)^-$, for all $t \in I$.

Analogously with admissible maps between 3-manifolds (replace the word "surface" in the above defintions by "3-manifold").

Every admissible homotopy equivalence $f\colon (F,\underline{\underline{f}}) \to (F',\underline{\underline{f}}')$ can be considered as an admissible $(\emptyset,\emptyset)$-homeomorphism, and our aim is to deform $f$ admissibly so that the regions where it is a homeomorphism are as large as possible. This problem will be considered in the next paragraph. As a result we obtain the existence of a unique "obstruction surface" for a homotopy equivalence (see 30.15).

In fact, this result is a consequence of a general property of homotopies of homotopy equivalences which we establish in §30. This fact will still be somewhat improved in §31. There we shall investigate homotopy equivalences between I-bundles, a situation similar to that described in §26 for homeomorphisms. As a result we obtain a theorem, similar to that of §26, which was needed in §23.

To avoid circle reasoning we use in this chapter only the

statements from Part I (except for 28.3 used in the proof of 31.1).

## §30. Homotopy equivalence of surfaces

To begin with we first describe the situation which we shall consider throughout this paragraph. For this let $(F,\underline{\underline{f}})$, $(F',\underline{\underline{f}}')$ be connected, orientable surfaces and let $H: (F,\underline{\underline{f}}) \times I \to (F',\underline{\underline{f}}')$ be an admissible homotopy. We consider $(F,\underline{\underline{f}}) \times I$ as an admissible product I-bundle over $(F,\underline{\underline{f}})$. With this interpretation $F \times 0$ and $F \times 1$ are lids. Let $F_i$, $i = 0,1$ and $F_i'$ be essential surfaces in $F \times i$ and $F'$, respectively, and suppose that $(F' - F_0')^-$ is in very good position to $(F' - F_1')^-$. Furthermore, suppose that $f_i = H|F \times i: F \times I \to F'$ is an admissible homotopy equivalence which is an admissible $(F_i,F_i')$-homeomorphism. For convenience we finally also suppose that neither $(F,\bar{\underline{\underline{f}}})$ nor $(F',\bar{\underline{\underline{f}}}')$ is the square, annulus, or torus.

Given the above situation we shall prove,

<u>30.1 Theorem</u>. <u>There is an admissible homotopy</u> $H_t$, $t \in I$, <u>of</u> $H$ <u>and an admissible ambient isotopy</u> $\alpha_t$ <u>of</u> $(F,\underline{\underline{f}}) \times I$ <u>with</u>:

1. $S_i = (\alpha_1 \cdot H_1)^{-1}(\partial F_i' - \partial F')^-$ <u>is a system of vertical squares or annuli in</u> $(F,\bar{\underline{\underline{f}}}) \times I$, $i = 0$ <u>and</u> $i = 1$,
2. $S_i \cap F \times i = (\partial F_i - \partial(F \times i))^-$, $i = 0$ <u>and</u> $i = 1$,
3. $H_t|F \times i$, $t \in I$, $i = 0,1$ <u>is an admissible</u> $(F_i,F_i')$-<u>homotopy</u>.

<u>Remark</u>. From this theorem we deduce later (see 30.15) the existence of the obstruction surface for homotopy equivalences between surfaces.

Before starting the proof of 30.1, we should first establish the following fact about admissible $(G\ G')$-homeomorphisms. Here we refer to the notations given in the beginning of this chapter.

<u>30.2 Lemma</u>. <u>If</u> $f: (F,\underline{\underline{f}}) \to (F',\underline{\underline{f}}')$ <u>is an essential map which is an admissible</u> $(G\ G')$-<u>homeomorphism, then</u> $f|\tilde{F}: (\tilde{F},\tilde{\underline{\underline{f}}}) \to (\tilde{F}',\tilde{\underline{\underline{f}}}')$ <u>is also an essential map</u>.

<u>Proof</u>. Let $k$ be any admissible singular curve in $(F,\tilde{\underline{\underline{f}}})$ such that

$f \circ k$ is inessential in $(\tilde{F}', \tilde{\underline{\underline{f}}}')$. Then we have to show that $k$ is inessential in $(\tilde{F}, \tilde{\underline{\underline{f}}})$.

Suppose that no end-point of $k$ lies in $(\partial G - \partial F)^{-}$. Then $f \circ k$ is an inessential singular curve in $(F', \underline{\underline{f}}')$. Hence $k$ is inessential in $(F, \underline{\underline{f}})$ since $f$ is an essential map, and so inessential in $(\tilde{F}, \tilde{\underline{\underline{f}}})$ for $G$ is an essential surface in $(F, \bar{\underline{\underline{f}}})$.

Suppose that precisely one end-point $x$ of $k$ lies in $(\partial G - \partial F)^{-}$. Then $f \circ k$ can be admissibly deformed (rel $x$) in $(F', \underline{\underline{f}}')$ into a component $t'$ of $(\partial G' - \partial F')^{-}$ since $f \circ k$ is inessential in $(\tilde{F}', \tilde{\underline{\underline{f}}}')$. Hence $t'$ is an arc as $k$ is. Now, $f|G\colon G \to G'$ is a homeomorphism and so $x$ lies in an arc $t$ of $(\partial G - \partial F)^{-}$. Thus $x$ splits $t$ into two arcs. One of them, together with $k$, defines an admissible singular arc $k^*$ in $(F, \bar{\underline{\underline{f}}})$ such that $f \circ k^*$ is an inessential singular arc in $(F', \underline{\underline{f}}')$ whose end-points lie in <u>one</u> side of $(F', \underline{\underline{f}}')$. Since $f$ is essential, this implies that $k$ is inessential in $(\tilde{F}, \tilde{\underline{\underline{f}}})$.

Finally, suppose that two end-points of $k$ lie in $(\partial G - \partial F)^{-}$. $f \circ k$ can be deformed (rel $\partial k$) into one component $t'$ of $(\partial G' - \partial F')^{-}$ since it is inessential in $(\tilde{F}', \tilde{\underline{\underline{f}}}')$. Since $f|G\colon G \to G'$ is a homeomorphism, this implies that the end-points of $k$ lie in <u>one</u> component $t$ of $(\partial G - \partial F)^{-}$. Since $f|t\colon t \to t'$ is a homeomorphism, it is easily seen that there is a singular arc $s$ in $t$ which joins the end-points of $k$ and such that $f \circ (s * k) \simeq 0$. Hence $s * k \simeq 0$ since $f$ is essential, and so $k$ is inessential in $(\tilde{F}, \tilde{\underline{\underline{f}}})$.

q.e.d.

We divide the actual proof of 30.1 into different steps. The crucial step is the proof of the next proposition.

<u>30.3 Proposition.</u> <u>Let the situation be given as described in the beginning. Then there is an admissible homotopy</u> $H_t$, $t \in I$, <u>of</u> $H$ <u>with:</u>

1. <u>The intersection of</u> $H_1^{-1}((\partial F_0' \cup \partial F_1') - \partial F')^{-}$ <u>with any free side</u> $r \times I$ <u>of</u> $(F, \underline{\underline{f}}) \times I$ <u>is a system of essential curves in</u> $r \times I$
2. $H_t | F \times i$, $t \in I$ $i = 0,1$, <u>is an admissible</u> $(F_i, F_i')$-

homotopy.

Proof. Define $R \times I$ to be the union of all free sides of $(F,\underline{\underline{f}}) \times I$, and denote by $h$ the restriction of $H$ to $R \times I$. Furthermore, let $\underline{\underline{r}}$ be the set of all components of $\partial R \times I$, and observe that $h: (R \times I, \underline{\underline{r}}) \to (F', \underline{\underline{f}}')$ is an admissible map. Finally, recall that $f_i$, $i = 1,2$, is an admissible homotopy equivalence etc. defined by $f_i = H|F \times i$.

Without loss of generality, we may suppose that $H$ is admissibly deformed, so that, in addition, $f_i$ is transverse with respect to $(\partial F'_{i+1} - \partial F')^-$ (indices mod 2). This implies that in particular $f_i^{-1}(\partial F'_{i+1} - \partial F')^-$ is a system of admissible curves in $(F,\bar{\underline{\underline{f}}}) \times i$. Moreover, note that every admissible homotopy of $h$ which is constant on $R \times \partial I$ can be extended to an admissible homotopy of $H$ with 2 of 30.3.

Observe that $h|R \times \partial I$ is transverse with respect to $((\partial F'_0 \cup \partial F'_1) - \partial F')^-$. Thus we may suppose that $h$ is admissibly deformed (rel $R \times \partial I$) so that $h^{-1}(\partial F'_i - \partial F')^-$ consists of admissible curved in $(R \times I, \bar{\underline{\underline{r}}})$, for $i = 0$ and $i = 1$. Define

$$K_i = h^{-1}(\partial F'_i - \partial F')^- .$$

Then $K_0$ and $K_1$ are transverse since $F'_0$ and $F'_1$ are in a good position.

Let $\alpha(h)$ be the number of points of $K_0 \cap K_1$, let $\beta(h)$ be the number of points of $(R \times \partial I) \cap (K_0 \cup K_1)$, and let $\gamma(h)$ be the sum of the numbers of components of $K_0$ and $K_1$. Suppose that $H$ is admissibly deformed, using an admissible homotopy $H_t$, $t \in I$, which satisfies 2 of 30.3, such that $(\alpha(h),\beta(h),\gamma(h))$ is as small as possible with respect to the lexicographical order. We claim that $H_t$ is the required homotopy, i.e. that $H_1$ satisfies 1 of 30.3.

To show this we first investigate the components of $R \times I - K_i$, $i = 0,1$, and of $R \times I - (K_0 \cup K_1)$. If $D$ is the closure of any component of $R \times I - K_i$ or of $R \times I - (K_0 \cup K_1)$ define

$$\underline{\underline{d}} = \{\text{components of } D \cap K_i \text{ and } D \cap b, \text{ where } b \in \bar{\underline{\underline{r}}}\}, \text{ or}$$

$$\underline{\underline{d}} = \{\text{components of } D \cap K_0,\ D \cap K_1, \text{ and } D \cap b, \text{ where } b \in \bar{\underline{\underline{r}}}\},$$

respectively.

For convenience we finally introduce the following conventions on our notations: the components of $(\partial F_0' - \partial F')^-$ and of their preimages under $f_0$, $f_1$ and $h$ will be denoted by Latin symbols, while the components of $(\partial F_1' - \partial F')^-$ and of their preimages under $f_0$, $f_1$, and $h$ will be denoted by Greek symbols.

30.4 Assertion. *Let $D$ be the closure of a component of $R \times I - K_i$, $i = 0,1$, which meets $R \times i$. Then $(D, \underline{\underline{d}})$ is not a j-faced disc, $j = 2$ or $j = 3$.*

Assume the converse. Define $r = D \cap R \times i$. Then, by the choice of $D$, $r$ is an arc in $R \times i$ such that at least one end-point $x$ of $r$ lies in $K_i$, and so in $(\partial F_i - \partial F)^-$. Moreover, $r$ can be considered as a free side of $(\tilde{F}, \tilde{\underline{\underline{f}}})$, where $(\tilde{F}, \tilde{\underline{\underline{f}}})$ is defined to be the surface obtained from $(F, \underline{\underline{f}})$ by splitting at $(\partial F_i - \partial F)^-$. The existence of the map $h|D$ shows that $h|r$ is inessential in $(\tilde{F}', \tilde{\underline{\underline{f}}}')$, where $(\tilde{F}', \tilde{\underline{\underline{f}}}')$ denotes the surface obtained from $(F', \underline{\underline{f}}')$ by splitting at $(\partial F_i' - \partial F')^-$. Hence, by 30.2, $r$ is inessential in $(\tilde{F}, \tilde{\underline{\underline{f}}})$. This implies that the component of $(\partial F_i - \partial F)^-$, which contains $x$, is inessential in $(F, \bar{\underline{\underline{f}}})$. But this contradicts the fact that $F_i$ is essential in $(F, \bar{\underline{\underline{f}}})$.

30.5 Assertion. *Let $D$ be the closure of a component of $R \times I - (K_0 \cup K_1)$. Suppose that $D$ lies in $h^{-1}F_i'$ if $D \cap R \times i \neq \emptyset$, $i = 0$ or $i = 1$. Then $(D, \underline{\underline{d}})$ is not a j-faced disc, $1 \leq j \leq 3$.*

Assume the converse. Consider the map $h|D\colon D \to F'$. We claim that $h(\partial D) \subset \partial F_0' \cup \partial F_1' \cup \partial F'$. This is clear if $D \cap R \times \partial I = \emptyset$. If $D \cap R \times \partial I \neq \emptyset$, notice that $D \cap R \times \partial I \subset D \cap R \times i$, $i = 0$ or $i = 1$, and that $D \cap R \times i \subset h^{-1}F_i'$, by our suppositions on $D$. Hence $h(D \cap R \times i) \subset \partial F_i'$ since $H|F \times i = f_i$ is an admissible $(F_i, F_i')$-homeomorphism. If $h|\partial D$ is contractible in $\partial F_0' \cup \partial F_1' \cup \partial F'$, we get a contradiction to our minimality condition on $(\alpha(h), \beta(h), \gamma(h))$, and if $h|\partial D$ is not contractible in $\partial F_0' \cup \partial F_1' \cup \partial F'$, we get either a contradiction to 11.1, i.e. to the fact that $F_0'$ and $F_1'$ are in a good position, or to the fact that $F_0'$ and $F_1'$ are essential surfaces in $(F', \bar{\underline{\underline{f}}}')$.

<u>30.6 Assertion</u>. <u>Let</u> D <u>be the closure of a component of</u> $R \times I - (K_0 \cup K_1)$ <u>which is also a component of</u> $R \times I - K_i$, $i = 0$ <u>or</u> $i = 1$. <u>Then</u> $(D, \underline{\underline{d}})$ <u>is not a</u> j-<u>faced disc</u>, $1 \leq j \leq 3$.

Assume the converse. Then, by 30.4 and 30.5, D must meet $R \times i + 1$ (indices mod 2). Let $r = D \cap R \times i + 1$. If r is contained in $F_{i+1}$, then $h(r) \subset \partial F'$ and $h|r$ is an embedding, since $f_{i+1} = H|F \times i + 1$ is an admissible $(F_{i+1}, F_{i+1}')$-homeomorphism. In this case, $h|\partial D \subset (\partial F_i' - \partial F')^- \cup \partial F'$ and $h|\partial D$ is not contractible in $(\partial F_i' - \partial F')^- \cup \partial F'$. Thus we get a contradiction to the fact that $F_i'$ is an essential surface in $(F', \bar{\underline{\underline{f}}}')$. Therefore $r \subset F \times i + 1 - F_{i+1}$, and so $h(r) \subset F' - F_{i+1}'$. The existence of the map $h|D$ shows that $h|r$ can be admissibly deformed (rel $r \cap K_i$) in $(F' - F_{i+1}')^-$ into $(\partial F_i' - \partial F')^-$. Certainly, this homotopy can be extended to an admissible homotopy of H which staisfies 2 of 30.3 and which pulls h into a map $\hat{h}$ so that

$$\hat{h}^{-1}((\partial F_0' \cup \partial F_1') - \partial F')^- = ((K_0 \cup K_1) - U(r))^- \cup \hat{r},$$

where $U(r)$ is a regular neighborhood of r in $R \times I$ and where $\hat{r}$ is a copy of r contained in $(\partial U(r) - R \times \partial I)^-$. This contradicts our minimality condition on $(\alpha(h), \beta(h), \gamma(h))$.

<u>30.7 Assertion</u>. <u>Let</u> D <u>be the closure of a component of</u> $R \times I - (K_0 \cup K_1)$ <u>which meets</u> $R \times i$ <u>and which does not meet</u> $R \times i + 1$, <u>for</u> $i = 0$ <u>or</u> $i = 1$ <u>(indices mod</u> 2). <u>Suppose that</u> D <u>lies in</u> $h^{-1}F_i'$ <u>and that</u> $D \cap R \times i$ <u>is not a component of</u> $R \times i$.
<u>Then the side of</u> $(D, \underline{\underline{d}})$ <u>which does not meet</u> $R \times i$ <u>must lie in</u> $K_i$, <u>if</u> $(D, \underline{\underline{d}})$ <u>is a square</u>.

Assume the converse. Then $(D, \underline{\underline{d}})$ is a square such that one side $r_0$ of $(D, \underline{\underline{d}})$ lies in $R \times 0$, say, and that the side $\rho_0$ of $(D, \underline{\underline{d}})$

opposite to $r_0$ is contained in $K_1$. Let $x_1$, $x_2$ and $z_1$, $z_2$ be the end-points of $r_0$ and $\rho_0$, respectively. Since $r_0$ is not a component of $R \times 0$, it follows that $x_1$, say, lies in a component $k_1$ of $K_0$. $x_2$ lies either in $K_0$ or in $\partial R \times \partial I$. We only deal with the case that $x_2$ lies in a component $k_2$ of $K_0$, for the argument in the other case is analogous.

Now, we consider the surface $(F', \underline{\underline{f}}')$. Notice that D lies in $h^{-1}F_0'$, and let $G_0'$ be the component of $F_0'$ which contains $h(D)$. Since $f_0$ is an admissible $(F_0, F_0')$-homeomorphism, there is a component $r_0'$ of $G_0' \cap \partial F'$ such that $h|r_0 = f_0|r_0 \colon r_0 \to r_0'$ is a homeomorphism. In particular, the end-points $x_1$ and $x_2$ of $r_0$ are mapped under h onto the end-points $x_1'$ and $x_2'$ resp. of $r_0'$. Let $t_j'$, $j = 1,2$, be the component of $(\partial F_0' - \partial F')^-$ which contains $h(k_j)$, and let $\tau'$ be the component of $(\partial F_1' - \partial F')^-$ which contains $h(\rho_0)$. $h(\partial D) \subset \partial F_0' \cup \partial F_1' \cup \partial F'$ and, since $h(x_1) \neq h(x_2)$, it follows that $h|\partial D$ is not contractible in $\partial F_0' \cup \partial F_1' \cup \partial F'$. Thus the existence of the map $h|D$ shows that $\tau'$ separates a disc $D'$ from $G_0'$ such that $D' \cap \partial F_0' = r_0'$. If we define

$$\underline{\underline{d}}' = \{\text{components of } D' \cap \partial F',\ D' \cap (\partial F_0' - \partial F')^-,$$
$$\text{and } D' \cap (\partial F_1' - \partial F')^-\},$$

then $(D', \underline{\underline{d}}')$ is a square.

Consider $R \times 1 \cap h^{-1}(t_1' \cup t_2')$. Since $h^{-1}(t_1' \cup t_2')$ is a system of admissible curves in $(R \times I, \underline{\underline{r}})$, $R \times 1 \cap h^{-1}(t_1' \cup t_2')$ is a system of points. Define

$$P = \{p | p \in R \times 1 \cap h^{-1}(t_1' \cup t_2') \text{ and } h(p) \in D'\}, \text{ and}$$

$$Q = \{(p_1, p_2) \in P \times P | \text{there is a component of } R \times 1 \cap h^{-1}G_0'$$
$$\text{which contains } p_1 \text{ and } p_2\}.$$

A point of P is called an <u>essential point</u> if it lies in an arc of $h^{-1}(t_1' \cup t_2')$ which is essential in $(R \times I, \underline{\underline{r}})$. A pair $(p_1, p_2) \in Q$ is called an <u>essential pair</u> if $p_1$ and $p_2$ are essential points.

Let $(p_1,p_2)$ be any element of $Q$, and let $r_1$ be the component of $R \times 1 \cap h^{-1}G_0'$ which contains $p_1$ and $p_2$. Then we claim that $p_1$ and $p_2$ are mapped under $h$ into different components of $D' \cap (t_1' \cup t_2')$. For notice that otherwise $h|r_1$ can be deformed (rel $\partial r_1$) in $F' - F_1'$ into $(\partial F_0' - \partial F')^-$ since $f_1(r_1)$ lies in the disc $D'$. Therefore there is an admissible homotopy of $h$ which diminishes $\beta(h)$ without enlarging $\alpha(h)$. Moreover, this homotopy can be chosen so that it extends to an admissible homotopy of $H$ with 2 or 30.3. But this contradicts our minimality condition on $(\alpha(h)\ \beta(h),\gamma(h))$.

<u>Case 1. There is no element of</u> $Q$ <u>which is an essential pair</u>.

We show that Case 1 leads to contradictions.

Recall that $f_1\colon (F,\underline{\underline{f}}) \times 1 \to (F',\underline{\underline{f}}')$ is an admissible homotopy equivalence which is an admissible $(F_1,F_1')$-homeomorphism, and that $\tau'$ is a component of $(\partial F_1' - \partial F')^-$. Thus, in particular, the preimage $\tau = f_1^{-1}\tau'$ is precisely one curve, and, moreover, $f_1|\tau\colon \tau \to \tau'$ is a homeomorphism. Especially, $R \times 1 \cap h^{-1}\tau'$ $= R \times 1 \cap f_1^{-1}\tau'$ consists of at most two points.

(A) <u>Suppose that</u> card $(Q) = 0$.

To get a contradiction, it suffices to construct an element $(p_1,p_2)$ of $Q$. This can be done in such a way that $p_1$ can be joined in $f_1^{-1}t_1'$ with $\tau$. To begin with note that $f_1|\tau\colon \tau \to \tau'$ is a homeomorphism and that $t_1' \cap \tau' \neq \emptyset$. This implies that $\tau \cap f_1^{-1}t_1' \neq \emptyset$. Let $y$ be the point of $\tau \cap f_1^{-1}t_1'$ with $f_1(y) = h(z_1)$, and let $t_1$ be the component of $f_1^{-1}t_1'$ which contains $y$. Since $t_1'$ is an arc and since $f_1|F_1\colon F_1 \to F_1'$ is a homeomorphism, it follows that $t_1$ is an arc. $y$ splits $t_1$ into two arcs. One of them, say $t_1^*$, is mapped under $f_1$ entirely into $t_1' \cap D'$. Now $r_0$ lies in a free side of $(F,\underline{\underline{f}}) \times 0$ (see beginning of 30.7), $f_0(r_0) = r_0'$, and $f_0$ is an admissible $(F_0,F_0')$-homeomorphism. This implies that $r_0'$, and so $x_1'$, lies in a free side of $(F',\underline{\underline{f}}')$. Hence the end-point $p_1$ of $t_1^*$ different from $y$ lies in a free side of $(F,\underline{\underline{f}}) \times 1$, i.e. in $R \times 1$. $D'$ is a disc, and $f_1(R \times 1) \cap (D' \cap \partial F_1') = \emptyset$ since $f_1$ is an admissible $(F_1,F_1')$-homeomorphism. Thus it follows immediately the existence of a point $p_2 \in R \times 1$ with $(p_1,p_2) \in Q$.

(B) Suppose that card $(Q) = 1$.

Let $(p_1,p_2)$ be the element of $Q$. Since we are in Case 1, $p_1$ say, lies in an arc $k$ of $h^{-1}(t_1' \cup t_2')$, i.e. of $K_0$, which is inessential in $(R \times I,\bar{\underline{\underline{r}}})$. Then $k$ separates an admissible disc $(D^*,\underline{\underline{e}}^*)$ from $(R \times I,\bar{\underline{\underline{r}}})$ so that $(D^*,\underline{\underline{e}}^*)$ is a 2- or 3-faced disc. Define $d = D^* \cap R \times 1$. By 30.6 it follows that $k \cap K_1 \neq \emptyset$. This implies the existence of a component $\mu$ of $D^* \cap h^{-1}\tau'$ with $\mu \cap k \neq \emptyset$. By 30.5, it follows that the arcs of $K_1 \cap D^*$ join $k$ with $d$. Thus, in particular, $\mu$ joins $k$ with a point $q$ of $d$. Notice that $q$ is a point of $R \times 1 \cap h^{-1}(\partial F_1' - \partial F')^-$ and that $R \times 1 \cap h^{-1}(\partial F_1' - \partial F')^- = R \times 1 \cap f_1^{-1}(\partial F_1' - \partial F')^-$. Hence it follows from the fact that $f_1$ is an admissible $(F_1\ F_1')$-homeomorphism that $q$ is mapped under $f_1$ into a point of $\tau' \cap \partial F'$, say $q'$. In particular, $\tau'$ cannot be a closed curve. Therefore the point $h(z_1)$ splits $\tau'$ into two arcs $\tau_1'$ and $\tau_2'$, and $t_1'$ into two arcs $t_{11}'$ and $t_{12}'$. Let the indices be chosen so that $\tau_1'$ contains $q'$ and that $t_{11}'$ contains $x_1'$.

Define $w' = \tau_1' \cup t_{11}'$. Then $w'$ is an arc, and it follows that the end-points of $w'$ lie in free sides of $(F',\underline{\underline{f}}')$. This is already proved for $x_1'$ and for $q'$ notice that $q$ lies in a free side of $(F,\underline{\underline{f}}) \times 1$ and that $f_1$ is an admissible $(F_1,F_1')$-homeomorphism. Since $f_1$ is an admissible $(F_1,F_1')$-homeomorphism, it follows that $q = \partial F \times 1 \cap f_1^{-1}\tau_1'$. Moreover, it follows that $p_1 = \partial F \times 1 \cap f_1^{-1}t_{11}'$ $= R \times 1 \cap h^{-1}t_{11}'$ since we are in (B). Thus $p_1$ and $q$ are the only points of $\partial F \times 1 \cap f_1^{-1}w'$.

Now, $\mu$ separates a disc $D_1$ from $D^*$ which contains $p_1$ and $q$. The existence of the map $h|D_1$ shows that $f_1|d_1$ can be deformed (rel $\partial d_1$) into $w'$, where $d_1$ is the arc in $d = D^* \cap R \times 1$ which joins $p_1$ with $q$. Thus $f_1$ can be admissibly deformed so that afterwards $\partial F \cap f_1^{-1}w' = \emptyset$. Then $f_1^{-1}w'$ consists of closed curves. The restriction of $f_1$ to each such closed curve is contractible in the arc

$w'$. Now, $f_1$ is an essential map, in fact an admissible homotopy equivalence. Thus each curve of $f_1^{-1}w'$ is contractible in $F \times 1$. Hence, applying the transversality lemma to $f_1$, it follows that $f_1$ can be deformed (rel $\partial F \times 1$) so that afterwards $f_1^{-1}w' = \emptyset$. But $f_1\colon (F,\underline{\underline{f}}) \times 1 \to (F',\underline{\underline{f}}')$ is an admissible homotopy equivalence. Therefore it follows from 16.1 (take the product with the interval) that $w'$ is either admissibly parallel in $(F',\bar{\underline{\underline{f}}}')$ to a free side of $(F',\underline{\underline{f}}')$, or $w'$ is inessential in $(F',\bar{\underline{\underline{f}}}')$. The first case is impossible since the end-points of $w'$ lie both in free sides of $(F',\underline{\underline{f}}')$, and the latter is impossible since $F_0'$ and $F_1'$ are in good position.

(C) <u>Suppose that</u> card $(Q) = 2$.

Let $(p_1,p_2)$ and $(p_3,p_4)$ be the elements of $Q$, and suppose that $p_1$ and $p_3$ are mapped under $f_1$ into the same component of $D' \cap t_1'$. Consider the map $f_1$. Recall that $f_1|\tau\colon \tau \to \tau'$ is a homeomorphism. Hence precisely one point of $\tau$, say $y$, is mapped under $f_1$ into $h(z_1)$. By an argument of (A), we may suppose that $p_1$ can be joined in $f_1^{-1}t_1'$ with $\tau$, i.e. with $y$. Thus it follows that $p_3$ cannot be joined in $f_1^{-1}t_1'$ with $\tau$. This means that the component $t_1$ of $f_1^{-1}t_1'$ which contains $p_3$ is mapped under $f_1$ entirely into $D' \cap t_1'$. This implies that the end-point $p_5$ of $t_1$ which is different from $p_3$ is contained in $R \times 1 \cap f_1^{-1}(t_1' \cup t_2') = R \times 1 \cap h^{-1}(t_1' \cup t_2')$. Certainly, $p_5$ is different from $p_3$, and different from $p_1$ since $t_1 \cap \tau = \emptyset$. Moreover, there must be a point $p_6$ of $R \times 1 \cap h^{-1}(t_1' \cup t_2')$ such that $(p_5,p_6) \in Q$ (recall the properties of the disc $D'$). Thus we have found an element of $Q$ which is different from $(p_1,p_2)$ and $(p_3,p_4)$. But this is a contradiction since we are in (C).

(D) <u>Suppose that</u> card $(Q) \geq 3$.

Let $(p_1,p_1^*),\ldots,(p_n,p_n^*)$ be all the elements of $Q$. Since we are in Case 1, we may suppose that every $p_j$ lies in a component $s_j$ of $h^{-1}(t_1' \cup t_2')$ which is inessential in $(R \times I,\bar{\underline{\underline{r}}})$, $1 \leq j \leq n$. Hence $s_j$ separates an i-faced disc $D_j$ from $(R \times I,\bar{\underline{\underline{r}}})$, $i = 2,3$. Moreover, by an argument of (B), there must be a component of $D_j \cap h^{-1}\tau'$

which joins $s_j$ with $R \times 1 \cap D_j$.

Define $G = (R \times I - h^{-1}F_1')^-$. Then $p_1,\ldots,p_n \in G$, for otherwise we get a contradiction to 30.5. Denote by $\hat{D}_j$ the component of $G \cap D_j$ which contains $p_j$. Then $\hat{D}_j$ can be considered as a 3-faced disc such that one side lies in $h^{-1}\tau'$ and one in $R \times 1$. In particular, $\hat{D}_j$ contains a point of $R \times 1 \cap h^{-1}\tau'$. Since we are in (D), there are at least three different points $p_1$, $p_2$, $p_3$. On the other hand, recall from the beginning of Case 1 that $R \times 1 \cap h^{-1}\tau'$ consists of at most two points. Thus we may suppose that $\hat{D}_2 \subset \hat{D}_1$. Now $\hat{D}_2 \neq \hat{D}_1$ and so $p_2^* \in \hat{D}_1$. Moreover, $p_1$, $p_2$ and $p_2^*$ can be joined in $h^{-1}(t_1' \cup t_2')$, without crossing $h^{-1}\tau'$, with pairwise different points $x_1$, $x_2$, and $x_3$ of $h^{-1}\tau'$, resp. Now, $p_1$ and $p_2$ (or $p_1$ and $p_2^*$) are mapped into the same component of $(t_1' \cup t_2') \cap D'$. Hence it follows that $h(x_1)$ and $h(x_2)$ lie in the same point $z'$ of $\tau' \cap (t_1' \cup t_2')$.

The arc $\mu$ in $h^{-1}\tau'$ which joins $x_1$ with $x_2$ is mapped under $h$ into $\tau'$ with $h(\partial\mu) = z'$. Since $Q \neq \emptyset$, it follows that $R \times 1 \cap h^{-1}\tau' \neq \emptyset$ and so, from the beginning of Case 1, that $\tau'$ is an arc. Hence $h|\mu$ can be contracted (rel $\partial\mu$) in $\tau'$ into $z'$. Therefore there is a homotopy of $h$, constant outside a regular neighborhood of $\mu$, which diminishes $\alpha(h)$. Moreover, this homotopy can certainly be extended to an admissible homotopy of $H$ with 2 of 30.3. But this contradicts our minimality condition on $(\alpha(h),\beta(h),\gamma(h))$.

<u>Case 2. There is at least one element of $Q$ which is an essential pair.</u>

Let $(p_1,p_2)$ be an essential pair of $Q$. Then, by the very definition, there is a component $r_1$ of $R \times 1 \cap h^{-1}G_0'$ which contains $p_1$ and $p_2$, and, moreover, $p_1$ and $p_2$ lie in components of $h^{-1}(\partial G_0' - \partial F')$ which are essential in $(R \times I, \underline{\underline{r}})$. Let $A$ be the component of $h^{-1}G_0'$ which contains $r_1$. Then, applying 30.4, it follows that each component of $(\partial A - \partial(R \times I))^-$ is an essential curve in $(R \times I, \underline{\underline{r}})$. This implies that $A$ is an inner square in $(R \times I, \underline{\underline{r}})$.

Since $h(r_1)$ lies in the disc $D'$, $h|r_1$ can be deformed (rel $\partial r_1$) in $D'$ into $\partial G_0'$. Then $h(\partial A) \subset \partial G_0'$. $h|\partial A$ is not contractible in $\partial G_0'$, for otherwise $A \cap R \times 0 = D \cap R \times 0$ since $f_0$ is an admissible $(F_0,F_0')$-homeomorphism, and the restriction of $h$ to any component of $(\partial A - \partial(R \times I))^-$ can be deformed (rel boundary) in $(\partial G_0' - \partial F')^-$ into $D'$; so we get a contradiction to our minimality condition on $(\alpha(h),\beta(h),\gamma(h))$. Thus the existence of the map $h|A\colon A \to G_0'$ shows that $G_0'$ must be an inner square in $(F',\underline{\underline{f}}')$ (recall that $G_0'$ is essential in $(F',\bar{\underline{\underline{f}}}'))$.

$G_0' \cap (\partial F_1' - \partial F')^- \neq \emptyset$, and each component of $G_0' \cap (\partial F_1' - \partial F')$ is an arc which joins the two components of $(\partial G_0' - \partial F')^-$. Hence $(\partial F_1' - \partial F')^-$ splits $G_0'$ into squares. Two of them meet $\partial F'$. $D'$ is one of them, and denote by $D^*$ the other one.

Now, recall that we used $D'$ in the definition of the set $Q$. We copy this definition, and define a set $Q^*$ using $D^*$. Then, as in Case 1, it follows the existence of at least one element $(p_1^*,p_2^*)$ of $Q^*$ which is an essential pair. Denote by $A^*$ the component of $h^{-1}G_0'$ which contains $p_1^*$ and $p_2^*$. Then, as above, $A^*$ must be an inner square in $(R \times I,\bar{\underline{\underline{r}}})$.

$A \cap R \times 1$ is mapped under $h$ into $D'$. Since $f_0$ is an admissible $(F_0,F_0')$-homeomorphism, $A \cap R \times 0$ is mapped under $h$ into $G_0' \cap \partial F'$, i.e. either into $D'$ or into $D^*$. It follows from our minimality condition on $(\alpha(h),\beta(h),\gamma(h))$ that $h(A \cap R \times 0)$ lies in $D^*$ (see above). Analogously, $h(A^* \cap R \times 0) \subset D'$ and $h(A^* \cap R \times 1) \subset D^*$.

Consider $f_0\colon (F,\underline{\underline{f}}) \times 0 \to (F',\underline{\underline{f}}')$. Since $f_0$ is an admissible $(F_0,F_0')$-homeomorphism and since $G_0'$ is an inner square in $(F',\bar{\underline{\underline{f}}}')$, it follows that $G_0 = f_0^{-1}G_0'$ is an inner square in $(F,\bar{\underline{\underline{f}}}) \times 0$, and that $f_0|G_0\colon G_0 \to G_0'$ is a homeomorphism. In particular, $G_0 \cap \partial F \times 0 = (A \cup A^*) \cap R \times 0$.

There is an admissible $(F_0,F_0')$-homotopy, constant outside of a regular neighborhood of $G_0$, which pulls $f_0$ into an admissible $(F_0,F_0')$-homeomorphism $\hat{f}_0$ with $\hat{f}_0(A \cap R \times 0) \subset D'$ and $\hat{f}_0(A^* \cap R \times 0) \subset D^*$. Extend this homotopy to an admissible homotopy of $H$ into $\hat{H}$, which satisfies 2 of 30.3 and which is constant outside of a regular neighborhood $U(G_0)$ of $G_0$ in $F \times I$. Define

$\hat{h} = \hat{H}|R \times I$. After a small general position deformation of $\hat{h}$, constant outside of $U(G_0) \cap R \times I$, $\hat{h}^{-1}(\partial F_i' - \partial F')^-$, $i = 0,1$, is a system $\hat{K}_i$ of curves. Now $\alpha(\hat{h}) > \alpha(h)$. But notice that all the points of $\hat{K}_0 \cap \hat{K}_1$ which do not lie in $K_0 \cap K_1$ are contained in $A \cup A^*$.

$\hat{h}^{-1}(\partial F_0' - \partial F')^- = h(\partial F_0' - \partial F')^-$. However, by definition of $\hat{h}$, $\hat{h}(A \cap R \times \partial I) \subset D'$ and $\hat{h}(A^* \cap R \times \partial I) \subset D^*$. Hence the points of $\hat{K}_0 \cap \hat{K}_1$ which are contained in $A \cup A^*$ can be removed in the obvious way, using an admissible homotopy of $\hat{h}$, constant outside of a regular neighborhood of $A \cup A^*$. Thus $\alpha(h)$ can be diminished, and we get a contradiction to our minimality condition on $(\alpha(h),\beta(h),\gamma(h))$.

Thus, in any case, we get contradictions, and this proves 30.7.

With 30.4-30.7 we have all the tools available to prove 30.3.

An easy consequence of 30.4-30.7 is, that every component of $K_0$ and $K_1$ is an essential curve in $(R \times I, \bar{\underline{\underline{r}}})$. Hence, in order to prove 30.3, it remains to show that $K_0 \cap K_1 = \emptyset$.

Assume the converse. Then there is a curve $k_1$ of $K_0$ with $k_1 \cap K_1 \neq \emptyset$. Let $t'$ be the component of $(\partial F_0' - \partial F')^-$ which contains $h(k_1)$. Then, by our choice of $k_1$, $t' \cap (\partial F_1' - \partial F')^- \neq \emptyset$.

$k_1$ is an essential curve in $(R \times I, \bar{\underline{\underline{r}}})$. Hence it joins either two components of $\partial R \times I$ or of $R \times \partial I$. Applying 30.5 and 30.7, we see that $k_1$ must join $R \times 0$ with $R \times 1$. The same with the curves of $K_1$. Moreover, $k_1 \cap K_1$ consists of precisely one point. Therefore, applying 30.5, it follows that one end-point $y$ of $k_1$ lies in $R \times 0 \cap h^{-1}F_0' \cap h^{-1}F_1' = R \times 0 \cap h^{-1}(F_0' \cap F_1')$. Now, $h|R \times 0 = f_0|R \times 0$ and $f_0$ is an admissible $(F_0,F_0')$-homeomorphism. Thus $h(y)$ lies in $\partial F'$ (in fact in $F_0' \cap \partial F'$), and so in an end-point $y'$ of $t'$.

$y'$ lies in $F_1' \cap \partial F'$ since $y \in h^{-1}F_1'$. Thus there is a point $x_1$ of $R \times 1 \cap h^{-1}F_1'$ with $h(x_1) = y'$ since $f_1$ is an admissible $(F_1,F_1')$-homeomorphism. Since $t'$ is a component of $(\partial F_0' - \partial F')^-$, there is a component $k_2$ of $h^{-1}(\partial F_0' - \partial F')^-$ which contains $x_1$. Since $k_1 \cap K_1 \neq \emptyset$, it follows from our minimality condition on

$(\alpha(h),\beta(h),\gamma(h))$ that $h|k_1$ cannot be contracted (rel $\partial k_1$) in t' into y'. This implies that $k_1 \neq k_2$ since $h(x_1) = h(y) = y'$.

Let $x_2$ be the end-point of $k_2$ different from $x_1$. y and $x_2$ both lie in $R \times 0 \cap h^{-1}F_0'$, $h|R \times 0 = f_0|R \times 0$ and $f_0|F_0: F_0 \to F_0'$ is a homeomorphism. Hence it follows that $h(x_2) \neq h(y) = y'$. Moreover, $h(x_2)$ lies in an end-point of t' since $f_0$ is an admissible $(F_0,F_0')$-homeomorphism. Thus h maps the end-points of $k_2$ into two different end-points of t'. This implies that $k_2 \cap K_1 \neq \emptyset$ since $t' \cap (\partial F_1' - \partial F') \neq \emptyset$. Hence, as above, we see that $k_2 \cap K_1$ consists of precisely one point.

Since $k_2 \cap K_1$ is precisely one point, it follows from 30.4-30.7 that there is a 3-faced disc $(D,\underline{\underline{d}})$ whose sides lie in $R \times 1$, $K_1$, and $k_2$, and which is, more precisely, the closure of a component of $R \times I - (K_0 \cup K_1)$. Moreover, it contains $x_1$. Hence it follows that $D \subset h^{-1}F_1'$ since $x_1 \in h^{-1}F_0' \cap h^{-1}F_1'$. But this contradicts 30.5.

Thus our assumption $k_1 \cap K_1 \neq \emptyset$ must be wrong, and this proves 30.3. q.e.d.

<u>30.8 Corollary</u>. <u>Let the situation be given as described in the beginning of</u> §30. <u>Then there is an admissible homotopy</u> $H_t$, $t \in I$, <u>of</u> H <u>which satisfies</u> 2 <u>of</u> 30.3 <u>and so that for every free side</u> $r \times I$ <u>of</u> $(F,\underline{\underline{f}}) \times I$ <u>the system</u>

$$(r \times I) \cap H_1^{-1}((\partial F_0' \cup \partial F_1') - \partial F')^-$$

<u>consists of</u> (<u>pairwise disjoint</u>) <u>arcs joining</u> $r \times 0$ <u>with</u> $r \times 1$.

<u>Proof</u>. Assume the contrary. We use the same notations as given in the beginning of the proof 30.3. Then, by 30.3, we may suppose that H is admissibly deformed so that $H^{-1}((\partial F_0' \cup \partial F_1') - \partial F')^-$ intersects $R \times I$ in a system of curves which are essential in $(R \times I,\bar{\underline{\underline{r}}})$. In particular, $K_0 \cap K_1 = \emptyset$. Suppose that H is admissibly deformed, using an admissible homotopy with 2 of 30.3, so that the above holds and that, in addition, the number of curves of $K_0 \cup K_1$ is as small as possible. (Recall that $K_i = R \times I \cap H^{-1}(\partial F_i' - \partial F')^-$).

By our assumption, there is a component r of R (R is

the union of free sides of $(R \times I, \underline{\underline{r}}))$ such that $(r \times I) \cap K_0$, say, is a non-empty system of curves which do not meet $r \times \partial I$.

<u>30.9 Assertion</u>. <u>Let</u> $A'$ <u>be a component of</u> $F_i'$ <u>or of</u> $(F' - F_i')^-$, $i = 0$ <u>or</u> $i = 1$. <u>Suppose that</u> $A'$ <u>is an inner square or annulus in</u> $(F', \bar{\underline{\underline{f}}}')$ <u>which contains a free side of</u> $(F', \underline{\underline{f}}')$.

<u>Then</u> $r \times I \cap K_i = \emptyset$ <u>if</u> $h(r \times i + 1) \subset A'$ <u>(indices mod</u> 2).

Assume the converse. Then $h(r \times i + 1) \subset A'$ and $r \times I \cap K_i \neq \emptyset$. Hence $h(r \times i) \subset F' - A'$, for otherwise $h(r \times \partial I) \subset A'$ and then $h|r \times I$ can be admissibly contracted (rel $r \times \partial I$) into $A'$ which contradicts our minimality condition on $K_0 \cup K_1$. $A'$ is a component of $F_i'$ or of $(F' - F_i')^-$ and $f_i$ is an admissible $(F_i, F_i')$-homeomorphism. Hence there must be a component $A$ of $F_i$ or of $(F \times i - F_i)^-$ with $f_i(A) \subset A'$. By 30.2, $f_i|A\colon A \to A'$ is an essential map. Hence it follows that $A$ is an inner square or annulus resp. in $(F, \bar{\underline{\underline{f}}}) \times i$ which contains a free side $s \times i$ of $(F, \underline{\underline{f}}) \times i$, since $A'$ is such a surface in $(F', \underline{\underline{f}}')$. Now $h(r \times i) \subset F' - A'$ and $h(s \times i) \subset A'$. Therefore $r \neq s$. The existence of the map $h|r \times I$ shows that $f_i|r \times i$ can be admissibly deformed into $A'$, and so it follows that a multiple of $f_i|r \times i$ is admissibly homotopic in $(F', \underline{\underline{f}}')$ to a multiple of $f_i|s \times i$. On the other hand, recall that $f_i$ is an admissible homotopy equivalence. Hence there is an admissible homotopy inverse $g_i$ of $f_i$. In particular, $g_i f_i|r \times i$ and $g_i f_i|s \times i$ are admissibly homotopic, and admissibly homotopic to $\mathrm{id}|r \times i$ and $\mathrm{id}|s \times i$, respectively. Thus two different free sides of $(F, \underline{\underline{f}})$ are admissibly homotopic in $(F, \underline{\underline{f}})$, and so $(F, \bar{\underline{\underline{f}}})$ must be a square or annulus, which contradicts our suppositions on $(F, \underline{\underline{f}})$. Hence we have proved 30.9.

Now we split the proof of 30.8 into two cases:

<u>Case 1</u>. $r \times i \cap K_1 = \emptyset$.

In this case either $r \times I \subset h^{-1}F_1'$ or $r \times I \subset h^{-1}(F' - F_1')^-$. If $r \times I \subset h^{-1}F_1'$, notice that $r \times I \cap K_0 \neq \emptyset$ and that $K_0$

splits $r \times I$ into a system of squares or annuli. By our choice of $r$, one of them, say $C$, contains $r \times 1$. $r \times 1$ is a free side of $(F,\underline{\underline{f}}) \times 1$ which lies in $h^{-1}F_1'$, and so in $f_1^{-1}F_1'$. Since $f_1$ is an admissible $(F_1,F_1')$-homeomorphism, $r \times 1$ is mapped under $f_1$, and so under $h$, into a free side of $(F',\underline{\underline{f}}')$. Thus the existence of the map $h|C$ shows that $h(r \times 1)$ lies in a component of $F_0'$ or of $(F' - F_0')^-$ which is an inner square or annulus in $(F',\bar{\underline{\underline{f}}}')$ and which contains a free side of $(F',\underline{\underline{f}}')$. Thus, by 30.9, $r \times I \cap K_0 = \emptyset$, which contradicts our choice of $r$.

If $r \times I \subset h^{-1}(F' - F_1')^-$, then $h|r \times I$ can be considered as a homotopy of $f_1|r \times 1$ which can be extended to an admissible $(F_1,F_1')$-homotopy of $f_1$, which is constant outside of a regular neighborhood of $r \times 1$. This homotopy of $f_1$ can be extended to an admissible homotopy of $H$, with 2 of 30.3, which pulls $H$ into $\hat{H}$ so that $\hat{H}|r \times 0 = \hat{H}|r \times 1$. Then all the components of $r \times I \cap \hat{H}^{-1}(\partial F_0' - \partial F')^-$ can be removed, using an admissible homotopy of $\hat{H}$ which is constant on $F \times \partial I$ and outside a regular neighborhood of $r \times I$ in $F \times I$. But this contradicts our minimality condition on $K_0 \cap K_1$.

<u>Case 2</u>. $r \times I \cap K_1 \neq \emptyset$.

$K_1$ splits $r \times I$ into a system of squares or annuli resp. By our choice of $r$ and since $K_0 \cap K_1 = \emptyset$, it follows that one of them, say $C$, contains $r \times 1$. Let $k$ be the component of $r \times I \cap K_1$ which lies in $C$, and let $k'$ be the component of $(\partial F_1' - \partial F')^-$ which contains $h(k)$. Since $f_1$ is an admissible $(F_1,F_1')$-homeomorphism, $t = f_1^{-1}k'$ is an essential curve in $(F,\bar{\underline{\underline{f}}})$ and $f_1|t\colon t \to k'$ is a homeomorphism. Hence the existence of the map $h|C$ shows that a multiple of $f_1|r \times 1$ is admissibly homotopic to a multiple of $f_1|t$. Since $f_1$ is an admissible homotopy equivalence, it follows that $r \times 1$ is admissibly homotopic in $(F,\underline{\underline{f}}) \times 1$ to $t$, and so $t$ separates an inner square or annulus $A$ from $(F,\underline{\underline{f}}) \times 1$ which contains $r \times 1$.

Let $U(A)$ be a regular neighborhood of $A$ in $(F,\bar{\underline{\underline{f}}}) \times 1$. Define an admissible isotopy $\alpha_t$, $t \in I$, of $id_{F\times 1}$ which is constant outside of $U(A)$ and which contracts $U(A)$ into $U(A) - A$. Then $f_1\alpha_t$, $t \in I$, is an admissible homotopy of $f_1$ which pulls $f_1$ into $\hat{f}_1$ so that $\hat{f}_1^{-1}k' = f_1^{-1}k' - t$. Since $t = f_1^{-1}k'$, we have $\hat{f}_1^{-1}k' = \emptyset$. Hence, applying 16.1 (take the product with the interval), it follows that $k'$ is admissibly parallel to a free side of $(F', \underline{\underline{f}}')$. This means that $k'$ separates an inner square or annulus $A'$ from $(F', \bar{\underline{\underline{f}}}')$ which contains a free side of $(F', \underline{\underline{f}}')$.

$A'$ is a component of $F_1'$ or of $(F' - F_1')^-$ and $f_1$ is an admissible $(F_1, F_1')$-homeomorphism. Hence there must be a component $A^*$ of $F_1$ or of $(F \times 1 - F_1)^-$ with $f_1(A^*) \subset A'$. By 30.2, $f_1|A^*: A^* \to A'$ is an essential map. This implies that $A^*$ is an inner square or annulus in $(F, \bar{\underline{\underline{f}}}) \times 1$ which contains a free side of $(F, \underline{\underline{f}}) \times 1$. Since $A$ and $A^*$ both contain $t$, $A$ must be equal to $A^*$, for otherwise $(F, \bar{\underline{\underline{f}}})$ is a square or annulus. Thus, in particular, $h(r \times 1) \subset A'$.

Thus we may suppose that $h(r \times 1)$ lies in a free side of $(F', \underline{\underline{f}}')$. Now recall that $r \times I \cap K_0 \neq \emptyset$. $K_0$ splits $r \times I$ into a system of squares or annuli resp. By our choice of $r$, one of them, say $B$, contains $r \times 1$. Then the existence of the map $h|B$ shows that $h(r \times 1)$ lies in a component of $F_0'$ or of $(F' - F_0')^-$ which is an inner square or annulus in $(F', \bar{\underline{\underline{f}}}')$ and which contains a free side of $(F', \underline{\underline{f}}')$. Thus by 30.9, $r \times I \cap K_0 = \emptyset$, which contradicts our choice of $r$. q.e.d.

As an immediate consequence of 30.8 we obtain:

<u>30.10 Corollary</u>. <u>Let the situation be given as described in the beginning of</u> §30. <u>Then there is an admissible homotopy</u> $H_t$, $t \in I$, <u>of</u> $H$ <u>which satisfies</u> 2 <u>of</u> 30.3 <u>and so that for every side</u> $k$ <u>of</u> $(F, \bar{\underline{\underline{f}}})$ <u>the complex</u>

$$H_1^{-1}((\partial F_0' \cup \partial F_1') - \partial F')^-$$

<u>intersects</u> $k \times I$ <u>in</u> (<u>pairwise disjoint</u>) <u>arcs joining</u> $k \times 0$ <u>with</u> $k \times 1$.

So far we have seen that the restriction $H|\partial(F \times I)$ can be normalized. It remains to show that the normalization can be extended to the interior of $F \times I$. One step in this direction is the following lemma:

<u>30.11 Lemma</u>. <u>Let the situation be given as described in the beginning of §30. Suppose that the map</u> $H$ <u>is transverse with respect to</u> $(\partial F_0' - \partial F')^-$ <u>and</u> $(\partial F_1' - \partial F')^-$.

<u>Then</u> $H$ <u>can be deformed (rel</u> $\partial(F \times I)$<u>) so that afterwards each component of</u> $H^{-1}(\partial F_0' - \partial F')^-$ <u>and</u> $H^{-1}(\partial F_1' - \partial F')^-$ <u>is either a 2-sphere or an incompressible surface</u>.

<u>Proof</u>. Define $S_i = H^{-1}(\partial F_i' - \partial F')^-$. $H$ is transverse with respect to $(\partial F_i' - \partial F')^-$. This means that there are neighborhoods $S_i \times I'$ of $S_i$, and $(\partial F_i' - \partial F') \times I'$ of $(\partial F_i' - \partial F')^-$, (here $I'$ denotes the unit interval) with $S_i = S_i \times \frac{1}{2}$, and $(\partial F_i' - \partial F') = (\partial F_i' - \partial F')^- \times \frac{1}{2}$, such that $H(x,y) = (H(x),y)$, for all $x \in S_i$ and $y \in I'$. Then $S_0$ and $S_1$ are admissible surfaces in $(F,\underline{\underline{\bar{f}}}) \times I$, and without loss of generality $S_0$ is transverse to $S_1$ (with respect to the product structures $S_0 \times I$ and $S_1 \times I$).

Let $D$ be any disc in $F \times I$ with $D \cap S_0 = \partial D$ and $D \cap S_1 = \emptyset$. Fix a neighborhood $D \times I'$ of $D$ with $D = D \times \frac{1}{2}$ and $(D \times I') \cap S_0 = \partial D \times I'$. Recall that $F_0'$ is essential and $F'$ is aspherical. Hence it follows the existence of a deformation which pulls $H|D$ first (rel $\partial D$) into $(\partial F_0' - \partial F')^-$ and then out of $(\partial F_0' - \partial F')^-$. If the latter is done in the right direction and if the deformation is chosen so that it does not meet $(\partial F_1' - \partial F')^-$ (see suppositions on $D$), it can be extended to a homotopy $H_t$, $t \in I$, of $H$ with the following properties: $H_t$ is constant outside of a regular neighborhood of $D \times I'$, $H_1^{-1}(\partial F_1' - \partial F')^- = S_1$, and $H_1^{-1}(\partial F_0' - \partial F') = (H^{-1}(\partial F_0' - \partial F')^- - D \times I') \cup D \times \partial I$.

Assume there is a component of $S_0$, say, which is compressible, but not a 2-sphere. Then we find a disc $D$ with $D \cap S_0 = \partial D$ which does not bound a disc in $S_0$ and which intersects $S_1$ in a system of curves. To prove 30.11, it suffices to show that $H$ is homotopic (rel $\partial(F \times I)$) to a map $H_1$ with $H_1^{-1}(\partial F_1' - \partial F')^- = S_1$ and $H_1^{-1}(\partial F_0' - \partial F')^- = (H^{-1}(\partial F_0' - \partial F')^- - D \times I')^- \cup D \times \partial I'$. By what we have seen so far, we may suppose that $D \cap S_1$ consists of arcs. Let

$U$ be a regular neighborhood of $D \cap S_1$ in $D \times I'$. Then $(D - U)^-$ consists of discs $D^*$ with $D^* \cap (U \cup S_0) = \partial D^*$ and $D^* \cap S_1 = \emptyset$. Hence by what we have seen above, it remains to show that $H$ is homotopic (rel $\partial(F \times I)$) to $H_1$ with $H_1^{-1}(\partial F_1' - \partial F')^- = S_1$ and $H_1^{-1}(\partial F_0' - \partial F')^- = (H^{-1}(\partial F_0' - \partial F')^- - U)^- \cup (\partial U - S_0)^-$. For this we have to show that, for every arc $k$ of $D \cap S_1$, the restriction $H|k$ can be deformed (rel $\partial k$) in $(\partial F_1' - \partial F')^-$ into $(\partial F_0' - \partial F')^-$. Let $z_1$ and $z_2$ be the two end-points of $k$, and observe that they are mapped under $H$ into points of $(\partial F_0' - \partial F')^- \cap (\partial F_1' - \partial F')$. Since $F_0'$ and $F_1'$ are in good position, the existence of the map $H|D$ shows that $H(z_1) = H(z_2)$. Hence the required deformation for $H|k$ exists, for otherwise $H(k)$ lies in a closed curve of $(\partial F_1' - \partial F')^-$ which meets $(\partial F_0' - \partial F')^-$ in one point, which is impossible since $(\partial F_0' - \partial F')^-$ is separating. q.e.d.

<u>30.12 Proposition</u>. <u>Let the situation be given as described in the beginning of §30. Then there is an admissible homotopy of</u> $H$ <u>which satisfies</u> 2 <u>of</u> 30.3 <u>and which pulls</u> $H$ <u>such that afterwards</u> $S_i = H^{-1}(\partial F_i' - \partial F')^-$ <u>consists of essential squares and annuli in</u> $(F,\underline{\underline{f}}) \times I$. $i = 0$ <u>and</u> $i = 1$, <u>and that each component of</u> $S_0 \cap S_1$ <u>is an arc which joins</u> $F \times 0$ <u>with</u> $F \times 1$.

<u>Proof</u>. By the transversality lemma and by 30.10 and 30.11, we may suppose that $H$ is admissibly deformed so that

(i) $k \times I \cap (S_0 \cup S_1)$ is a system of arcs which join $k \times 0$ with $k \times 1$, for any side $k$ of $(F,\underline{\underline{f}})$,

(ii) each component of $S_0$ and $S_1$ is either a 2-sphere or an incompressible surface, and

(iii) $S_0$ and $S_1$ are transverse.

Denote by $\alpha(H)$ the number of all components of $S_0 \cap S_1$, by $\beta(H)$ the sum of the numbers of components of $S_0$ and $S_1$, and by $\gamma(H)$ the sum of the numbers of components of $S_0 \cap F \times \partial I$ and $S_1 \cap F \times \partial I$. Suppose $H$ is admissibly deformed, using an admissible homotopy with 2 of 30.3 such that the above holds, and that, in addition, $(\alpha(H),\beta(H),\gamma(H))$ is as small as possible with respect to the lexicographical order. We claim that $H$ then satisfies the conclusion of 30.12.

Of course, H induces an injection of the fundamental groups. Thus it follows from (ii) that any component G of $S_i$, $i = 0,1$ is homeomorphic to a disc, a 2-sphere, or an annulus. More precisely, it holds the following:

30.13 Assertion. 1. If $G \cap F \times i \neq \emptyset$, then G is an essential square or annulus in $(F,\bar{\underline{f}}) \times I$.

2. If $G \cap F \times i = \emptyset$ then G is either an admissible 1-faced disc with $\partial G \subset (F \times i+1) - F_{i+1}$ (indices mod 2), or an admissible, incompressible annulus with $\partial G \subset (F \times i+1) - F_{i+1}$, or a 2-sphere.

For the following we may suppose, without loss of generality, that G is a component of $S_0$.

Proof to 1 of 30.13. Since $G \cap (F \times 0) \neq \emptyset$, G cannot be a 1-faced disc. For otherwise it follows that $\partial G$ is contractible in $F \times 0$. But $G \cap (F \times 0)$ is a component of $(\partial F_0 - \partial F)^-$, and so we get a contradiction to the fact that $F_0$ is an essential surface in $F \times 0$.

G is mapped under H into a component of $(\partial F_0' - \partial F')^-$ and $H|F_0\colon F_0 \to F_0'$ is a homeomorphism. This implies that $G \cap (F \times 0)$ is connected. Since G is homeomorphic to a disc or an annulus, it is easily checked, using (i) above, that G must be an admissible square or annulus in $(F,\bar{\underline{f}}) \times I$, and that $G \cap F \times i \neq \emptyset$, for $i = 0$ and $i = 1$. That G is in fact essential follows then since $G \cap (F \times 0)$ is an essential curve in $F \times 0$.

Proof to 2 of 30.13. Suppose $G \cap F \times 0 = \emptyset$, and recall that G is homeomorphic to a disc, a 2-sphere, or an annulus. If G is a 2-sphere, we are done. So, suppose the converse. Then, using (i) and (ii), it is easily checked that G is either an admissible 1-faced disc with $\partial G \subset F \times 1$, or an admissible annulus with $\partial G \subset F \times 1$. Thus it remains to show that $\partial G \subset (F \times 1) - F_1$.

Assume the converse. If at least one component of $G \cap (F \times 1)$ lies in $F_1$ recall that $H|F \times 1$ is an $(F_1,F_1')$-homeomorphism, and so we get a contradiction either to the fact that $F_0'$ is essential or that $G$ is mapped under $H$ into one component of $(\partial F_0' - \partial F')^-$. Thus, by our assumption, $\partial G \cap (\partial F_1 - \partial F)^- \neq \emptyset$. This means that there is a component $t$ of $G \cap S_1$ which is an arc whose end-points lie in $F_1$. $t$ is mapped under $H$ into one point of $(\partial F_0' - \partial F')^- \cap (\partial F_1' - \partial F')^-$. In particular, $H(\partial t)$ is one point--which contradicts the supposition that $H|F_1\colon F_1 \to F_1'$ is a homeomorphism. This proves 30.13.

<u>30.14 Assertion</u>. <u>Every component of</u> $S_0 \cap S_1$ <u>is an arc which joins</u> $F \times 0$ <u>with</u> $F \times 1$.

Let $k$ be a component of $S_0 \cap S_1$. If $k$ is an arc, it follows from (i) that the end-points of $k$ lie in $F \times \partial I$, and furthermore they cannot both lie in $F \times 0$ or $F \times 1$ (see the end of the above proof). If $k$ is a closed curve, observe that $k$ is mapped under $H$ into a point of $(\partial F_0' - \partial F')^- \cap (\partial F_1' - \partial F')^-$ and so $k$ is contractible in $F \times I$ since $H$ induces an injection of the fundamental groups. By (ii) and the loop-theorem, it follows that $k$ bounds a disc $D$ in $S_0$. By an argument of the proof of 30.11, the number $\alpha(H)$ of intersection curves of $S_0 \cap S_1$ can be diminished, using an admissible homotopy of $H$ with 2 of 30.3. This contradicts our suppositions on $H$, and so 30.14 is proved.

To complete the proof of 30.12, it remains to show that there is no component of $S_i$, $i = 0,1$, as described in 2 of 30.13. Assume the converse and let $G$ be a component of $S_0$, say, as described in 2 of 30.13.

If $G$ is a 2-sphere, it bounds a ball $E$. $G \cap S_1 = \emptyset$ since, by 30.14, every component of $S_0 \cap S_1$ is an arc. Since $(\partial F_0' - \partial F')^-$ and $F'$ are aspherical, $H|E$ can be deformed (rel $\partial E$) into $(\partial F_0' - \partial F')^-$. It follows that $G$ can be removed using an admissible homotopy of $H$ with 2 of 30.3, i.e. we obtain a contradiction to our minimality condition on $(\alpha(H), \beta(H), \gamma(H))$.

If $G$ is a 1-faced disc with $\partial G \subset (F \times 1) - F_1$, then $\partial G$ bounds a disc $G^*$ in $(F \times 1) - F_1$. Since $F_0'$ is essential and $F' - F_1'$ is aspherical, it follows that $H|G^*$ can be deformed (rel $\partial G^*$) into

$(\partial F_0' - \partial F')^-$. Hence we find an admissible homotopy of $H$ with 2 of 30.3, which diminishes $\gamma(H)$ without enlarging $\alpha(H)$ or $\beta(H)$. Again this is a contradiction.

If finally $G$ is an admissible, incompressible annulus with $\partial G \subset (F \times 1) - F_1$, the two boundary curves of $G$ are freely homotopic in $F \times 1$. Hence, by Nielsen's theorem, $\partial G$ bounds an annulus $G^*$ in $F \times 1$. $G$ is mapped under $H$ into one component of $(\partial F_0' - \partial F')^-$. In particular, $H(\partial G^*)$ is one component of $(\partial F_0' - \partial F')$. Hence and since $F'$ is not a torus and $H|F \times 1$ is an admissible $(F_1, F_1')$-homeomorphism, it follows that $G^*$ does not contain any component of $F_1$ and that $H|G^*$ can be deformed (rel $\partial G^*$) into $(\partial F_0' - \partial F')^-$ (Nielsen's theorem). Therefore we again find an admissible homotopy of $H$ with 2 of 30.3 which diminishes $\gamma(H)$ without enlarging $\alpha(H)$ or $\beta(H)$.

So in all cases we get a contradiction. q.e.d.

Using 30.12 we can now complete the proof of 30.1.

Proof of 30.1. Let $H$ be admissibly deformed, using a homotopy $H_t$, $t \in I$, with 2 of 30.1, such that $S_i = H^{-1}(\partial F_i' - \partial F')^-$ satisfies the conclusion of 30.12.

Then, in particular, $S_0$ consists of essential squares and annuli. $(F, \underline{\underline{f}})$ is neither a square nor an annulus. Hence, by 5.6, there is an admissible ambient isotopy $\beta_t$, $t \in I$, which pulls $S_0$ into a vertical system.

Finally let $\tilde{M}$ be the manifold obtained from $F \times I$ by splitting along $\beta_1(S_0)$. By our choice of $\beta_t$, $\tilde{M}$ is in a canonical way again an admissible product I-bundle. $\tilde{S}_1 = \beta_1(S_1) \cap \tilde{M}$ is a system of admissible squares or annuli in $\tilde{M}$ which cannot be admissibly isotoped in $\tilde{M}$ into a horizontal system since $S_0 \cap S_1$ consists of arcs joining $F \times 0$ with $F \times 1$. Hence, by 5.6 and 10.1, $\tilde{S}_1$ can be admissibly isotoped in $\tilde{M}$ into a vertical system. Of course this isotopy can be extended to an admissible ambient isotopy $\gamma_t$, $t \in I$, of $F \times I$ which preserves $\beta_1(S_0)$ and which pulls $\beta_1(S_1)$ into a vertical system.

The product $\alpha_t = \gamma_t \beta_t$ is the required ambient isotopy. q.e.d.

As an application of 30.1 we obtain the existence of the obstruction surface for homotopy equivalences between (orientable) surfaces:

30.15 Corollary. *For each admissible homotopy equivalence* $f\colon (F,\underline{\underline{f}}) \to (F',\underline{\underline{f}}')$ *there is an essential surface* $F'_f$ *in* $(F',\bar{\underline{\underline{f}}}')$ *with the following properties*:

1. $f$ *can be admissibly deformed into an admissible* $((F - f^{-1}F'_f)^-,\ (F' - F'_f)^-)$*-homeomorphism*.
2. $F'_f$ *can be admissibly isotoped in* $(F',\bar{\underline{\underline{f}}}')$ *into every essential surface with* 1.
3. $F'_f$ *minus a component of* $F'_f$ *does not satisfy* 1.

Remark. 2 and 3 of 30.15 imply that the surface $F'_f$ is unique, up to admissible ambient isotopy. The surface $F'_f$ associated with $f$ is called the *obstruction surface of* $f$. It intuitively measures how "bad" the homotopy equivalence $f$ is (e.g. note that $f$ is a homeomorphism if $F'_f$ is empty).

Proof. Let an admissible homotopy equivalence $f\colon (F,\underline{\underline{f}}) \to (F',\underline{\underline{f}}')$ be given. An essential surface $G'$ in $(F',\bar{\underline{\underline{f}}}')$ is called *good* (with respect to $f$) if

1. $f$ can be admissibly deformed into a map $g$ which is an admissible $(g^{-1}G',G')$-homeomorphism,
2. no component of $(F' - G')^-$ is an inner square or annulus in $(F',\bar{\underline{\underline{f}}}')$ which meets a component of $G'$ which is also an inner square or annulus.

The inclusion defines a partial-order relation on the set $\underline{\underline{F}}'$ of all admissible isotopy classes of good surfaces in $(F',\bar{\underline{\underline{f}}}')$. By 9.3, any well-ordered subset of $\underline{\underline{F}}'$ has a supremum, and so, by Zorn's lemma, there are maxima in $\underline{\underline{F}}'$. Let $F'_0$ be one such maximum.

It remains to prove that every good surface in $F'$ can be admissibly isotoped into $F'_0$, for then $F'_f = (F' - F'_0)^-$ is the required surface.

Let $F'_1$ be another good surface and isotop it admissibly so that is is in a very good position with respect to $F'_0$. Since $F'_0$, $F'_1$

are good, $f$ can be admissibly deformed to maps $f_i$, $i = 0,1$, which are admissible $(f_i^{-1}F_i', F_i')$-homeomorphisms. Of course, there is an admissible homotopy which deforms $f_0$ into $f_1$. Hence, by 30.1, it follows that there is an admissible ambient isotopy $\alpha_t$, $t \in I$, of $(F,\underline{\underline{f}})$ and an admissible $(f_0^{-1}F_0', F_0')$-homotopy which pulls $f_0$ into a map $g$ which is an admissible $(\alpha_1(f_1^{-1}F_1'), F_1')$-homeomorphism. Denote by $G$, $G'$ the essential unions of $f_0^{-1}F_0'$, $\alpha_1(f_1^{-1}F_1')$ and $F_0'$, $F_1'$, respectively. Then, by our choice of $g$, $g$ is also a $(G,G')$-homeomorphism. If we remove trivial component we obtain a good surface from $G'$ which contains $F_0'$ and $F_1'$, up to admissible isotopy. Hence, by our choice of $F_0'$, $G'$ and so $F_1'$ can be admissibly isotoped into $F_0'$. q.e.d.

## §31. Homotopy equivalences of product I-bundles

Given two product I-bundles $(X,\underline{\underline{x}})$ and $(X',\underline{\underline{x}}')$, not over the square, annulus or torus, we shall consider essential maps $f: (X,\underline{\underline{x}}) \to (X',\underline{\underline{x}}')$. These maps should satisfy various properties. To describe them fix an admissible fibration of $(X,\underline{\underline{x}})$, i.e. fix an actual product structure $L \times I$ or equivalently a projection $p: X \to L$. The same with $(X',\underline{\underline{x}}')$. We further shall make free use of the following notations.

Denote by $F$ the union of the lids of $(X,\underline{\underline{x}})$, i.e. $F = (\partial X - p^{-1}\partial L)^{-}$, and let $\underline{\underline{f}}$ be the boundary-pattern of $F$ induced by $\underline{\underline{x}}$. Define $d: (F,\underline{\underline{f}}) \to (F,\underline{\underline{f}})$ as the fixpoint-free, admissible involution given by the reflections in the I-fibres. Furthermore let $G$ be an essential surface in $(F,\bar{\underline{\underline{f}}})$, and denote by $S_i$ the vertical surface in $(X,\bar{\underline{\underline{x}}})$ which intersects the lid $L \times i$ in the system $(\partial(G \cap (L \times i)) - \partial(L \times i))^{-}$, $i = 0,1$. The same with $(X',\underline{\underline{x}}')$.

Suppose, that $(F' - G')^{-}$ is in a very good position to $(F' - dG')^{-}$ and suppose that $f|F: F \to F'$ is an admissible homotopy equivalence which is an admissible $(G,G')$-homeomorphism (for definition see the beginning of this chapter).

Given the above situation we are going to prove the following result with the help of 30.1. A non-singular version of this (which is easier to prove) is described in 26.3.

<u>31.1 Proposition</u>. <u>It may be supposed that the surface</u> $G$ <u>is admissibly isotoped in</u> $(F,\bar{\underline{\underline{f}}})$ <u>such that</u> $(F - G)^{-}$ <u>is in a very good position to</u> $(F - dG)^{-}$ <u>and such that there is an admissible homotopy</u> $f_t$, $t \in I$, <u>of</u> $f$ <u>with the following properties</u>:

1. $f_1$ <u>is an admissible</u> $(p^{-1}pH, (p')^{-1}p'H')$-<u>homeomorphism, where</u> $H$, $H'$ <u>are the essential unions of</u> $G$, $dG$ <u>and</u> $G', dG'$, <u>respectively</u>.
2. $f_t|F: F \to F'$ <u>is an admissible</u> $(G,G')$-<u>homotopy</u>.

<u>In addition</u>. $f_t$ <u>may be chosen so that furthermore</u> $f_1^{-1}S_i' = S_i$.

Proof. The map $p'\cdot f$ can be interpreted as an admissible homotopy $L \times I \to L'$ as considered in §30. Hence, using 30.1, we may suppose that $G$ is admissibly isotoped in $F$, and that moreover $f$ is admissibly deformed, using a homotopy which satisfies 2 of 31.1, so that $S_i = f^{-1}S_i'$. Since $f|F$ is an admissible $(G,G')$-homeomorphism, it is then easily seen that $f$ can be admissibly deformed (rel $F$) into an admissible $(p^{-1}pH,(p')^{-1}p'H')$-homeomorphism, where $H$, $H'$ are essential unions of $G,dG$ and $G',dG'$, respectively. For this use the fact (see 28.3) that any essential map of a product I-bundle is admissibly homotopic to a fibre preserving map, using a homotopy constant on one lid. Thus it remains to prove that $(F - G)^-$ is in a very good position to $(F - dG)^-$.

Assume $(F - G)^-$ is not in a good position to $(F - dG)^-$. Then there is an embedding of a 2- or 3-faced disc $D$ into $L \times 0$, say, such that each side is mapped into $S_0$, $S_1$, or a side of $(F,\underline{\underline{f}})$. One side is mapped into a free side of $(F,\underline{\underline{f}})$, for otherwise $f$ maps $D$ into a similiar disc in $F'$ which is impossible since $(F' - G')^-$ is in a good position to $(F' - dG')^-$. Let $b_0$, $b_1$ be the sides of $D$ which lie in $S_0$, resp. $S_1$. Denote by $z$ their common end-point. Then $b_0$ and $b_1$ join $z$ with a free side of $(F,\underline{\underline{f}})$, and let $z_0$, $z_1$ be the end-points of $b_0$, $b_1$, resp., different from $z$. Define $z' = f(z)$. Then, in particular, $z'$ is a point of $(L' \times 0) \cap (S_0' \cap S_1')$ since $S_i = f^{-1}S_i'$ (it is to be understood that $f(L \times 0) \subset (L' \times 0))$. Furthermore, $b_i$, $i = 0,1$, is mapped under $f$ into an arc $b_i'$ of $S_i' \cap (L' \times 0)$ which joins $z'$ with a free side of $(F',\underline{\underline{f}}')$. This is clear for $b_0$ since $b_0$ lies in $(\partial G - \partial F)^-$ and $f|F$ is an admissible $(G,G')$-homeomorphism. To see that it also holds for $b_1 = \underline{b}_1 \times 0$, recall that both $S_0$ and $S_1$ are vertical and equal to $f^{-1}S_0'$, resp. $f^{-1}S_1'$, and apply the above argument to $b_1 \times 1$. Now, define $b' = b_0' \cup b_1'$. $(f|L \times 0)^{-1}b'$ is a system of curves in $L \times 0$. Furthermore $(f|L \times 0)^{-1}b'$ meets $\partial L \times 0$ only in two points, namely $z_0$ and $z_1$. To see the latter observe that, for every point $y$ of $(f|L \times 0)^{-1}b'$ contained in $\partial L \times 0$, either $y \times 0$ or $y \times 1$ lies in $G$ and that only one such point is mapped under $f$ to $z_0$, resp. $z_1$, since $f$ is an admissible $(G,G')$-homeomorphism. The existence of the map $f|D$ shows that $f|L \times 0$ can be admissibly deformed so that

$(f|L \times 0)^{-1}b'$ consists of closed curves. Furthermore, since $f|L \times 0$ is essential, $f|L \times 0$ can then be admissibly deformed so that $(f|L \times 0)^{-1}b$ is empty (transversality lemma). But $f|L \times 0$: $L \times 0 \to L' \times 0$ is an admissible homotopy equivalence, so it follows from 16.1 (take the product of $L$ and $L'$ with an interval) that the arc $b'$ is either admissibly parallel to a free side of $(F',\underline{\underline{f}}')$ or it is inessential in $(F',\bar{\underline{\underline{f}}}')$. The first case is impossible since the end-points of $b'$ lie in free sides of $(F',\underline{\underline{f}}')$, and the latter case is impossible since $(F' - G')$ and $(F' - dG')^-$ are in a good position. So we have proved that $(F - G)^-$ and $(F - dG)^-$ are in a good position.

Assume finally that $(F - G)^-$ is not in a very good position to $(F - dG)^-$. Since $(F - G)^-$ is in a good position to $(F - dG)^-$, there is then an inner square or annulus $A$ in $(F,\bar{\underline{\underline{f}}})$ which lies in $G = (F - (F - G)^-)^-$ and in $dG$ such that one component of $(\partial A - \partial F)^-$ is contained in $S_0$ and the other one in $S_1$. Since $f|F$: $F \to F'$ is a $(G,G')$-homeomorphism it follows that $f$ maps $A$ to a similar inner square or annulus in $(F',\bar{\underline{\underline{f}}}')$. But this contradicts the fact that $(F' - G')^-$ is in a very good position to $(F' - dG')^-$. q.e.d.

Our aim is to improve 31.1 somewhat using an appropriate homotopy inverse of $f$. For this we need

<u>31.2 Lemma</u>. <u>Let</u> $(F,\underline{\underline{f}})$ <u>be a connected surface, and suppose that</u> $(F,\bar{\underline{\underline{f}}})$ <u>is not a square, annulus, or torus. Let</u> $G$ <u>be an essential surface in</u> $(F,\bar{\underline{\underline{f}}})$. <u>Let</u> $h$: $(F,\underline{\underline{f}}) \to (F,\underline{\underline{f}})$ <u>be an admissible map which is admissibly homotopic to the identity. Suppose that</u> $h$ <u>is an admissible</u> $(G,G)$-<u>homeomorphism</u>.

<u>Then</u> $h^{-1}k = k$, <u>for every component</u> $k$ <u>of</u> $(\partial G - \partial F)^-$.

<u>In addition: If</u> $(\tilde{F},\tilde{\underline{\underline{f}}})$ <u>is the surface obtained from</u> $(F,\underline{\underline{f}})$ <u>by splitting at</u> $(\partial G - \partial F)^-$, <u>then</u> $h|\tilde{F}$: $(\tilde{F},\tilde{\underline{\underline{f}}}) \to (\tilde{F},\tilde{\underline{\underline{f}}})$ <u>is an admissible map which is admissibly homotopic to the identity</u>.

<u>Proof</u>. Assume the contrary. Let $F \times I \to F$ be an admissible homotopy which pulls $h$ into the identity. Since $(F - G)^-$ is in a very good

position to $U(F - G)^-$ , where $U(F - G)^-$ is a regular neighborhood, this homotopy is a homotopy as considered in §30. Hence, by 30.1, we may suppose that $H$ is admissibly deformed (rel $F \times \partial I$) so that $S = H^{-1}(\partial G - \partial F)^-$ consists of essential squares or annuli. $S$ joins $(\partial G - \partial F) \times 0$ with $(\partial G - \partial F)^- \times 1$. By our assumption, there is a component $A$ of $S$ which joins a component $k_1 \times 0$ of $(\partial G - \partial F) \times 0$ with a component $k_2 \times 1$ of $(\partial G - \partial F)^- \times 1$ such that $k_1 \times 0 \neq k_2 \times 0$.

Denote by $T_i$, $i \geq 1$, the component of $(\partial G - \partial F)^- \times I$ which contains the component $k_i$ of $(\partial G - \partial F)^-$, and define $M_{ij}$ to be the manifold obtained from $F \times I$ by splitting along $T_i \cup T_j$. The existence of $A$ shows that one component of $M_{12}$ is the product I-bundle over the square or annulus. Since $(F, \underline{\underline{f}})$ is neither the square nor the annulus, it easily follows that $A$ can be admissibly isotoped (rel $A \cap (F \times \partial I)$) into the forementioned component of $M_{12}$. By the same argument there is no component of $S$ which joins $k_1 \times 1$ with $k_2 \times 0$. Hence considering the component of $S$ containing $k_1 \times 1$, we find a component $k_3$ of $(\partial G - \partial F)^-$ different from $k_1$ and $k_2$. Repeating this argument we get an infinite number of components of $(\partial G - \partial F)^-$ which is the required contradiction. q.e.d.

<u>31.3 Corollary</u>. <u>Let us give the situation of</u> 31.1,<u>and suppose that</u> $G$ <u>and</u> $f$ <u>are deformed so that they satisfy the conclusion of</u> 31.1. <u>Furthermore</u>, <u>let</u> $g: X' \to X$ <u>be an essential map such that</u> $g|F': F' \to F$ <u>is an admissible homotopy inverse of</u> $f|F$ <u>and which is an admissible</u> $(G',G)$-<u>homeomorphism</u>. <u>Then the following holds</u>:

1. $g$ <u>can be admissibly deformed so that afterwards</u> $g$ <u>is an admissible</u> $((p')^{-1}p'H', p^{-1}pH)$-<u>homeomorphism</u>.
2. $f|\tilde{X}: \tilde{X} \to \tilde{X}'$ <u>is an admissible homotopy equivalence with admissible homotopy inverse</u> $g|\tilde{X}'$, <u>where</u> $\tilde{X}$ <u>and</u> $\tilde{X}'$ <u>denote the manifolds obtained from</u> $X$ <u>resp</u>. $X'$ <u>by splitting along</u> $(\partial p^{-1}pH - \partial X)^-$ <u>and</u> $(\partial (p')^{-1}p'H' - \partial X')^-$.

<u>Remark</u>. 1. For the definition of $H$ and $H'$ see 31.1.

2. 31.3 immediately leads to the 2-dimensional result 23.2.

Proof. Using 30.1, we may assume that $g$ is admissibly deformed so that $g^{-1}S_i$, $i = 0,1$ consists of essential squares or annuli and that $g^{-1}S_0 \cap g^{-1}S_1$ consists of arcs joining $L' \times 0$ with $L' \times 1$.

Hence the conclusion of 31.3 follows easily from 31.2, if there is an admissible ambient isotopy $\gamma_t$, $t \in I$, of $(X',\underline{\underline{x}}')$ with the following properties:

(i) $\gamma_t|F': F' \to F'$ is an admissible $(G',G')$-isotopy,

(ii) $\gamma_1(g^{-1}S_1) = S_1'$, and

(iii) $\gamma_1(g^{-1}S_0) = S_0'$.

Hence it suffices to find an admissible ambient isotopy with (i) and (ii), for the required ambient isotopy $\gamma_t$ then follows by an application of 5.6 to the manifold obtained from X' by splitting along $S_1'$.

Let A' be a component of $g^{-1}S_1$. Denote by a' and $\alpha_0'$ the sides of A' contained in the lids $L' \times 1$ resp. $L' \times 0$, and let $a_0'$ be the side of the vertical surface $a' \times I$ opposite to a'. Projecting A' down to $L' \times 0$, we get an admissible homotopy $h: t \times I \to L' \times 0$ of $\alpha_0'$ to $a_0'$. $h$ can be admissibly deformed so that $S = h^{-1}(\partial G' - \partial F')^-$ is either a system of arcs joining $t \times 0$ with $t \times 1$, or a system of essential curves which do not meet $t \times \partial I$ To see this recall that (1) G' is essential in F', (2) A' is a component of $g^{-1}S_1$, (3) $g$ is an admissible (G',G)-homeomorphism, and (4) $(F - G)^-$ is in a very good position to $(F - dG)^-$ (see 31.1). In particular, $\alpha_0'$ and $a_0'$ are sides of an inner square or annulus C' in $L' \times 0$. Furthermore it is easily seen (apply Nielsen's theorem) that $C' \cap (\partial G' - \partial F')^-$ is a system of essential curves with the properties of S. Hence it remains to show that C' does not contain a component of $(\partial G' - \partial F')^-$. For then we certainly find an admissible ambient isotopy with (i) above and which pulls A' into $S_1'$. More precisely, this isotopy can be chosen to be constant on all components of $S_1'$ which do not meet A', and the existence of the ambient isotopy $\gamma_t$ follows inductively.

So assume that C' contains a component $\beta_0'$ of $(\partial G' - \partial F')^-$. Let $\beta'$ be the side of $\beta_0' \times I$ opposite to $\beta_0'$, let B' be the component of $g^{-1}S_0$ containing $\beta_0'$, and let b' be the side of B' opposite to $\beta_0'$.

The component $G_1'$ of G' containing a' cannot contain $\beta'$ since $(F' - G')^-$ is in a very good position to $(F' - dG')^-$. A' and

$B'$ are components of $g^{-1}S_1$ resp. $g^{-1}S_0$. Hence it follows that $G_1'$ cannot contain $b'$ since $(F - G)^-$ is in a very good position to $(F - dG)^-$ (see 31.1). Thus $G_1'$ is an inner square or annulus in the lid $L' \times 1$.

Without loss of generality we may suppose that the component $A'$ of $g^{-1}S_1$ is chosen so that the inner square or annulus determined by $a'$ and $b'$ does not contain any other component of $(\partial G' - \partial F')^-$. In particular, $G_1'$ must lie in the inner square or annulus determined by $\beta'$ and $a'$. So we may keep in mind that there is no admissible homotopy in $L' \times 1$ which moves $\beta'$ into $a'$ without crossing $G_1'$ (recall that by supposition $L' \times 1$ is neither square nor annulus).

Now consider the two disjoint components $A$ and $B$ of $S_1$ resp. $S_0$ which contain $g(A')$ resp. $g(B')$. Denote by $\alpha_i, \beta_i$, $i = 0,1$, the sides of $A$ resp. $B$ contained in the lid $L \times i$. Recall that $f^{-1}S_0' = S_0$ and that, by 31.2, $gf(\beta_0) = \beta_0$. Hence $B$ is mapped under $f$ into $\beta_0' \times I$. Moreover, $\alpha_0$ and $\beta_0$ determine an inner square or annulus which does not contain any component of $(\partial G - \partial F)^-$ since $\alpha_0'$ and $\beta_0'$ do. The same is certainly true for $\alpha_1$ and $\beta_1$, and let $C$ be the inner square or annulus determined by $\alpha_1$ and $\beta_1$. Now, $f(\alpha_1) = fg(a') = a'$, by 31.2, and $f(\beta_1) = \beta'$ since $gf(\beta_0') = \beta_0$ and so $f(B) \subset \beta_0' \times I$. Therefore $f|C$ defines an admissible homotopy of $\beta'$ to $a'$ which does not cross $G_1'$. This is the required contradiction.

q.e.d.

## Chapter XII: Geometric properties of 3-manifold groups.

Recall from §28 the definition of an exceptional curve. In this chapter we study the influence of the exceptional curves on the fundamental group of 3-manifolds. For this we use the concept of characteristic submanifolds as developed in Part I and II. In particular, we shall give a geometric criterion for the fundamental group of a Haken 3-manifold to be an R-group (see 32.4), and another proof of Shalen's result [Sh 1] that such groups have no element which is infinitely divisible. From the latter we deduce (see 32.9) that the centralizer of any element of a 3-manifold group as above is carried geometrically (see also [Ja 1], [JS 1]).

### §32. The influence of exceptional curves on 3-manifold groups

We begin this paragraph with proving that a closed curve in the boundary of a 3-manifold is divisible (i.e. freely homotopic to a non-trivial multiple of another curve) only by geometric reasons.

32.1 Lemma. *Let $(M,\underline{\underline{m}})$ be an irreducible 3-manifold whose complete boundary-pattern is useful and non-empty. Let $F$ be a bound side of $(M,\underline{\underline{m}})$, and let $k: S^1 \to F$ be an essential singular curve in $F$. Suppose that $k$ is freely homotopic in $M$ to an n-multiple, $n \geq 2$, of some singular curve $\ell: S^1 \to M$, but not freely homotopic in $F$ to a non-trivial multiple of some singular curve.*

*Then there are curves $k'$, $\ell'$ freely homotopic to $k$, $\ell$ in $F$, $M$, respectively, such that either 1 or 2 holds:*

1. *$n = 2$, and there is an essential I-bundle $(X,\underline{\underline{x}})$ in $(M,\underline{\underline{m}})$ which contains an essential singular and vertical Möbius band $g: B \to X$ such that $g|\partial B = k'$ and that $\ell'$ is the restriction of $g$ to the core of $B$.*
2. *$n \neq 2$, and there is an essential solid torus $(X,\underline{\underline{x}})$ in $(M,\underline{\underline{m}})$ which contains $k'$ and $\ell'$, such that $k'$ has circulation number $n$ with respect to $X$.*

Proof. Let $(W,\underline{\underline{w}})$ be a solid torus with fixed admissible fibration as Seifert fibre space, and with precisely one bound side and one free side. Denote by B and C the bound and the free side resp., and let b, c and w be the cores of B, C and W, respectively. Suppose that b has circulation number n with respect to W.

By our suppositions, there is an admissible map $f: (W,\underline{\underline{w}}) \to (M,\underline{\underline{m}})$ with $k = f|b$ and $\ell = f|w$.

We assert that $f|C$ is essential in $(M,\underline{\underline{m}})$. For this note that k is indivisible in F. This means, in particular, that f cannot be deformed (rel B) into F. Hence our assertion follows since $\underline{\underline{m}}$ is a useful boundary-pattern of M and since M is aspherical.

Thus, by 12.5, $f|C$ can be admissibly deformed into the characteristic submanifold V of $(M,\underline{\underline{m}})$, and so into an essential I-bundle or Seifert fibre space, say $(X,\underline{\underline{x}})$. Furthermore, by 4.4 and 5.7, it follows that f can be admissibly deformed so that $C \subset f^{-1}X$ and that $f^{-1}(\partial V - \partial M)^-$ consists of essential vertical annuli in $(W,\underline{\underline{w}})$. Suppose that f is admissibly deformed so that the above holds and that, in addition, the number of components of $f^{-1}(\partial V - \partial M)^-$ is minimal.

We assert that $f^{-1}(\partial V - \partial M)^- = \emptyset$. Assume the converse. Then $f^{-1}(\partial V - \partial M)^-$ consists of essential annuli in $(W,\underline{\underline{w}})$ which are admissibly parallel to C (to see this consider the fibre projection of $f^{-1}(\partial V - \partial M)^-$ onto the orbit surface). This means that C lies in a component $W_1$ of $f^{-1}V$ which is a regular neighborhood of C. Let $W_2$ be the component of $(W - f^{-1}V)^-$ which meets $W_1$ and fix an essential annulus in $W_2$ which is admissibly parallel to a component of $W_2 \cap \partial W$. Then, by our minimality condition on $f^{-1}(\partial V - \partial M)^-$, it follows that $f|A$ is an essential singular annulus in $(M',\underline{\underline{m}}')$, where $(M',\underline{\underline{m}}')$ denotes the manifold obtained from $(M,\underline{\underline{m}})$ by splitting at $(\partial V - \partial M)^-$. Hence it follows from 12.6 and 10.4 that $f(W_2)$ is contained in a component Y of $(M',\underline{\underline{m}}')$ which is an I-bundle over the annulus. Now, $f|W_1$ can be admissibly contracted $(\mathrm{rel}(\partial W_1 - \partial W)^-)$ into Y and $f|W_2$ can be admissibly deformed out of Y into V. Therefore the number of components of $f^{-1}(\partial V - \partial M)^-$ can be diminished which is a contradiction.

Thus we may suppose that f(W) lies entirely in X, and 32.1 follows easily if X is a Seifert fibre space (see 28.3). If X is an I-bundle, observe that $f|C$ is essential in $(X,\underline{\underline{x}})$ since it is essential in $(M,\underline{\underline{m}})$. By our choice of f, X cannot be the I-bundle over the annulus, and if it is the one over the Möbius band we are done. In the other case we may suppose, by the argument of Case 3 of 5.10, that $f|C$ is vertical in X. If we denote by s some side of C and by $p: X \to G$ the projection onto the base, then $p\cdot f|s$ is homotopic either to $p\cdot f|w$ or to $(p\cdot f|w)^2$. Now, $s \simeq w^n$ in W, with $n \geq 2$. Therefore either $p.f|w$ or $(p\cdot f|w)^2$ is homotopic in G to $(p.f|w)^n$. For surfaces (orientable or not) this is only possible if n = 2. Since $f|C$ is vertical, f can then be admissibly deformed (rel C) in X so that the fibration of X lifts via f to an admissible fibration of $(W,\underline{\underline{w}})$ as I-bundle over the Möbius band and this implies 32.1 immediately. q.e.d.

The following is a "key-lemma" for the proof of 32.6.

32.2 Lemma. Let $(M,\emptyset)$ be an irreducible 3-manifold. Let F be an essential surface in $(M,\emptyset)$, with $F \cap \partial M = \partial F$. Let W be a solid torus. Denote by w the core of W, and let t be a simple closed curve in $\partial W$ with circulation number $\geq 1$ (with respect to w). Let $f: (W,\emptyset) \to (M,\emptyset)$ be an essential map with $f(t) \subset F$, and suppose that $f|w$ can be deformed into F.

Then f can be deformed (rel t) into F.

Proof. Let A be a regular neighborhood of t in $\partial W$, define $\underline{\underline{w}} = \{A\}$, and fix an admissible fibration of $(W,\underline{\underline{w}})$ as Seifert fibre space. By 4.4 and 5.7, we may suppose that f is deformed (rel A) so that $f^{-1}F - A$ is a system of essential and vertical annuli in $(W,\bar{\underline{\underline{w}}})$. Hence, by an induction, it suffices to prove 32.2 in the case that $f^{-1}F = A$.

f can be considered as an admissible map $(W,\underline{\underline{w}}) \to (\tilde{M},\tilde{\underline{\underline{m}}})$ which induces an injection of the fundamental groups, where $(\tilde{M},\tilde{\underline{\underline{m}}})$ denotes the manifold obtained from $(M,\emptyset)$ by splitting along F. Let

$G$ be the free side of $(W,\underline{\underline{w}})$. Then 32.2 follows immediately from the following assertion since $\tilde{M}$ is an aspherical manifold with useful boundary-pattern.

<u>32.3 Assertion</u>. $f|G$ <u>is inessential in</u> $(\tilde{M},\tilde{\underline{\underline{m}}})$.

Assume the converse. Then $f|G$ can be admissibly deformed into a component $X$ of the characteristic submanifold of $\tilde{M}$ (see 12.5). By an argument of 32.1, we may suppose that $f\colon W \to \tilde{M}$ is admissibly deformed so that even $f(W) \subset X$

<u>Case 1</u>. $X$ <u>is an I-bundle</u>.

Recall that $w$ can be deformed into $F$. Since $F$ is essential, this homotopy can be chosen within $\tilde{M}$ and since $X$ is essential, it can be chosen within $X$. Let $W^* = W - \mathring{U}(w)$, where $U(w)$ is a regular neighborhood of $w$, and define $\underline{\underline{w}}^* = \{A_1, A_2\}$, where $A_1 = A$ and $A_2 = \partial U(w)$ Then, by what we have seen so far, $f$ can be deformed (rel $A_1$) so that $g = f|W^*\colon (W^*,\underline{\underline{w}}^*) \to (X,\underline{\underline{x}})$ is an admissible map (which maps the bound sides into lids of $X$). Since $f|w$ is not contractible, there must be a vertical square or annulus in $(X,\bar{\underline{\underline{x}}})$ so that $f|w$ cannot be deformed out of $C$ (in $X$). This means that $g$ can be admissibly deformed so that $S = g^{-1}C$ is a non-empty horizontal surface in $W^*$ (see 4.4 and 5.6). Let $B$ be a vertical annulus $W^*$ which joins $A_1$ with $A_2$, i.e. an essential vertical annulus. Then $B$ must meet $S$ in arcs joining $A_1$ with $A_2$ since $S$ is horizontal. Let $c_1$, $c_2$ be the sides of $C$ contained in the lids of $X$, and observe that the bound sides of $S$ are mapped under $f$ into these sides, $c_1$ and $c_2$. Hence, by a "counting argument" we easily see that the restriction of $f$ either to a free side of $S$ or to an arc of $B \cap S$ is inessential in $C$, and so in $\tilde{M}$. In the first case we are done. In the second case it follows as usual that $f|B$ can be deformed (rel $\partial B$) in $X$ into a lid of $X$, and so into $F$. But observe that $B \cup U(w)$ is an admissible deformation retract of $W$, and so assertion 32.3 follows also in this case.

Case 2. X is a Seifert fibre space.

First of all, we may suppose that X is a solid torus, for otherwise, using essential and vertical annuli, we split X into a system of solid tori, and apply the argument below successively to components of this system.

$(W,\underline{\underline{w}})$ is a solid torus which carries an admissible fibration as Seifert fibre space. By our suppositions on the circulation number of t, it follows that, without loss of generality, the core w is an exceptional fibre of W. Moreover, recall that $(W,\underline{\underline{w}})$ has precisely one bound side, namely A, and precisely one free side, namely G. By our assumption, $f|G$ is an essential map into $(X,\underline{\underline{x}})$. Hence, by the additional remark of 28.1, we may suppose that $f: (W,\underline{\underline{w}}) \to (X,\underline{\underline{x}})$ is admissibly deformed into a fibre preserving map, that $(X,\underline{\underline{x}})$ has an exceptional fibre w', and that $f|w$ is an s-multiple of w', with $s \geq 1$. Finally, let m be the order of the exceptional fibre w' in X, then m is equal to the circulation number of any side of $(X,\underline{\underline{x}})$ with respect to X (note that any side of $(X,\underline{\underline{x}})$ is an annulus, for F is essential, and that a solid torus has at most one exceptional fibre [Wa 1]).

Recall that $f|w$ can be deformed into F. Since F and X are essential, this homotopy can be chosen within X. Hence s must be a multiple of m. Let D be an essential and horizontal disc in X, i.e. a meridian disc. This exists, for, by our suppositions, X is a solid torus. By 4.4 and 5.7, again, f can be admissibly deformed in X so that afterwards $f^{-1}D$ is a system of discs which are essential and horizontal in $(W,\underline{\underline{w}})$. Let $D_1$ be one of them. Let $\alpha$ be a simple arc in $\partial D_1$ with $\alpha \cap t = \partial\alpha$ (this arc exists since, by supposition the circulation number of t is at least two). Of course, the end-points of $\alpha$ can be joined in t by an arc k. $D_1$ splits k into arcs $k_1,\ldots,k_q$, $q \geq 1$. Now recall that m is the circulation number of the sides of X (with respect to X). Moreover, $f|w$ is an s-multiple of w' and s itself is a multiple of m. It is easily checked (by counting) that these facts imply that $f(\partial k_i)$ is one point, for all $1 \leq i \leq q$. Hence also $f(\partial k)$, and so $f(\partial\alpha)$, is one point. Thus $f|\alpha$ can be deformed (rel $\partial\alpha$) in the disc $D$ into one point, and this implies, by our choice of $\alpha$, that $f|G$ is inessential in $(X,\underline{\underline{x}})$. q.e.d.

It is a well-known fact that a Klein bottle $K$ can be decomposed into two Möbius bands $K_1$ and $K_2$, and that this decomposition is unique, up to an ambient isotopy. Denote by $k_1$ and $k_2$ the cores of $K_1$ and $K_2$, respectively. Now, let $F$ be any surface (orientable or not) and let $x$ and $y$ be two non-trivial elements of $\pi_1 F$. Suppose that $x^m = y^n$, $m,n \geq 2$, and that there is no $z \in \pi_1 F$ with $z^\alpha = x$ and $z^\beta = y$. Then it follows that $F = K$ and that $k_1$ and $k_2$ induce elements of $\pi_1 K$ which are conjugate to $x$ and $y$ (a proof for this can be modelled after that of 32.4 and 32.5, but we leave this as an exercise for the reader). Thus the algebraic relation $x^m = y^n$ in $\pi_1 F$ can hold just for geometric reasons. The aim is to establish a similiar phenomenon also for 3-manifolds. For this recall from §28 the definition of an exceptional curve for Haken 3-manifolds, and observe that the system of all exceptional curves of a Haken 3-manifold is unique, up to ambient isotopy.

<u>32.4 Proposition.</u> <u>Let</u> $(M,\emptyset)$ <u>be a Haken</u> 3-<u>manifold whose completed boundary-pattern is useful</u>. <u>Suppose</u> $M$ <u>is not one of the exceptions</u> 5.1.1-5.1.5 <u>and</u> 5.7.4.

<u>Then the following two statements are equivalent</u>:

1. $\pi_1 M$ <u>is an</u> R-<u>group</u>, <u>i.e.</u> $x^n = y^n$ <u>implies</u> $x = y$, <u>for all</u> $x,y \in \pi_1 M$.
2. $(M,\emptyset)$ <u>contains no exceptional curve and no Klein bottle</u>.

<u>Remark</u>. See [Mu 1] for results in this direction for knot groups.

For later use we give the following version of 32.4.

<u>32.5 Proposition</u>. <u>Let</u> $(M,\emptyset)$ <u>be given as in</u> 32.4. <u>Let</u> $k_1, k_2 \colon S^1 \to M$ <u>be two essential singular curves in</u> $(M,\emptyset)$ (<u>without base points</u>). <u>Suppose that</u> $k_1^m \simeq k_2^n$, <u>with</u> $m,n \geq 2$, <u>and that there is no curve</u> $b_1$ <u>with</u> $b_1^\alpha \simeq k_1$ <u>and</u> $b_1^{\alpha'} \simeq k_2$, $\alpha,\alpha' \geq 1$. <u>Suppose that</u>, <u>in addition</u>,

*either* 1 *or* 2 *holds*:

1. M *contains no Klein bottle*, *or*
2. *there is a curve* $k_3$ *with* $k_3^p \simeq k_1^m \simeq k_2^n$, $p \geq 2$, *but no curves* $b_2$, $b_3$, *with* $b_2^\beta \simeq k_1$ *and* $b_2^{\beta'} \simeq k_3$, *or* $b_3^\gamma \simeq k_2$ *and* $b_3^{\gamma'} \simeq k_3$, $\beta, \beta', \gamma, \gamma' \geq 1$.

*Then* $k_i$, $i = 1,2$, *can be deformed into* $t_i$, *where* $t_1$ *and* $t_2$ *are two different exceptional curves of* $(M,\emptyset)$.

*Proof of 32.4 and 32.5*. As a start we prove that 1 of 32.4 implies 2 of 32.4. For this assume there is an essential Seifert fibre space X in $(M,\emptyset)$ with at least one exceptional fibre t, of order n, say. Since X cannot be one of the exceptions of 5.1.4 or 5.1.5, it follows the existence of at least one essential and vertical torus $T = S^1 \times S^1$ in X (see 5.4). We may suppose that $(0,0) \in S^1 \times S^1$ is the base point of $\pi_1 M$, and we denote by $k_1$ and $k_2$ the based loops $S^1 \times 0$ and $0 \times S^1$ in M, resp., as well as the corresponding elements in $\pi_1 M$. Moreover, we may suppose that the product structure of T is chosen such that $k_2 = 0 \times S^1$ is an ordinary fibre of X. Since t is an exceptional fibre of order n, it induces an element $x_1$ of $\pi_1 M$ with $x_1^n = k_2$. Now, define $x_2 = k_1 \cdot x_1 \cdot k_1^{-1} \in \pi_1 M$. Then, of course, $x_1^n = k_2 = k_1 \cdot k_2 \cdot k_1^{-1} = k_1 \cdot x_1^n \cdot k_1^{-1}$ $= (k_1 \cdot x_1 \cdot k_1^{-1})^n = x_2^n$. Thus to prove that $\pi_1 M$ is not an R-group, it suffices to show that $x_1 \neq x_2$ in $\pi_1 M$. Assume the converse. Then $x_1 x_2^{-1} = 1$ in $\pi_1 X$, for $(\partial X - \partial M)^-$ is essential in $(M,\emptyset)$. Thus, by definition of $x_2$, $k_1$ must lie in the centralizer of $x_1$ with respect to $\pi_1 X$. The centralizer of an element of the fundamental group of a Seifert fibre space induced by an exceptional fibre is isomorphic to $\mathbb{Z}$ (to see this geometrically use a vertical hierarchy). Thus $k_1$ can be (freely) deformed in X into t. We extend this deformation to a homotopy of the inclusion map $T \to M$. Then, by our supposition on $k_2$, it follows that $k_2$ can be pulled into t, by a homotopy which is constant on the base point (0,0). Hence $T \to M$ can be deformed into t, for T is a torus. This means that T is inessential in X (loop-theorem) which contradicts our choice of T, and so $x_1 \neq x_2$ in $\pi_1 M$.

By what we have seen so far $\pi_1 M$ is not an R-group if $M$ has an exceptional curve. If $M$ contains a Klein bottle, $\pi_1 M$ is also not an R-group. This can be proved in the same way. Indeed, a regular neighborhood $U$ of the Klein bottle is a Seifert fibre space. In face, $U$ admits two Seifert fibrations. One of these has two exceptional fibres and as orbit surface the disc, and the other one is the $S^1$-bundle structure over the Möbius band. These two Seifert fibrations are transversal in $\partial U$. In particular, this implies that the complement of $U$ in $M$ cannot be a solid torus since $M$ is not one of the exceptions of 5.1.4, or 5.1.5. Hence by the irreducibility of $M$, $U$ is essential, and so by the argument above, $\pi_1 M$ is not an R-group.

To prove 32.4, we still have to show that $\pi_1 M$ is an R-group if $(M,\emptyset)$ contains neither an exceptional curve nor a Klein bottle. This can be proved along the line of arguments leading to 32.5. Thus we shall prove the two assertions simultaneously.

For this let $V_1$ be a solid torus and $A_1$ an annulus in $\partial V_1$ with circulation number $m$, and let $V_2$ be a solid torus and $A_2$ an annulus in $\partial V_2$ with circulation number $n$ (with respect to the solid tori). Denote by $W$ the manifold obtained from $V_1$ and $V_2$ by identifying $A_1$ and $A_2$ (under a homeomorphism). Then we may fix a fibration of $W$ as a Seifert fibre space such that the cores $s_1$ and $s_2$ of $V_1$ and $V_2$ are the exceptional fibres (if we are in the proof of 32.4, let $m = n$, fix a base point $z$ in $A_1 = A_2$, and deform $s_i$ in $V_i$ into a loop with base point $z$).

By the suppositions, there is a map $f\colon W \to M$ with $f|s_j = k_j$, $j = 1,2$ (resp. $[f|s_j] = x_j$ in $\pi_1 M$).

Fix a hierarchy $(M,\emptyset) = (M_1,\underline{\underline{m}}_1),\ldots,(M_n,\underline{\underline{m}}_n)$, $n \geq 1$. It is to be understood that the hierarchy surface $F_j$ in $M_j$, $1 \leq j \leq n-1$, is an essential surface which is chosen as a square, annulus, or torus whenever possible (for the existence of a hierarchy recall that $M$ is Haken 3-manifold and see 4.3). Since $f|s_j$, $j = 1,2$, is not contractible, there must be an integer $i$, $1 \leq i \leq n$, such that $f(W) \subset M_i$ and that $f$ cannot be deformed out of $F_i$.

By 4.4 and 5.6, $f$ can be deformed in $M_i$ so that afterwards $f^{-1}F_i$ is an essential surface in $(W,\bar{\emptyset})$ which is either vertical or horizontal. Suppose that $f$ is deformed so that the above holds, and that, in addition, the number of components of $f^{-1}F_i$ is minimal (also we may suppose that the base point $z$ lies in $f^{-1}F_i$).

Case 1. Suppose that $f^{-1}F_i$ is vertical in $W$.

Let $p\colon W \to G$ be the fibre projection. Then $G$ is a disc with precisely two exceptional points, and $p(f^{-1}F_i)$ is a system of pairwise parallel arcs which separate the two exceptional points (note that $p(f^{-1}F_i)$ cannot contain any closed curve, for $f$ cannot be deformed out of $F_i$).

$\pi_1 M$ has no torsion element [Ep 1, 8.4]. Hence it follows easily from 4.7.1 and our minimality condition on $f^{-1}F_i$ that $f|\partial W$ is an essential singular torus in $M_i$, and so in $M$. Hence, by 12.5, $f|\partial W$ can be deformed into an essential Seifert fibre space $X$ in $M$. Moreover we may suppose that $f$ lies in an essential Seifert fibre space $Z$. To see the latter deform $f$ so that $f^{-1}\partial X$ consists of essential, vertical tori. Then $f^{-1}X = W$, or there are essential vertical annuli $A$ in $(W - f^{-1}X)^-$. If $f|A$ is inessential in $(M - X)^-$ $f|A$ can be deformed (rel $\partial A$) into $X$, if not, it can be admissibly deformed in $(M - X)^-$ into an essential Seifert fibre space (by 12.5). Since every essential Seifert fibre space in $(M - X)^-$ is also essential in $M$, our claim follows immediately (recall our definition of $W$).

In this situation the proof of 1 of 32.4 is straightforward: deform $f$ in $Z$ into the complement of an ordinary fibre (observe that $f$ can be deformed out of an ordinary fibre contained in an essential vertical annulus or torus). If $Z$ does not contain any Klein bottle or exceptional fibre, the complement of an ordinary fibre of $Z$ is a product $S^1$-bundle over a surface with boundary, i.e. $\pi_1 Z$ is a free group. So $x_1 = x_2$ in $\pi_1 Z$, and so in $\pi_1 M$.

To prove 32.5 observe that, by our choice of $F_i$, it may be supposed that $F_i$ is an essential, vertical annulus or torus in the Seifert fibre space $Z$. Let $\tilde{Z}$ and $\tilde{W}$ be the Seifert fibre spaces obtained from $Z$ and $W$ by splitting along $F_i$ and $f^{-1}F_i$. Then $\tilde{W}$ consists of solid tori and it follows that $f$ can be deformed into a fibre preserving map which maps the exceptional fibres of $W$ onto exceptional fibres $t_1$ and $t_2$ of $Z$. To see this apply the additional remarks of 28.1 and 28.3 to the map $f|\tilde{W}: \tilde{W} \to \tilde{Z}$ and recall the description of $f^{-1}F_i$ in the beginning of Case 1. $t_1$ and $t_2$ are different since there is no curve $b_1$ with $b_1^{\alpha} \simeq k_1$ and $b_1^{\alpha'} \simeq k_2$, $\alpha,\alpha' \geq 1$.

This proves 32.4 and 32.5 in Case 1.

Case 2. Suppose that $f^{-1}F_i$ is horizontal in $W$.

Let $G_1,\ldots,G_q$, $q \geq 1$, be the components of $f^{-1}F_i$, and suppose that the base point $z$ of $\pi_1 W$ is contained in $G_1$. Let $(\tilde{W},\tilde{\emptyset})$ be the manifold obtained from $(W,\emptyset)$ by splitting at $f^{-1}F_i$. Since we are in Case 2, each component $W_j$, $1 \leq j \leq q$, of $\tilde{W}$ is a product I-bundle $G_j \times I$. Since $\tilde{W} \cap \partial W$ consists of annuli, it follows from our minimality condition on $f^{-1}F_i$ that the restriction of $f$ to any essential arc in $\tilde{W} \cap \partial W$, and so in $\tilde{W}$, is essential in $(M_{i+1},\underline{\underline{m}}_{i+1})$.

Every surface $G_j$, $1 \leq j \leq q$, has a non-empty boundary and hence a bouquet $R_j$ of 1-spheres as deformation retract. We may suppose that the base point $z$ is the centre point of $R_1$.

If, for every loop $k$ of $R_1$, the restriction $f|k$ is contractible in $F_i$, it follows that $f|R_1$ can be deformed (rel z) in $F_i$ into one point. By the asphericity of $M_i$, this implies that $f$ can be deformed (rel z) into a loop in $M_i$. Then we cannot be in the proof of 32.5, and 32.4 follows immediately.

Therefore we may suppose that there is an essential curve $k: S^1 \to G_1$ such that $f \circ k$ is essential in $F_i$. Since $W$ is a Seifert fibre space, we easily find an essential and vertical singular torus $g: S^1 \times S^1 \to W$ such that $g|S^1 \times 0$ is a multiple of $k$. Since $f^{-1}F_i$ is horizontal, $g^{-1}(f^{-1}F_i)$ is a system of closed curves which split $S^1 \times S^1$ into a system $\tilde{A}$ of annuli and $g|\tilde{A}$ must be essential in $(\tilde{W},\tilde{\emptyset})$. Recall that $\tilde{W}$ consists of product I-bundles and that the restriction of $f$ to any essential arc in $\tilde{W}$ is essential in $(M_{i+1},\underline{\underline{m}}_{i+1})$. By our choice of $g$, this implies that $f \circ g$ is an essential singular torus in $(M_i,\underline{\underline{m}}_i)$.

Hence, by 12.5, the characteristic submanifold of $(M_i,\underline{\underline{m}}_i)$

is non-empty. By our choice of $F_i$, this implies that $F_i$ is an essential square, annulus, or torus in $(M_i, \underline{m}_i)$. Moreover, the existence of the singular curve $f \circ k$ shows that $F_i$ is not a square.

We assert that $f|G_1$ can be deformed in $F_i$ into a simple closed curve. Assume the converse. Then, in particular, $F_i$ cannot be an annulus. Since $F_i$ is a torus, it follows from our assumption the existence of at least two loops $c_1$, $c_2$ of $R_1$ such that $f|c_1$ and $f|c_2$ cannot be freely deformed in $F_i$ into the same curve (of course this is not true for an arbitrary surface $F_i$). Each one of the loops $c_1$ and $c_2$ is a boundary curve of two vertical annuli in $\tilde{W}$. The restrictions of $f$ to these annuli are essential singular annuli in $M_{i+1}$, and so, by 12.5, they can be admissibly deformed into components of the characteristic submanifold of $M_{i+1}$. By our choice of the $c_j$'s and since $F_i$ is a torus, these components have to be I-bundles which contain the sides of $M_{i+1}$ which are copies of $F_i$. It follows that $M_i$, and so $M$, is one of the exceptions of 5.7.4, which is the required contradiction.

Recall that each component $W_j$, $1 \leq j \leq g$, of $\tilde{W}$ is a product I-bundle, that $M_{i+1}$ is aspherical and that $F_i$ is an essential surface in $M_i$. Define $f_j = f|W_j$ and denote by $T_j$ an annulus. Since $f|G_1$ can be deformed in $F_i$ into a simple closed, essential curve, a moments reflection shows that there are two maps $g_j$ and $h_j$ which make the following diagram commutative, up to admissible homotopy:

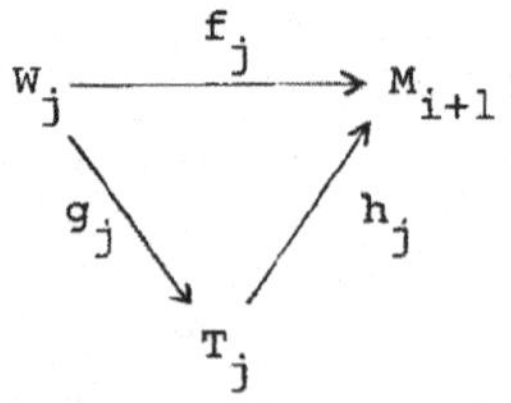


Since the restriction of $f$ to any essential arc of $\tilde{W}$ is essential in $M_{i+1}$, $h_j$ is an essential singular annulus in $M_{i+1}$. It can be admissibly deformed in $M_{i+1}$ into a Seifert fibre space $X_j$ of the characteristic submanifold, by 12.5. Since $M$ is not the exception 5.7.4, it follows from 5.10 and 5.11 that $h_j$ can be admissibly deformed in $X_j$ into a vertical map. Hence, attaching the copies of

$F_i$, we obtain from the $X_j$'s a Seifert fibre space $X$ in $M_j$. Identifying the boundary curves of the $T_j$'s in the right way we obtain a surface $T$ such that the quotient maps $g$ and $h$ make the following diagram commutative, up to homotopy:

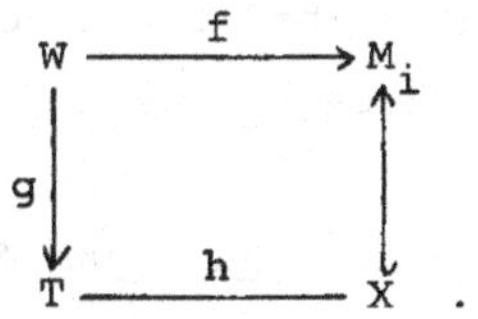


By construction, $T$ is either a torus or Klein bottle. If $T$ is a torus we cannot be in the proof of 32.5, and 32.4 follows immediately, for $\pi_1(S^1 \times S^1)$ is an R-group.

If $T$ is a Klein bottle, we obtain, by cutting and pasting along the singular curves of $h$, an embedding of a Klein bottle in $M$. In particular we cannot be in the proof of 32.4. Hence, by the suppositions of 32.5, there is a curve $k_3$ with $k_3^l \simeq k_1^m \simeq k_2^n$, $p \geq 2$, as described in 32.5. Let $V_3$ be a solid torus, $s_3$ the core of $V_3$, and $A_3$ an annulus in $\partial V_3$ with circulation number $p$, with respect to $V_3$. Let $W^*$ be the manifold obtained from $W$ and $V_3$ by identifying $A_3$ with a fibre-parallel annulus in $\partial W$. Then, by the construction, $f$ extends to a map $f^*: W^* \to M$ with $f^*|s_j = k_j$, $j = 1,2,3$. In the same way as for $f$ it follows also for $f^*$ that there is a torus or Klein bottle $T^*$ and that there are maps $g^*$ and $h^*$ such that the following diagram is commutative, up to homotopy:

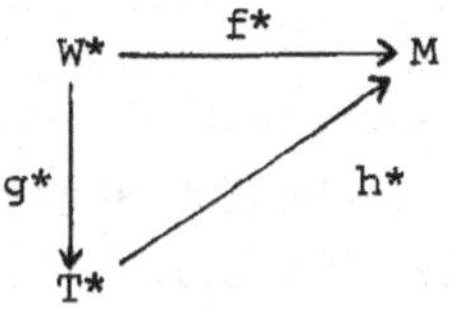


But this is impossible, for there are no curves $b_2, b_3$ as described in 32.5.2. Hence we have the required contradiction to our assumption that $T$ is a Klein bottle. q.e.d.

The following proposition is a version of a theorem first proved by P. Shalen [Sh 1]. Another proof can be found in [Ja 1]. Here we offer a third, complete geometric proof based on 32.2 and 32.5. For the statement of the proposition recall that a curve is <u>divisible</u> if it is freely homotopic to a non-trivial multiple of another curve. Finally we remark that the conclusion of 32.6 for surfaces can be proved along the same line of arguments.

<u>32.6 Proposition</u>. <u>Let</u> $(M,\emptyset)$ <u>be a Haken 3-manifold whose completed boundary-pattern is useful</u>. <u>Suppose</u> $M$ <u>is not one of the exceptions</u> 5.1.1-5.1.5, <u>and</u> 5.7.4. <u>Let</u> $k$ <u>be an essential closed curve in</u> $M$.

<u>Then there is only a finite number of curves</u> $t$ <u>in</u> $M$, <u>up to homotopy, for which</u> $t^n \simeq k$, <u>for some integer</u> $n \geq 2$.

<u>Moreover, the number of homotopy classes of indivisible curves</u> $t$ <u>with</u> $t^n \simeq k$, <u>for some integer</u> $n \geq 2$, <u>cannot be larger than the number of exceptional curves of</u> $(M,\emptyset)$ <u>plus</u> 2.

<u>32.7 Corollary</u> [Shalen]. $\pi_1 M$ <u>has no element which is infinitely divisible</u>.

<u>Proof of 32.6</u>. Define $S(k)$ to be the set of homotopy classes of all the curves $t$ in $M$ with $t^n \simeq k$, for some integer $n \geq 2$. Let $T(k) = \{s \in S(k) | s \text{ is not divisible}\} = \{s \in S(k) | s \simeq b^\alpha$, $b$ curve in $M$, implies $\alpha = 1\}$. Since $(M,\emptyset)$ has only finitely many exceptional curves, it follows from 32.5 that $T(k)$ must be a finite set (more precisely, the cardinality of $T(k)$ cannot be larger than the number of exceptional curves of $(M,\emptyset)$ plus 2).

Thus we still have to show that $k$ cannot be a "rational curve". For this let $k = k_1, k_2, \dots$ be any sequence of essential closed curves in $(M,\emptyset)$ such that, for all $j \geq 1$, the curve $k_j$ is freely homotopic in $M$ to $k_{j+1}^{n_j}$, for some integer $n_j \geq 2$. Then we have to show that this sequence must be finite. The idea is to introduce a "length" for curves in $M$ and then to show that the curves $k_1, k_2, \dots$ become shorter and shorter.

To make this idea work we choose a hierarchy for

$(M_1,\underline{\underline{m}}_1) = (M,\emptyset)$:

$(M_i,\underline{\underline{m}}_i)$, $F_i$ essential in $(M_i,\underline{\underline{m}}_i)$, and $(M_{i+1},\underline{\underline{m}}_{i+1})$ obtained from $(M_i,\underline{\underline{m}}_i)$ by splitting at $F_i$, $1 \leq i < n$.

If $k: S^1 \to M$ is an essential curve which is in <u>good position to the hierarchy</u>, i.e. in such a position that the intersection of $k$ with any $M_i$, $1 \leq i \leq n+1$, is a system of essential curves in $(M_i,\underline{\underline{m}}_i)$, then we may denote

$$\ell(k) = \text{number of components of } k^{-1}M_{n+1}.$$

The <u>length of</u> $k$ is defined to be min $\ell(k')$, where the minimum is taken over all curves $k'$ homotopic to $k$ which are in good position to the hierarchy.

Without loss of generality we may suppose that the curves of the sequence $k_1,k_2,\ldots$ are deformed so that they are in a good position to the hierarchy and that furthermore the number of components of $k_i^{-1}M_{n+1}$ is equal to the length of $k_i$, for all $i = 1,2,\ldots$ . Let $W$ denote a solid torus with $w$ its core. Recall that $k_p$ is freely homotopic to $k_q^m$, for some $m \geq 2$, provided $k_p$ and $k_q$, $p < q$, are curves from the sequence $k_1,k_2,\ldots$ . This means that there is a simple closed curve $w_{pq}$ in $\partial W$ such that

1. $w_{pq}$ has circulation number $m$ with respect to $W$, and
2. there is a map $f_{pq}: W \to M$ with $f_{pq}|w = k_q$ and $f_{pq}|w_{pq} = k_p$.

For any $1 \leq i \leq n$, there are only finitely many curves of $k_1,k_2,\ldots$ which are closed in $M_i$ and which can be deformed into $F_i$. Otherwise it follows from 32.2 the existence of a curve which is rational in the surface $F_i$. Forgetting finitely many curves of $k_1,k_2,\ldots$ we may hence suppose that no curve of $k_1,k_2,\ldots$, which is closed in $M_i$, can be deformed into $F_i$, $1 \leq i \leq n$.

We claim that the length of $k_2$ is strictly smaller than that of $k_1$. To see this let $j$ be the index such that $k_2$ is closed in $M_j$ and intersects $F_j$. Consider the map $f_{12}: W \to M$ as defined above. Since $k_2 \subset M_j$ and since $k_1$ is in a good position to the hierarchy,

$f_{12}$ can be deformed (rel $w$ and $w_{12}$) so that afterwards $f_{12}^{-1}F_i$, $1 \leq i \leq j-1$ is either empty or vertical (with respect to the Seifert fibration induced by $w_{12}$). But the latter is excluded, by our supposition that no curve of $k_1, k_2, \ldots$, which is closed in $M_i$, can be deformed into $F_i$. Thus we may suppose that $f_{12}(W) \subset M_j$. Hence $f_{12}$ can be deformed (rel $w$ and $w_{12}$) in $M_j$ so that afterwards $f_{12}^{-1}F_j$ consists of horizontal meridian discs. In particular, $k_1$ must intersect $F_j$. Indeed, it intersects $F_j$ in strictly more points than $k_2$ since the circulation number of $w_{12}$ is strictly larger than one. Then our claim follows without difficulty.

In the same way it follows that the length of $k_3$ is strictly smaller than that of $k_2$, etc. Hence the sequence $k_1, k_2, \ldots$ must be finite since the lengths have certainly a lower bound. q.e.d.

Let $(X,\emptyset)$ be a Seifert fibre space, but not one of the exceptions 5.1.3-5.1.5, and let $(M,\emptyset)$ be a Haken 3-manifold.

If the orbit surface of $X$ is orientable, the center of $\pi_1 X$ is non-trivial (namely a free group generated by an ordinary fibre). Conversely, if $\pi_1 M$ has a non-trivial center, then $M$ is a Seifert fibre space whose orbit surface is orientable [Wa 3 ].

We are going to deduce a similiar geometric property from the non-triviality of centralizers. So, let $\alpha$ be any element of a group $\pi$ (e.g. $\pi = \pi_1 X$ or $\pi_1 M$). Then it is convenient first to introduce the so-called twisted centralizer of $\alpha$ in $\pi$. More precisely, we define

$\mathfrak{Z}_t(\alpha,\pi) = \{b \in \pi \mid \alpha b \alpha^{-1} b^{-1} = 1 \text{ or } \alpha b \alpha b^{-1} = 1\}$
to be the <u>twisted</u> <u>centralizer</u> of $\alpha$ in $\pi$, and

$\mathfrak{Z}(\alpha,\pi) = \{b \in \pi \mid \alpha b \alpha^{-1} b^{-1} = 1\}$ to be the <u>centralizer</u> of $\alpha$ in $\pi$.

The following lemma describes centralizers and twisted centralizers of an element $\alpha$ of the fundamental group of the Seifert fibre space $X$ (we denote by $\alpha$ also any based loop in $X$ inducing $\alpha$). The proof of it will be left to the reader. Hint: note that every element of $\mathcal{B}(\alpha,\pi_1 X)$ and $\mathcal{B}_t(\alpha,\pi_1 X)$ defines a singular torus or Klein bottle in X, and that hence the techniques of §5 and §7 are available.

<u>32.8 Lemma.</u> X <u>is a Seifert fibre space, but not one of</u> 5.1.4-5.1.5.

1. <u>Suppose</u> $\alpha$ <u>cannot be freely deformed into an ordinary fibre of</u> X, <u>for no fibration of</u> X. <u>Then</u>

$$\mathcal{B}(\alpha,\pi_1 X) \cong \begin{cases} \mathbb{Z}, \text{ if } \alpha \text{ can be freely deformed into an} \\ \quad \text{exceptional fibre.} \\ \mathbb{Z} \oplus \mathbb{Z}, \text{ otherwise} \end{cases}$$

2. <u>Suppose</u> $\alpha$ <u>can be freely deformed into an ordinary fibre of</u> X, <u>for some fibration of</u> X. <u>Then</u> $\mathcal{B}_t(\alpha,\pi_1 X) \cong \pi_1 X$, <u>and</u>

$$\mathcal{B}(\alpha,\pi_1 X) \cong \begin{cases} \mathcal{B}_t(\alpha,\pi_1 X), \text{ if the orbit surface of } X \\ \qquad\qquad \text{is orientable} \\ \text{subgroup of index two in } \mathcal{B}_t(\alpha,\pi_1 X), \text{ otherwise.} \end{cases}$$

<u>32.9 Proposition.</u> <u>Let</u> (M,∅) <u>be a Haken</u> 3-<u>manifold whose completed boundary-pattern is useful</u>.

<u>Suppose</u> M <u>is not one of the exceptions</u> 5.1.4-5.1.5. <u>Let</u> $\alpha \in \pi_1 M$ <u>and suppose that</u> $\mathcal{B}(\alpha,\pi_1 M) \not\cong \mathbb{Z}$ or $\mathbb{Z} \oplus \mathbb{Z}$.

<u>Then there is an essential Seifert fibre space</u> X <u>in</u> (M,∅), <u>and a</u> 1- <u>or</u> 2-<u>sheeted covering map</u> $q\colon \tilde{X} \to X$ <u>such that</u>

$$\mathcal{B}(\alpha,\pi_1 M) \cong \pi_1 \tilde{X}.$$

<u>Remark</u>. For other proofs of this proposition or related results see also [Ja 1] [Fe 4], [JS 1]

<u>Proof</u>. From 32.6 and our suppositions on $\mathcal{B}(\alpha,\pi_1 M)$ we deduce that there is a curve b in M such that $\alpha b \alpha^{-1} b^{-1}$ is nullhomotopic and that this homotopy defines an essential singular torus f in (M,∅). Hence, by 12.5, the characteristic submanifold V of (M,∅) is not

empty, and $f$ can be deformed into a component $X$ of $V$. $X$ must be a Seifert fibre space, for $X \cap \partial M = \emptyset$. Hence we may suppose that the curve $\alpha$, and so the base point of $\pi_1 M$, lies in $X$. Moreover, it is easily seen that $X$ may be chosen so that, for every $c \in \mathcal{B}_t(\alpha,\pi_1 M)$ or $\mathcal{B}(\alpha,\pi_1 M)$, $c$ is homotopic (rel base point) to a curve $c'$ in $X$ (apply 12.5, 13.2, 12.6, 10.4 and note that $V$ is a full F-manifold). Let the association $\varphi_1: \mathcal{B}(\alpha,\pi_1 M) \to \mathcal{B}(\alpha,\pi_1 X)$ and $\varphi_2: \mathcal{B}_t(\alpha,\pi_1 M) \to \mathcal{B}_t(\alpha,\pi_1 X)$ be given by $\varphi_i(c) = c'$, $i = 1,2$. Since $(\partial X - \partial M)^-$ is an essential surface in $(M,\emptyset)$, it is easily checked that $\varphi_i$, $i = 1,2$, is in fact an isomorphism. In particular, by 32.8 and since $\mathcal{B}(\alpha,\pi_1 M) \not\cong \mathbb{Z}$ or $\mathbb{Z} \oplus \mathbb{Z}$, $\alpha$ must be freely homotopic in $X$ to an ordinary fibre of $X$. Hence 32.9 follows from 32.8 (recall the very special ways defining the isomorphism $\varphi_1$ and $\varphi_2$). q.e.d.

## References


[Al 1] Alexander, J.W.: On the subdivision of 3-space by a polyhedron. Proc. Nat. Acad. Sc. 10, 6-8 (1924).

[Bi 1] Birman, J.: Braids, links and mapping class groups. Ann. of Math. Studies 82 (1974).

[Bo 1] Boehme, H.: Fast genügend große irreducible 3-dimensionale Mannigfaltigkeiten. Inv. Math. 17, 303-316 (1972).

[BZ 1] Burde, G., Zieschang, H.: Eine Kennzeichnung des Torusknoten. Math. Ann. 167, 169-176 (1966).

[CF 1] Cannon, J.W., Feustel, C.D.: Essential embeddings of annuli and Möbius bands in 3-manifolds. Trans. A.M.S. 215, 219-239 (1976).

[De 1] Dehn, M.: Die Gruppe der Abbildungsklassen. Acta math. 69, 135-206 (1938).

[Ep 1] Epstein, D.B.A.: Projective planes in 3-manifolds. Proc. Lond. Math. Soc. (3) 11, 469-484 (1961).

[Ep 2] Epstein, D.B.A.: Curves on 2-manifolds and isotopies. Acta math 115, 83-107 (1966).

[Fe 3] Feustel, C.D.: On the torus theorem and its applications. Trans. A.M.S. 217, 1-43 (1976).

[Fe 4] Feustel, C.D.: On realizing centralizers of certain elements in the fundamental group of a 3-manifold. Proc. A.M.S. 55, 213-216 (1976).

[Fe 5] Feustel, C.D.: On the torus theorem for closed 3-manifolds. Trans. A.M.S. 217, 45-57 (1976).

[Fn 1] Fenchel, W.: Estensioni di gruppi discontinui e transformazioni periodiche delle superficie. Ati accad. Naz. Lincei Rend. Cl. Sci. Fis. Mat. Nat. Ser. 8, 5, 326-329 (1948).

[Gr 1] Gramain, A.: Rapport sur la theorie classique des noeuds. Seminaire Bourbaki, 28e année n°485 (1975/76).

[Ha 1] Haken, W.: Theorie der Normalflächen. Acta math. 105, 245-375 (1961).

[Ha 2] Haken, W.: Über das Homöomorphieproblem der 3-Mannigfaltigkeiten I. Math. Z. 80, 89-120 (1962).

[Ha 3] Haken, W.: Some results on surfaces in 3-manifolds. Studies in Modern Topology, Prentice Hall (1968).

[HT 1] Hatcher, A., Thurston, W.: A presentation for the mapping class group of a closed orientable surface. Preprint.

[He 1] Hemion, G : On the classification of homeomorphisms of 2-manifolds and the classification of 3-manifolds. Acta math. 142, 123-155 (1979).

[Hp 1] Hempel, J.: 3-manifolds. Ann. of Math. Studies 86, Princeton University Press (1976).

[Hp 2] Hempel, J.: unpublished.

[Hu 1] Hudson, J.F.P.: Piecewise linear topology. W.A. Benjamin, New York, Amsterdam (1969).

[Ja 1] Jaco, W.: Roots, relations and centralizers in three-manifold groups. Geometric Topology, L.C. Glaser and T.B. Rushing, eds. Springer Lecture notes 438,283-309 (1975).

[JS 1] Jaco, W., Shalen, P.B.: Seifert fibered spaces in 3-manifolds. Preprint.

[JS 2] Jaco, W., Shalen, P.B.: A new decomposition theorem for irreducible sufficiently-large 3-manifolds. Proc. Symp. Pure Math. 32 (1978).

[Jo 1] Johannson, K.: Êquivalences d'homotopie des variétés de dimension 3. C.R. Acad. Sci., Paris 281, Serie A, 1009-1010 (1975).

[Jo 2] Johannson, K.: Homotopy equivalences of knot spaces. Preprint.

[Jo 3] Johannson, K.: Homotopy equivalences of 3-manifolds with boundary. Preprint.

[Jo 4] Johannson, K.: On the mapping class group of simple 3-manifolds, to appear.

[Jo 5] Johannson, K.: On exotic homotopy equivalences of 3-manifolds, to appear.

[Kn 1] Kneser, H.: Geschlossene Flächen und dreidimensionale Mannigfaltigkeiten. Jahresb. d. Deut. Math. Verein. 38, 248-260 (1929).

[La 1] Laudenbach, F.: Topologie de la dimension troi-homotopie et isotopie. Asterisque 12, Soc. Math. France (1974).

[Li 1] Lickorish, W.B.R.: A representation of orientable combinatorial 3-manifolds. Ann. of Math. 76 (3), 531-540 (1962).

[Li 2] Lickorish, W.B.R.: Homeomorphisms of non-orientable two-manifolds. Proc. Camb. Phil. Soc. 59, 307-317 (1963).

[Li 3] Lickorish, W.B.R.: A finite set of generators for the homeotopy group of a 2-manifold. Proc. Camb. Phil. Soc. 60, 769-778 (1964).

[Li 4] Lickorish, W.B.R.: On the homeomorphisms of a non-orientable surface. Proc. Camb. Phil. Soc. 61, 61-64 (1965).

[Ma 1] Macbeath, A.M.: On a theorem by J. Nielsen. Quart. J. Math. 13 (2), 235-236 (1962).

[ML 1] MacLane, S.: Homology. Springer, Berlin (1963).

[Me 1] Melvin, Bordism of diffeomorphisms. Preprint.

[Mu 1] Murasugi, K.: On the divisibility of knot groups. Pacific

J. Math. 52, 491-503 (1974).

[Ne 1] Neumann, W.D.: Equivariant Witt rings. Bonner Math. Schriften 100 (1977).

[Ni 1] Nielsen, J.: Die Isomorphismengruppe der freien Gruppen. Math. Ann. 91, 169-209 (1924).

[Ni 2] Nielsen, J.: Untersuchungen zur Topologie der geschlossenen zweiseitigen Flächen II. Acta math. 53, 1-76 (1929).

[Ni 3] Nielsen, J.: Die Struktur periodischer Transformationen von Flächen. Mat. Fys. Medd. Danske Vid. Selsk. 15, 1-27 (1937).

[Ni 4] Nielsen, J.: Abbildungsklassen endlicher Ordnung. Acta math. 75, 23-115 (1943).

[Ni 5] Nielsen, J.: Surface transformation classes of algebraically finite type. Math. fys. Meddelelser Kgl. Danske Vidensk. Selsk. XXI, 2 (1944).

[Or 1] Orlik, P.: Seifert manifolds. Springer Lecture notes 291 (1972).

[OVZ 1] Orlik, P., Vogt, E., Zieschang, H.: Zur Topologie gefaserter dreidimensionaler Mannigfaltigkeiten. Topology 6, 49-64 (1967).

[Pa 1] Papakyriakopoulos, C.D.: On Dehn's Lemma and the asphericity of knots. Ann. of Math. 66, 1-26 (1957).

[Pa 2] Papakyriakopoulos, C.D.: On solid tori. Proc. London Math. Soc (3) 7, 281-299 (1959).

[Sch 1] Schubert, H.: Knoten und Vollringe. Acta math. 90, 131-286 (1953).

[Sc 1] Scott, G.P.: On sufficiently large 3-manifolds. Quart. J. of Math. 23, 159-172 (1972).

[Sc 2] Scott, G.P.: On the annulus and torus theorem. Preprint.

[Se 1] Seifert, H.: Topologie dreidimensionaler gefaseter Raume. Acta math. 60, 147-238 (1933).

[Sh 1] Shalen, P.B.: Infinitely divisible elements in 3-manifold groups, in: Knots,groups, and 3-manifolds, Neuwirth (ed.) Ann. of Math. Studies, 293-335 (1975).

[Si 1] Simon, J.: An algebraic classification of knots in $S^3$. Ann. of Math. 97, 1-13 (1973).

[Si 2] Simon, J.: Roots and centralizers of peripheral elements in knot groups. Math. Ann. 222, 205-209 (1976).

[St 1] Stallings, J.: On the loop theorem. Ann. of Math. 72, 12-19 (1960).

[St 2] Stallings, J.: On fibering certain 3-manifolds, in: Topology of 3-manifolds (ed. M.K. Fort), Englewood Cliffs, Prentice Hall (1962).

[St 3] Stallings, J.: Group theory and three-dimensional

manifolds. Yale Univ. Press (1971).

[St 4] Stallings, J.: On the recursiveness of sets of presentations of 3-manifold groups. Fundamenta Math. 51, 191-194 (1962).

[ST 1] Seifert, H., Threlfall, W.: Lehrbuch der Topologie, Teubner, Leipzig (1934).

[Sw 1] Swarup, A.: On a theorem of Johannson. Preprint.

[Sw 2] Swarup, A.: Deforming homotopy equivalences. Preprint.

[Th 1] Thurston, W.: On the geometry and dynamics of diffeomorphisms of surfaces I, to appear.

[Wa 1] Waldhausen, F.: Eine Klasse von 3-dimensionalen Mannigfaltigkeiten I. II, Inv. math. 3, 308-333 (1967), 4, 87-117 (1967).

[Wa 2] Waldhausen, F.: Eine Verallgemeinerung des Schleifensatzes. Topology 6, 501-504 (1967).

[Wa 3] Waldhausen, F.: Gruppen mit Zentrum und 3-dimensionale Mannigfaltigkeiten. Topology 6, 505-517 (1967).

[Wa 4] Waldhausen, F.: On irreducible 3-manifolds which are sufficiently large. Ann. of Math. 87, 56-88 (1968).

[Wa 5] Waldhausen, F.: The word problem in fundamental groups of sufficiently large 3-manifolds. Ann. of Math. 88, 272-280 (1968).

[Wa 6] Waldhausen, F.: On the determination of some bounded 3-manifolds by their fundamental group alone. Proc. of the Intern. Symp. on Topology and its applications, Herceg-Novi, Yugoslavia, Beograd 1969, 331-332 (1969).

[Wa 7] Waldhausen, F.: Recent results on sufficiently large 3-manifolds. Proc. Symp. Pure Math. 32 (1978).

[Wh 1] Whitehead, J.H.C.: On 2-spheres in 3-manifolds. B.A.M.S. 64, 161-166 (1958).

[ZVC 1] Zieschang, H., Vogt, E., Coldewey, H.D.: Flächen und ebene diskontinuierliche Gruppen. Springer Lecture notes 122 (1970).

[Zi 1] Zieschang, H.: Lifting and projecting homeomorphisms. Archiv d. Math. 14, 416-421 (1973).

[Zim 1] Zimmermann, B.: Erweiterungen nichteuklidischer kristallographischer Gruppen. Math. Ann. 231, 187-192 (1977).

Index

Vol. 609: General Topology and Its Relations to Modern Analysis and Algebra IV. Proceedings 1976. Edited by J. Novák. XVIII, 225 pages. 1977.

Vol. 610: G. Jensen, Higher Order Contact of Submanifolds of Homogeneous Spaces. XII, 154 pages. 1977.

Vol. 611: M. Makkai and G. E. Reyes, First Order Categorical Logic. VIII, 301 pages. 1977.

Vol. 612: E. M. Kleinberg, Infinitary Combinatorics and the Axiom of Determinateness. VIII, 150 pages. 1977.

Vol. 613: E. Behrends et al., $L^p$-Structure in Real Banach Spaces. X, 108 pages. 1977.

Vol. 614: H. Yanagihara, Theory of Hopf Algebras Attached to Group Schemes. VIII, 308 pages. 1977.

Vol. 615: Turbulence Seminar, Proceedings 1976/77. Edited by P. Bernard and T. Ratiu. VI, 155 pages. 1977.

Vol. 616: Abelian Group Theory, 2nd New Mexico State University Conference, 1976. Proceedings. Edited by D. Arnold, R. Hunter and E. Walker. X, 423 pages. 1977.

Vol. 617: K. J. Devlin, The Axiom of Constructibility: A Guide for the Mathematician. VIII, 96 pages. 1977.

Vol. 618: I. I. Hirschman, Jr. and D. E. Hughes, Extreme Eigen Values of Toeplitz Operators. VI, 145 pages. 1977.

Vol. 619: Set Theory and Hierarchy Theory V, Bierutowice 1976. Edited by A. Lachlan, M. Srebrny, and A. Zarach. VIII, 358 pages. 1977.

Vol. 620: H. Popp, Moduli Theory and Classification Theory of Algebraic Varieties. VIII, 189 pages. 1977.

Vol. 621: Kauffman et al., The Deficiency Index Problem. VI, 112 pages. 1977.

Vol. 622: Combinatorial Mathematics V, Melbourne 1976. Proceedings. Edited by C. Little. VIII, 213 pages. 1977.

Vol. 623: I. Erdelyi and R. Lange, Spectral Decompositions on Banach Spaces. VIII, 122 pages. 1977.

Vol. 624: Y. Guivarc'h et al., Marches Aléatoires sur les Groupes de Lie. VIII, 292 pages. 1977.

Vol. 625: J. P. Alexander et al., Odd Order Group Actions and Witt Classification of Innerproducts. IV, 202 pages. 1977.

Vol. 626: Number Theory Day, New York 1976. Proceedings. Edited by M. B. Nathanson. VI, 241 pages. 1977.

Vol. 627: Modular Functions of One Variable VI, Bonn 1976. Proceedings. Edited by J.-P. Serre and D. B. Zagier. VI, 339 pages. 1977.

Vol. 628: H. J. Baues, Obstruction Theory on the Homotopy Classification of Maps. XII, 387 pages. 1977.

Vol. 629: W.A. Coppel, Dichotomies in Stability Theory. VI, 98 pages. 1978.

Vol. 630: Numerical Analysis, Proceedings, Biennial Conference, Dundee 1977. Edited by G. A. Watson. XII, 199 pages. 1978.

Vol. 631: Numerical Treatment of Differential Equations. Proceedings 1976. Edited by R. Bulirsch, R. D. Grigorieff, and J. Schröder. X, 219 pages. 1978.

Vol. 632: J.-F. Boutot, Schéma de Picard Local. X, 165 pages. 1978.

Vol. 633: N. R. Coleff and M. E. Herrera, Les Courants Résiduels Associés à une Forme Méromorphe. X, 211 pages. 1978.

Vol. 634: H. Kurke et al., Die Approximationseigenschaft lokaler Ringe. IV, 204 Seiten. 1978.

Vol. 635: T. Y. Lam, Serre's Conjecture. XVI, 227 pages. 1978.

Vol. 636: Journées de Statistique des Processus Stochastiques, Grenoble 1977, Proceedings. Edité par Didier Dacunha-Castelle et Bernard Van Cutsem. VII, 202 pages. 1978.

Vol. 637: W. B. Jurkat, Meromorphe Differentialgleichungen. VII, 194 Seiten. 1978.

Vol. 638: P. Shanahan, The Atiyah-Singer Index Theorem, An Introduction. V, 224 pages. 1978.

Vol. 639: N. Adasch et al., Topological Vector Spaces. V, 125 pages. 1978.

Vol. 640: J. L. Dupont, Curvature and Characteristic Classes. X, 175 pages. 1978.

Vol. 641: Séminaire d'Algèbre Paul Dubreil, Proceedings Paris 1976–1977. Edité par M. P. Malliavin. IV, 367 pages. 1978.

Vol. 642: Theory and Applications of Graphs, Proceedings, Michigan 1976. Edited by Y. Alavi and D. R. Lick. XIV, 635 pages. 1978.

Vol. 643: M. Davis, Multiaxial Actions on Manifolds. VI, 141 pages. 1978.

Vol. 644: Vector Space Measures and Applications I, Proceedings 1977. Edited by R. M. Aron and S. Dineen. VIII, 451 pages. 1978.

Vol. 645: Vector Space Measures and Applications II, Proceedings 1977. Edited by R. M. Aron and S. Dineen. VIII, 218 pages. 1978.

Vol. 646: O. Tammi, Extremum Problems for Bounded Univalent Functions. VIII, 313 pages. 1978.

Vol. 647: L. J. Ratliff, Jr., Chain Conjectures in Ring Theory. VIII, 133 pages. 1978.

Vol. 648: Nonlinear Partial Differential Equations and Applications, Proceedings, Indiana 1976–1977. Edited by J. M. Chadam. VI, 206 pages. 1978.

Vol. 649: Séminaire de Probabilités XII, Proceedings, Strasbourg, 1976–1977. Edité par C. Dellacherie, P. A. Meyer et M. Weil. VIII, 805 pages. 1978.

Vol. 650: C*-Algebras and Applications to Physics. Proceedings 1977. Edited by H. Araki and R. V. Kadison. V, 192 pages. 1978.

Vol. 651: P. W. Michor, Functors and Categories of Banach Spaces. VI, 99 pages. 1978.

Vol. 652: Differential Topology, Foliations and Gelfand-Fuks-Cohomology, Proceedings 1976. Edited by P. A. Schweitzer. XIV, 252 pages. 1978.

Vol. 653: Locally Interacting Systems and Their Application in Biology. Proceedings, 1976. Edited by R. L. Dobrushin, V. I. Kryukov and A. L. Toom. XI, 202 pages. 1978.

Vol. 654: J. P. Buhler, Icosahedral Golois Representations. III, 143 pages. 1978.

Vol. 655: R. Baeza, Quadratic Forms Over Semilocal Rings. VI, 199 pages. 1978.

Vol. 656: Probability Theory on Vector Spaces. Proceedings, 1977. Edited by A. Weron. VIII, 274 pages. 1978.

Vol. 657: Geometric Applications of Homotopy Theory I, Proceedings 1977. Edited by M. G. Barratt and M. E. Mahowald. VIII, 459 pages. 1978.

Vol. 658: Geometric Applications of Homotopy Theory II, Proceedings 1977. Edited by M. G. Barratt and M. E. Mahowald. VIII, 487 pages. 1978.

Vol. 659: Bruckner, Differentiation of Real Functions. X, 247 pages. 1978.

Vol. 660: Equations aux Dérivée Partielles. Proceedings, 1977. Edité par Pham The Lai. VI, 216 pages. 1978.

Vol. 661: P. T. Johnstone, R. Paré, R. D. Rosebrugh, D. Schumacher, R. J. Wood, and G. C. Wraith, Indexed Categories and Their Applications. VII, 260 pages. 1978.

Vol. 662: Akin, The Metric Theory of Banach Manifolds. XIX, 306 pages. 1978.

Vol. 663: J. F. Berglund, H. D. Junghenn, P. Milnes, Compact Right Topological Semigroups and Generalizations of Almost Periodicity. X, 243 pages. 1978.

Vol. 664: Algebraic and Geometric Topology, Proceedings, 1977. Edited by K. C. Millett. XI, 240 pages. 1978.

Vol. 665: Journées d'Analyse Non Linéaire. Proceedings, 1977. Edité par P. Bénilan et J. Robert. VIII, 256 pages. 1978.

Vol. 666: B. Beauzamy, Espaces d'Interpolation Réels: Topologie et Géometrie. X, 104 pages. 1978.

Vol. 667: J. Gilewicz, Approximants de Padé. XIV, 511 pages. 1978.

Vol. 668: The Structure of Attractors in Dynamical Systems. Proceedings, 1977. Edited by J. C. Martin, N. G. Markley and W. Perrizo. VI, 264 pages. 1978.

Vol. 669: Higher Set Theory. Proceedings, 1977. Edited by G. H. Müller and D. S. Scott. XII, 476 pages. 1978.

Vol. 670: Fonctions de Plusieurs Variables Complexes III, Proceedings, 1977. Edité par F. Norguet. XII, 394 pages. 1978.

Vol. 671: R. T. Smythe and J. C. Wierman, First-Passage Perculation on the Square Lattice. VIII, 196 pages. 1978.

Vol. 672: R. L. Taylor, Stochastic Convergence of Weighted Sums of Random Elements in Linear Spaces. VII, 216 pages. 1978.

Vol. 673: Algebraic Topology, Proceedings 1977. Edited by P. Hoffman, R. Piccinini and D. Sjerve. VI, 278 pages. 1978.

Vol. 674: Z. Fiedorowicz and S. Priddy, Homology of Classical Groups Over Finite Fields and Their Associated Infinite Loop Spaces. VI, 434 pages. 1978.

Vol. 675: J. Galambos and S. Kotz, Characterizations of Probability Distributions. VIII, 169 pages. 1978.

Vol. 676: Differential Geometrical Methods in Mathematical Physics II. Proceedings, 1977. Edited by K. Bleuler, H. R. Petry and A. Reetz. VI, 626 pages. 1978.

Vol. 677: Séminaire Bourbaki, vol. 1976/77, Exposés 489–506. IV, 264 pages. 1978.

Vol. 678: D. Dacunha-Castelle, H. Heyer et B. Roynette. Ecole d'Eté de Probabilités de Saint-Flour. VII-1977. Edité par P. L. Hennequin. IX, 379 pages. 1978.

Vol. 679: Numerical Treatment of Differential Equations in Applications, Proceedings, 1977. Edited by R. Ansorge and W. Törnig. IX, 163 pages. 1978.

Vol. 680: Mathematical Control Theory, Proceedings, 1977. Edited by W. A. Coppel. IX, 257 pages. 1978.

Vol. 681: Séminaire de Théorie du Potentiel Paris, No. 3, Directeurs: M. Brelot, G. Choquet et J. Deny. Rédacteurs: F. Hirsch et G. Mokobodzki. VII, 294 pages. 1978.

Vol. 682: G. D. James, The Representation Theory of the Symmetric Groups. V, 156 pages. 1978.

Vol. 683: Variétés Analytiques Compactes, Proceedings, 1977. Edité par Y. Hervier et A. Hirschowitz. V, 248 pages. 1978.

Vol. 684: E. E. Rosinger, Distributions and Nonlinear Partial Differential Equations. XI, 146 pages. 1978.

Vol. 685: Knot Theory, Proceedings, 1977. Edited by J. C. Hausmann. VII, 311 pages. 1978.

Vol. 686: Combinatorial Mathematics, Proceedings, 1977. Edited by D. A. Holton and J. Seberry. IX, 353 pages. 1978.

Vol. 687: Algebraic Geometry, Proceedings, 1977. Edited by L. D. Olson. V, 244 pages. 1978.

Vol. 688: J. Dydak and J. Segal, Shape Theory. VI, 150 pages. 1978.

Vol. 689: Cabal Seminar 76–77, Proceedings, 1976–77. Edited by A.S. Kechris and Y. N. Moschovakis. V, 282 pages. 1978.

Vol. 690: W. J. J. Rey, Robust Statistical Methods. VI, 128 pages. 1978.

Vol. 691: G. Viennot, Algèbres de Lie Libres et Monoïdes Libres. III, 124 pages. 1978.

Vol. 692: T. Husain and S. M. Khaleelulla, Barrelledness in Topological and Ordered Vector Spaces. IX, 258 pages. 1978.

Vol. 693: Hilbert Space Operators, Proceedings, 1977. Edited by J. M. Bachar Jr. and D. W. Hadwin. VIII, 184 pages. 1978.

Vol. 694: Séminaire Pierre Lelong – Henri Skoda (Analyse) Année 1976/77. VII, 334 pages. 1978.

Vol. 695: Measure Theory Applications to Stochastic Analysis, Proceedings, 1977. Edited by G. Kallianpur and D. Kölzow. XII, 261 pages. 1978.

Vol. 696: P. J. Feinsilver, Special Functions, Probability Semigroups, and Hamiltonian Flows. VI, 112 pages. 1978.

Vol. 697: Topics in Algebra, Proceedings, 1978. Edited by M. F. Newman. XI, 229 pages. 1978.

Vol. 698: E. Grosswald, Bessel Polynomials. XIV, 182 pages. 1978.

Vol. 699: R. E. Greene and H.-H. Wu, Function Theory on Manifolds Which Possess a Pole. III, 215 pages. 1979.

Vol. 700: Module Theory, Proceedings, 1977. Edited by C. Faith and S. Wiegand. X, 239 pages. 1979.

Vol. 701: Functional Analysis Methods in Numerical Analysis, Proceedings, 1977. Edited by M. Zuhair Nashed. VII, 333 pages. 1979.

Vol. 702: Yuri N. Bibikov, Local Theory of Nonlinear Analytic Ordinary Differential Equations. IX, 147 pages. 1979.

Vol. 703: Equadiff IV, Proceedings, 1977. Edited by J. Fábera. XIX, 441 pages. 1979.

Vol. 704: Computing Methods in Applied Sciences and Engineering, 1977, I. Proceedings, 1977. Edited by R. Glowinski and J. L. Lions. VI, 391 pages. 1979.

Vol. 705: O. Forster und K. Knorr, Konstruktion verseller Familien kompakter komplexer Räume. VII, 141 Seiten. 1979.

Vol. 706: Probability Measures on Groups, Proceedings, 1978. Edited by H. Heyer. XIII, 348 pages. 1979.

Vol. 707: R. Zielke, Discontinuous Čebyšev Systems. VI, 111 pages. 1979.

Vol. 708: J. P. Jouanolou, Equations de Pfaff algébriques. V, 255 pages. 1979.

Vol. 709: Probability in Banach Spaces II. Proceedings, 1978. Edited by A. Beck. V, 205 pages. 1979.

Vol. 710: Séminaire Bourbaki vol. 1977/78, Exposés 507–524. IV, 328 pages. 1979.

Vol. 711: Asymptotic Analysis. Edited by F. Verhulst. V, 240 pages. 1979.

Vol. 712: Equations Différentielles et Systèmes de Pfaff dans le Champ Complexe. Edité par R. Gérard et J.-P. Ramis. V, 364 pages. 1979.

Vol. 713: Séminaire de Théorie du Potentiel, Paris No. 4. Edité par F. Hirsch et G. Mokobodzki. VII, 281 pages. 1979.

Vol. 714: J. Jacod, Calcul Stochastique et Problèmes de Martingales. X, 539 pages. 1979.

Vol. 715: Inder Bir S. Passi, Group Rings and Their Augmentation Ideals. VI, 137 pages. 1979.

Vol. 716: M. A. Scheunert, The Theory of Lie Superalgebras. X, 271 pages. 1979.

Vol. 717: Grosser, Bidualräume und Vervollständigungen von Banachmoduln. III, 209 pages. 1979.

Vol. 718: J. Ferrante and C. W. Rackoff, The Computational Complexity of Logical Theories. X, 243 pages. 1979.

Vol. 719: Categorial Topology, Proceedings, 1978. Edited by H. Herrlich and G. Preuß. XII, 420 pages. 1979.

Vol. 720: E. Dubinsky, The Structure of Nuclear Fréchet Spaces. V, 187 pages. 1979.

Vol. 721: Séminaire de Probabilités XIII. Proceedings, Strasbourg, 1977/78. Edité par C. Dellacherie, P. A. Meyer et M. Weil. VII, 647 pages. 1979.

Vol. 722: Topology of Low-Dimensional Manifolds. Proceedings, 1977. Edited by R. Fenn. VI, 154 pages. 1979.

Vol. 723: W. Brandal, Commutative Rings whose Finitely Generated Modules Decompose. II, 116 pages. 1979.

Vol. 724: D. Griffeath, Additive and Cancellative Interacting Particle Systems. V, 108 pages. 1979.

Vol. 725: Algèbres d'Opérateurs. Proceedings, 1978. Edité par P. de la Harpe. VII, 309 pages. 1979.

Vol. 726: Y.-C. Wong, Schwartz Spaces, Nuclear Spaces and Tensor Products. VI, 418 pages. 1979.

Vol. 727: Y. Saito, Spectral Representations for Schrödinger Operators With Long-Range Potentials. V, 149 pages. 1979.

Vol. 728: Non-Commutative Harmonic Analysis. Proceedings, 1978. Edited by J. Carmona and M. Vergne. V, 244 pages. 1979.